Other Titles in This Series

192 **Tomek Bartoszyński and Marion Scheepers, Editors,** Set theory, 1996

191 **Tuong Ton-That, Kenneth I. Gross, Donald St. P. Richards, and Paul J. Sally, Jr., Editors,** Representation theory and harmonic analysis, 1995

190 **Mourad E. H. Ismail, M. Zuhair Nashed, Ahmed I. Zayed, and Ahmed F. Ghaleb, Editors,** Mathematical analysis, wavelets, and signal processing, 1995

189 **S. A. M. Marcantognini, G. A. Mendoza, M. D. Morán, A. Octavio, and W. O. Urbina, Editors,** Harmonic analysis and operator theory, 1995

188 **Alejandro Adem, R. James Milgram, and Douglas C. Ravenel, Editors,** Homotopy theory and its applications, 1995

187 **G. W. Brumfiel and H. M. Hilden,** $SL(2)$ representations of finitely presented groups, 1995

186 **Shreeram S. Abhyankar, Walter Feit, Michael D. Fried, Yasutaka Ihara, and Helmut Voelklein, Editors,** Recent developments in the inverse Galois problem, 1995

185 **Raúl E. Curto, Ronald G. Douglas, Joel D. Pincus, and Norberto Salinas, Editors,** Multivariable operator theory, 1995

184 **L. A. Bokut′, A. I. Kostrikin, and S. S. Kutateladze, Editors,** Second International Conference on Algebra, 1995

183 **William C. Connett, Marc-Olivier Gebuhrer, and Alan L. Schwartz, Editors,** Applications of hypergroups and related measure algebras, 1995

182 **Selman Akbulut, Editor,** Real algebraic geometry and topology, 1995

181 **Mila Cenkl and Haynes Miller, Editors,** The Čech Centennial, 1995

180 **David E. Keyes and Jinchao Xu, Editors,** Domain decomposition methods in scientific and engineering computing, 1994

179 **Yoshiaki Maeda, Hideki Omoro, and Alan Weinstein, Editors,** Symplectic geometry and quantization, 1994

178 **Hélène Barcelo and Gil Kalai, Editors,** Jerusalem Combinatorics '93, 1994

177 **Simon Gindikin, Roe Goodman, Frederick P. Greenleaf, and Paul J. Sally, Jr., Editors,** Representation theory and analysis on homogeneous spaces, 1994

176 **David Ballard,** Foundational aspects of "non"standard mathematics, 1994

175 **Paul J. Sally, Jr., Moshe Flato, James Lepowsky, Nicolai Reshetikhin, and Gregg J. Zuckerman, Editors,** Mathematical aspects of conformal and topological field theories and quantum groups, 1994

174 **Nancy Childress and John W. Jones, Editors,** Arithmetic geometry, 1994

173 **Robert Brooks, Carolyn Gordon, and Peter Perry, Editors,** Geometry of the spectrum, 1994

172 **Peter E. Kloeden and Kenneth J. Palmer, Editors,** Chaotic numerics, 1994

171 **Rüdiger Göbel, Paul Hill, and Wolfgang Liebert, Editors,** Abelian group theory and related topics, 1994

170 **John K. Beem and Krishan L. Duggal, Editors,** Differential geometry and mathematical physics, 1994

169 **William Abikoff, Joan S. Birman, and Kathryn Kuiken, Editors,** The mathematical legacy of Wilhelm Magnus, 1994

168 **Gary L. Mullen and Peter Jau-Shyong Shiue, Editors,** Finite fields: Theory, applications, and algorithms, 1994

167 **Robert S. Doran, Editor,** C^*-algebras: 1943–1993, 1994

166 **George E. Andrews, David M. Bressoud, and L. Alayne Parson, Editors,** The Rademacher legacy to mathematics, 1994

165 **Barry Mazur and Glenn Stevens, Editors,** p-adic monodromy and the Birch and Swinnerton-Dyer conjecture, 1994

(Continued in the back of this publication)

Set Theory

CONTEMPORARY MATHEMATICS

192

Set Theory

Annual Boise Extravaganza
in Set Theory (BEST) Conference
March 13–15, 1992
April 10–11, 1993
March 25–27, 1994
Boise State University, Boise, Idaho

Tomek Bartoszyński
Marion Scheepers
Editors

American Mathematical Society
Providence, Rhode Island

The annual Boise Extravaganza in Set Theory (BEST) conferences were held at Boise State University, Boise, Idaho, March 13–15, 1992, April 10–11, 1993, and March 25–27, 1994.

1991 *Mathematics Subject Classification*. Primary 03Exx;
Secondary 03E05, 03E15, 03E35, 90D44.

Library of Congress Cataloging-in-Publication Data

Boise Extravaganza in Set Theory Conference (1st : 1992 : Boise State University)

Set theory : annual Boise Extravaganza in Set Theory (BEST) Conference, March 13–15, 1992, April 10–11, 1993, March 25–27, 1994, Boise State University, Boise, Idaho / Tomek Bartoszyński, Marion Scheepers, editors.

p. cm. — (Contemporary mathematics, ISSN 0271-4132; v. 192)

Includes bibliographical references.

ISBN 0-8218-0306-9 (alk. paper)

1. Set theory—Congresses. I. Bartoszyński, Tomek, 1957– . II. Scheepers, Marion, 1957– . III. Boise Extravaganza in Set Theory Conference (2nd : 1993 : Boise State University) IV. Boise Extravaganza in Set Theory Conference (3rd : 1994 : Boise State University) V. Title. VI. Series: Contemporary mathematics (American Mathematical Society); v. 192.

QA248.B66 1992
511.3′22—dc20 95-34595
CIP

∞ The paper used in this book is acid-free and falls within the guidelines
established to ensure permanence and durability.
♻ Printed on recycled paper.

10 9 8 7 6 5 4 3 2 1 01 00 99 98 97 96

Contents

Preface

The Boise Extravaganza in Set Theory (better known as BEST) is an annual conference dedicated to Set Theory. The first meeting took place in March 1992. The papers presented in this volume are indicative of the scope and emphasis of the conference.

The conference would not have been possible without financial support from:

(1) Idaho State Board of Education

(2) Boise State University, in particular:

- (a) Department of Mathematics and Computer Science
- (b) Office of the Dean of the College of Arts and Sciences
- (c) Office of Research Administration

Tomek Bartoszyński and Marion Scheepers

List of Participants

(a) BEST 1 (March 13–15, 1992)

Tomek Bartoszyński *(Boise State University)*
Howard Becker *(University of South Carolina)*
Carl Darby *(Boise State University)*
Alex Feldman *(Boise State University)*
Steve Grantham *(Boise State University)*
Randall Holmes *(Boise State University)*
Winfried Just *(Ohio University, Athens)*
Claude Laflamme *(University of Calgary)*
Avner Landver *(University of Kansas, Lawrence)*
Arnold Miller *(University of Wisconsin, Madison)*
Janusz Pawlikowski *(Wrocław University)*
Marion Scheepers *(Boise State University)*
Juris Steprāns *(York University, Toronto)*
Stevo Todorcevic *(University of Toronto)*

(b) BEST 2 (April 10–11, 1993)

Rick Ball *(University of Denver)*
Tomek Bartoszyński *(Boise State University)*
James Baumgartner *(Dartmouth College)*
Andreas Blass *(University of Michigan, Ann Arbor)*
Krzysztof Ciesielski *(University of West Virginia, Morgantown)*
Carl Darby *(Boise State University)*
Mirna Dzamonja *(University of Wisconsin, Madison)*
Alex Feldman *(Boise State University)*
David Ferguson *(Boise State University)*

Bill Fleissner *(University of Kansas, Lawrence)*
Matthew Foreman *(Ohio State University, Columbus)*
Steve Grantham *(Boise State University)*
Randall Holmes *(Boise State University)*
Jakub Jasinski *(University of Pennsylvania, Scranton)*
Claude Laflamme *(University of Calgary)*
Judy Roitman *(University of Kansas, Lawrence)*
Marion Scheepers *(Boise State University)*
James Sharp *(Rutgers University, New Brunswick)*
Zoran Spasojevic *(University of Wisconsin, Madison)*
Jouko Vaananen *(University of Helsinki)*

(c) BEST 3 (March 25–27, 1994)

Tomek Bartoszyński *(Boise State University)*
Andreas Blass *(University of Michigan, Ann Arbor)*
Derrick DuBose *(University of Nevada, Las Vegas)*
Alex Feldman *(Boise State University)*
Matt Foreman *(UC Irvine)*
Thomas Forster *(Cambridge University)*
Fred Galvin *(University of Kansas, Lawrence)*
Steve Grantham *(Boise State University)*
Andras Hajnal *(Hungarian Academy of Sciences, Budapest)*
Randall Holmes *(Boise State University)*
Aleksander Kechris *(Caltech, Pasadena)*
Claude Laflamme *(University of Calgary)*
Marion Scheepers *(Boise State University)*
Chaz Schlindwein *(University of Nevada, Las Vegas)*
Zoran Spasojevic *(University of Wisconsin, Madison)*
Jindra Zapletal *(Penn State, University Park)*

Contemporary Mathematics
Volume **192**, 1996

Interval covers of a linearly ordered set

R. AHARONI, A. HAJNAL AND E.C. MILNER

ABSTRACT. We show that a κ-fold cover of a linearly ordered set by intervals is a union of κ disjoint covers.

Keywords: linearly ordered set, interval, cover.
AMS subject classification (1985): 06A05

1. Introduction

Let $\mathcal{F} = \{F_i : i \in I\}$ be a family of sets. $\mathcal{F}$ is a *cover* of S if $S \subseteq \bigcup \mathcal{F}$, and it is a *$\kappa$-fold cover* if

$$\forall x \in S \exists J \subseteq I \; (|J| = \kappa \wedge x \in \bigcap_{i \in J} F_i).$$

Two subfamilies $\mathcal{F}_0 = \langle F_i : i \in I_1 \rangle$, $\mathcal{F}_1 = \langle F_i : i \in I_1 \rangle$ of $\mathcal{F}$ are *disjoint* if the index sets I_0 and I_1 are disjoint subsets of I.

We are interested in the question whether a κ-fold cover can be factorized into κ pairwise disjoint covers. This is clearly false in general, but there are interesting questions when the base set S and the covering sets have some special structure. This type of question was investigated by Pach in [**2**], where he considered coverings of the plane by congruent copies of certain shapes. In fact, the particular problem we shall consider originated with him. We will show (Theorem 1.1) that there is such a factorization of $\mathcal{F}$ in the case when S is a linearly ordered set and the covering sets are intervals (convex subsets) of S. The theorem does not extend to partially ordered sets in any obvious way. For example, consider the the poset $S = \{a, a', b, b'\}$ on four points with order relations $a, a' < b, b'$. The family $\mathcal{F} = \{\{a, b\}, \{a, b'\}, \{a', b, b'\}, \{a'\}\}$ is a 2-fold cover of S by (connected) convex sets which does not factorize into two disjoint covers. It should be remarked that it is very simple to prove the theorem in the case when S is well-ordered, but the general case requires some care.

THEOREM 1.1. *Let κ be a cardinal number, $\langle S, \leq \rangle$ a linearly ordered set and let $\mathcal{F} = \langle F_i : i \in I \rangle$ be a κ-fold cover of S, where each F_i is an interval of*

Research of second author supported by Hungarian National Science Foundation OTKA grant #1908

Research of third author supported by NSERC grant #69-0982.

S. Then there are κ pairwise disjoint subsets I_ν of I such that each subfamily $\mathcal{F}_\nu = \langle F_i : i \in I_\nu \rangle$ $(\nu < \kappa)$ is a cover of S.

2. Preliminary remarks and some notation

We will say that a set $J \subseteq I$ is a *τ-fold cover* of $A \subseteq S$ if the corresponding family $\langle F_i : i \in J \rangle$ is a τ-fold cover of A. For $J \subseteq I$ and $x \in S$, we define $\mathcal{F}(J) = \bigcup_{i \in J} F_i$, and $J(x) = \{i \in J : x \in F_i\}$. Also, for sets $J \subseteq I$ and $A \subseteq S$, define

$$\theta(J, A) = \min\{\tau : \forall x \in A |J(x)| < \tau\}.$$

For intervals A, B of S,we write $A \leq_L B$ if $\forall b \in B \exists a \in A(a \leq b)$, and $A \leq_R B$ if $\forall a \in A \exists b \in B(a \leq b)$; also $A <_L B$ means that $A \leq_L B$ and $B \not\leq_L A$ etc. (Of course, $A \leq_L B$ is just the assertion that $\inf A \leq \inf B$ in the closure of S.) For $x \in S$, define $S(\geq x) = \{y \in S : y \geq x\}$, and $S(\leq x) = \{y \in S : y \leq x\}$.

Call an interval S' of S *good* if there is $\mathcal{F}' = \langle F_i : i \in I' \rangle$, a subfamily of $\mathcal{F}$, such that $\bigcup \mathcal{F}' = S'$, and there is a partition of $\mathcal{F}'$ into κ disjoint subfamilies each being a cover of S'. If S' is a good interval of S then we can reduce the problem to $S \backslash S'$, since each $(S \backslash S') \cap F_i$ is an interval of $S \backslash S'$. Thus, in order to prove the theorem, it will be enough by Zorn's lemma to show that each point of S belongs to a good subinterval of S.

3. The case $\kappa = k < \omega$

Let $\mathcal{F} = \langle F_i : i \in I \rangle$ be a k-fold interval cover of S, where $k < \omega$. As we already remarked, the theorem will follow if we show that each element $x \in S$ belongs to a good interval, and so without loss of generality we may assume that S contains no good *proper* subinterval. If the family $\mathcal{F}$ contains an increasing chain of intervals $F_{i_0} \subseteq F_{i_1} \subseteq \dots$, where the i_n $(n < \omega)$ are distinct elements of I, then their union $\bigcup_{n<\omega} F_{i_n}$ is good. Therefore, we may also assume that the family $\mathcal{F}$ contains no infinite increasing chain.

Call a sequence $\langle I_0, \dots, I_{k-1} \rangle$ of k subsets of I a *disjoint k-fold cover of* $A \subseteq S$ if the sets I_r $(r < k)$ are pairwise disjoint and $A \subseteq \bigcap_{r<k} \mathcal{F}(I_r)$. If $\langle I_0, \dots, I_{k-1} \rangle$ is any sequence of pairwise disjoint subsets of I and if $\langle I'_0, \dots I'_{k-1} \rangle$ is another such sequence of pairwise disjoint sets, we call it an *extension* of $\langle I_0, \dots, I_{k-1} \rangle$ if $I_r \subseteq I'_r$ holds for each $r < k$.

Definition: The sequence $\langle I_0, \dots, I_{k-1} \rangle$ of k subsets of I is *potentially good* if (i) the I_r $(r < k)$ are pairwise disjoint, (ii) $A_r = \mathcal{F}(I_r)$ is an interval of S $(r < k)$, (iii) the family of sets $\mathcal{F}^* = \langle F_i : i \in I \backslash \bigcup_{r<k} I_r \rangle \cup \langle A_0, \dots, A_{k-1} \rangle$ obtained from $\mathcal{F}$ by replacing the intervals F_i for $i \in I_r$ by the single interval A_r $(r < k)$, is also a k-fold cover of S, and (iv) either $A_r = \emptyset$ $(r < k-1)$ and $A_{k-1} \not\subset F_i$ for any $i \in I$ ($X \not\subset Y$ means that X is not a proper subset of Y) or $A = \bigcap_{r<k} A_r \not\subset F_i$ for any $i \in I$. For example, if F_i is a maximal member of $\mathcal{F}$, then the sequence $\langle \emptyset, \emptyset, \dots, \emptyset, \{i\} \rangle$ is potentially good.

The main idea for our proof of the theorem for the case of a k-fold cover ($k < \omega$) is the following lemma.

LEMMA 3.1. *Suppose $\mathcal{F} = \langle F_i : i \in I\rangle$ is a k-fold interval cover of S which contains no infinite increasing sequence. Suppose that I_r ($r < k$) are k pairwise disjoint subsets of I such that each $\mathcal{F}(I_r)$ is an interval of S ($r < k$), and that $A \subseteq \bigcap_{r<k} \mathcal{F}(I_r)$ is an interval which is not a proper subset of any F_i for $i \in I$. Then there are $I'_r \subseteq I_r$ ($r < k$) such that $\langle I'_0, \dots, I'_{k-1}\rangle$ is potentially good, $A \subseteq \bigcap_{i<k} \mathcal{F}(I'_r)$, and*

$$\{i \in I_r : F_i \subseteq A\} = \{i \in I'_r : F_i \subseteq A\}.$$

PROOF. Let T_0 be the set of those $r < k$ for which $A = F_{i_r}$ for some $i_r \in I_r$, and for each $r \in T_0$ define $I'_r = \{i \in I_r : F_i \subseteq A\}$.

If $r < k$ and $r \notin T_0$, we define

$$L_r = \{i \in I_r : F_i <_L A \wedge A \cap F_i \neq \emptyset\}, \quad R_r = \{i \in I_r : A <_R F_i \wedge A \cap F_i \neq \emptyset\}.$$

By our assumption, the sets L_r and R_r ($r < k, r \notin T_0$) are pairwise disjoint. We shall define subsets $L'_r \subseteq L_r$ and $R'_r \subseteq R_r$ as follows. Consider first the intervals F_i with $i \in R_r$. If $R_r = \emptyset$, we put $r \in T_{R,0}$ and define $R'_r = \emptyset$. If R_r contains an element, say $i_{R,r}$, such that $F_{i_{R,r}} \leq_L F_i$ for all $i \in R_r$, then we put $r \in T_{R,1}$ and define $R'_r = \{i_{R,r}\}$. Otherwise, we put $r \in T_{R,2}$. In this case, there is no $\leq_L$-minimal interval in $\mathcal{F}_r = \langle F_i : i \in R_r\rangle$, and so there is a sequence $i_\rho \in R_r$ ($\rho < \mu$), where μ is an infinite regular cardinal, such that

$$\dots <_L F_{i_\rho} <_L \dots <_L F_{i_1} <_L F_{i_0},$$

and for each $i \in R_r$ there is some $\rho < \mu$ such that $F_{i_\rho} <_L F_i$. Since there is no infinite increasing chain in $\mathcal{F}$, it follows, for example from the Erdős-Dushnik-Miller theorem [**1**], that we can further assume that

$$\dots <_R F_{i_\rho} <_R \dots <_R F_{i_1} <_R F_{i_0}.$$

In this case, we put $R'_r = \{i_\rho : k \leq \rho < \mu\}$. Note that, with this choice of the i_ρ when $r \in T_{R,2}$, we have that

$$M_r = \{i_\rho : \rho < k\} \subseteq I \setminus \Big(\bigcup_{s \neq r} I_s \cup \{i \in I_r : F_i \subseteq A\} \cup L_r \cup R'_r\Big). \tag{1}$$

This defines R'_r for all $r \notin T_0$, and L'_r is defined in an analogous way. Finally, put

$$I'_r = (I_r \setminus (L_r \cup R_r)) \cup (L'_r \cup R'_r).$$

It is clear that $A'_r = \mathcal{F}(I'_r)$ is an interval, $A'_r \supseteq A$ and $\{i \in I_r : F_i \subseteq A\} = \{i \in I'_r : F_i \subseteq A\}$ for all $r < k$. In order to show that $\langle I'_0, \dots, I'_{k-1}\rangle$ is potentially good, it remains only to show that $\mathcal{F}' = \langle F_i : i \in I \setminus \bigcup_{r<k} I'_r \rangle \cup \langle A'_0, \dots, A'_{k-1} \rangle$ is a k-fold cover of S.

If $x \in A$ then obviously x belongs to the k members A'_r $(r < k)$ of $\mathcal{F}'$. Suppose $A <_R \{x\}$. If $x \in A'_r$ for some $r \in T_{R,2}$, then $x \in F_i$ for each of the k members $i \in M_r$, and by (1) these belong to the family $\mathcal{F}'$. So we can assume that $x \notin \bigcup_{r \in T_{R,2}} A'_r$. Suppose that $x \in A'_r$ for t different values of $r \in T_{R,1}$. For each such r there is exactly one index $i \in \bigcup_{s<k} I'_s$ (namely $i = i_{R,r}$) such that $x \in F_i$. Therefore, there are $k - t$ different $i \in I \setminus \bigcup_{s<k} I'_s$ such that $x \in F_i$. Hence $\mathcal{F}'$ is a k-fold cover of S. □

We now conclude the proof of the theorem (for this case) as follows. We already remarked that we may assume that S contains no good proper subinterval, and also that $\mathcal{F}$ contains no infinite increasing chain, so we can apply Lemma 3.1. We also observed that there are trivial potentially good sequences, and so it will suffice to prove the following stronger assertion.

THEOREM 3.1. *Let k be a positive integer, $\langle S, \leq \rangle$ a linearly ordered set, $\mathcal{F} = \langle F_i : i \in I \rangle$ a k-fold interval cover of S which contains no infinite increasing sequence, and suppose that S contains no proper good subinterval. Then any potentially good sequence $\langle I_0, \dots, I_{k-1} \rangle$ can be extended to a disjoint k-fold cover $\langle I'_0, \dots, I'_{k-1} \rangle$ of S such that $\{i \in I'_r : F_i \subseteq \mathcal{F}(I_r)\} = I_r$.*

PROOF. We prove the theorem by induction on k. The result is obvious when $k = 1$ and so we assume that $k > 1$ and that Theorem 3.1 and also Theorem 1.1 hold when k is replaced by $k - 1$.

We first use the induction hypothesis to establish the following lemma.

LEMMA 3.2. *Let $\langle I_0, \dots, I_{k-1} \rangle$ be a potentially good sequence, $\mathcal{F}(I_r) = A_r$ $(r < k)$. Then there is a potentially good extension $\langle I'_r : r < k \rangle$ such that $\bigcup_{r<k} A_r \subseteq \bigcap_{r<k} \mathcal{F}(I'_r)$, and*

$$\{i \in I'_r : F_i \subseteq A_r\} = I_r. \tag{2}$$

PROOF. There are two cases to consider. Either (i) $A_r = \emptyset$ for $r < k - 1$ and A_{k-1} is not a proper subset of any F_i, or (ii) $A = \bigcap_{r<k} A_r$ is not a proper subset of any F_i.

Suppose (i) holds. Since $\mathcal{F}^* = \langle F_i \cap A_{k-1} : i \in I \setminus I_{k-1} \rangle$ is a $(k-1)$-fold cover of A_{k-1}, it follows from Theorem 1.1 for $k - 1$ that there are disjoint subsets $I''_0, \dots, I''_{k-2}$ of $I \setminus I_{k-1}$ such that $A_{k-1} \subseteq \mathcal{F}(I''_r)$ for $r < k - 1$. Put $I''_{k-1} = I_{k-1}$. Then, by Lemma 3.1, there are $I'_r \subseteq I''_r$ such that $\langle I'_0, \dots, I'_{k-1} \rangle$ is potentially good and $\{i \in I'_r : F_i \subseteq A_{k-1}\} = \{i \in I''_r : F_i \subseteq A_{k-1}\}$ for each $r < k$. In particular, this implies that $I'_{k-1} = I''_{k-1} = I_{k-1}$ so that $\langle I'_0, \dots, I'_{k-1} \rangle$

is a potentially good extension of $\langle I_0, \ldots, I_{k-1}\rangle$ such that (2) holds. This proves the lemma for this case.

Now suppose that (ii) holds. The proof is similar to the proof for the previous case, but we need to do it in two stages. For each $r < k$ define $B_r = \{x \in A_r : A \leq_L \{x\}\}$, $C_r = \{x \in A_r : \{x\} \leq_R A\}$, $B = \bigcup_{r<k} B_r$, $C = \bigcup_{r<k} C_r$. We first consider the right side B.

We can assume that our notation is such that $A \subseteq B_0 \subseteq \ldots \subseteq B_{k-1} = B$. Since $\langle I_0, \ldots, I_{k-2}\rangle$ is potentially good for the family $\mathcal{F}' = \langle F_i \cap B : i \in I \setminus I_{k-1}\rangle$ of subsets of B, it follows from the inductive assumption that there is a disjoint $(k-1)$-fold cover $\langle J_0, \ldots, J_{k-2}\rangle$ of B which extends $\langle I_0, \ldots, I_{k-2}\rangle$ and is such that

$$\{i \in J_r : F_i \cap B \leq_R B_r\} = I_r \tag{3}$$

holds for each $r < k-1$. In other words, the only new sets F_i with $i \in J_r \setminus I_r$ extend B_r (and hence A_r) to the right. Put $J_{k-1} = I_{k-1}$. Then (3) also holds for $r = k-1$.

Now by a similar argument on the left side, it follows that there is a disjoint k-fold cover $\langle J'_0, \ldots, J'_{k-1}\rangle$ of C which also extends $\langle I_0, \ldots, I_{k-1}\rangle$ and is such that

$$\{i \in J'_r : C_r \leq_L F_i\} = J_r. \tag{4}$$

It follows from (3) and (4) that

$$\{i \in J_r \cup J'_r : F_i \subseteq A_r\} = I_r. \tag{5}$$

Since $A \not\subseteq F_i$ for all $i \in I$, it follows that $J_r \cap J'_s = \emptyset$ for all $r, s < k$, and therefore, $\langle J_0 \cup J'_0, \ldots, J_{k-1} \cup J'_{k-1}\rangle$ is a disjoint k-fold cover of $B \cup C = \bigcup_{r<k} A_r$ which extends $\langle I_0, \ldots, I_{k-1}\rangle$ and satisfies (5).

Finally, it follows from Lemma 3.1, that there are sets $I'_r \subseteq J_r \cup J'_r$ $(r < k)$ such that $\langle I'_0, \ldots, I'_{k-1}\rangle$ is potentially good for the family $\mathcal{F}$, $B \cup C \subseteq \bigcap_{r<k} \mathcal{F}(J'_r)$, and

$$\{i \in I'_r : F_i \subseteq B \cup C\} = \{i \in J_r \cup J'_r : F_i \subseteq B \cup C\}. \tag{6}$$

The lemma follows since (5) and (6) together imply that (2) also holds. □

Theorem 3.1 follows from Lemma 3.2. Let $\langle I^0_0, \ldots, I^0_{k-1}\rangle$ be a given potentially good sequence. Let $n < \omega$ and suppose we have already defined a potentially good sequence $\langle I^n_0, \ldots, I^n_{k-1}\rangle$. Let $A^n_r = \mathcal{F}(I^n_r)$ $(r < k)$. By Lemma 3.2 there is a potentially good extension $\langle I^{n+1}_0, \ldots, I^{n+1}_{k-1}\rangle$ such that $\bigcup_{r<k} A^n_r \subseteq \bigcap_{r<k} \mathcal{F}(I^{n+1}_r)$. This inductively defines the sets I^n_r for all $n < \omega$

and $r < k$. The sets $I_r = \bigcup_{n<\omega} I_r^n$ $(r < k)$ are pairwise disjoint. Also, since $\mathcal{F}(I_s^n) \subseteq \mathcal{F}(I_r^{n+1})$ for all $r, s < k$, it follows that

$$\mathcal{F}(I_r) = \bigcup_{n<\omega} A_r^n = A$$

for some $A \subseteq S$. Therefore, A is a good set, and hence $A = S$. □

4. The case when κ is an infinite cardinal

As before, we assume that $\mathcal{F} = \langle F_i : i \in I \rangle$ is a κ-fold interval cover of the linearly ordered set S, where now κ is an infinite cardinal number. We will deduce the theorem from the following lemma. In the case when κ is regular we really only need this lemma for the case when $\tau = 1$, but we shall prove it without this restriction.

LEMMA 4.1 (MAIN LEMMA). *Let $\tau < \kappa$ be a regular cardinal number. Then there are pairwise disjoint subsets J, $\tilde{J}_\nu$ $(\nu < \kappa)$ of I and disjoint subsets $A, R \subseteq S$ such that*

(i) $S = A \cup R$,
(ii) $\theta(J \cup \bigcup_{\nu<\kappa} \tilde{J}_\nu, A) \leq \kappa$,
(iii) *J is a τ-fold cover of A,*
(iv) $R \subseteq \bigcap_{\nu<\kappa} \mathcal{F}(\tilde{J}_\nu)$.

Before proving the lemma we show that the theorem follows from it.

If κ is regular, we define $\tau_\xi = 1$ $(\xi < \kappa)$, and if κ is singular, we choose τ_ξ $(\xi < cf(\kappa))$ to be an increasing sequence of infinite regular cardinals cofinal in κ. We shall use the lemma to recursively define sets S^ξ, I^ξ, J^ξ, $\tilde{J}_\nu^\xi$ $(\nu < \kappa)$, A^ξ and R^ξ for $\xi < cf(\kappa)$ as follows. For $\xi = 0$, let $S^0 = S$, $I^0 = I$, and let J^0, $\tilde{J}_\nu^0$ $(\nu < \kappa)$, A^0 and R^0 be the sets given by the lemma when $\tau = \tau_0$. Let $0 < \xi < cf(\kappa)$ and suppose that we have already defined I^η, J^η, $\tilde{J}_\nu^\eta$ $(\nu < \kappa)$, S^η, A^η and R^η for $\eta < cf(\kappa)$, so that all the conditions of the lemma are satisfied. In other words, for each $\eta < \xi$ we assume that the family $\mathcal{F}^\eta = \langle F_i : i \in I^\eta \rangle$ is a κ-fold cover of S^η, J^η and $\tilde{J}_\nu^\eta$ $(\nu < \kappa)$ are pairwise disjoint subsets of I^η, A^η and R^η are disjoint subsets of S^η such that $(i)_\eta$ $S^\eta = A^\eta \cup R^\eta$, $(ii)_\eta$ $\theta(J^\eta \cup \bigcup_{\nu<\kappa} \tilde{J}_\nu^\eta, A^\eta) \leq \kappa$ $(iii)_\eta$ J^η is a τ_η-fold cover of A^η, and $(iv)_\eta$ $R^\eta \subseteq \bigcap_{\nu<\kappa} \mathcal{F}(\tilde{J}_\nu^\eta)$.

Let $S^\xi = \bigcap_{\eta<\xi} A^\eta$, $I^\xi = I \setminus \bigcup\{\tilde{J}_\nu^\eta \cup J^\eta : \nu < \kappa, \eta < \xi\}$. Then by the inductive assumptions, $\mathcal{F}^\xi = \langle F_i : i \in I^\xi \rangle$ is a κ-fold cover of S^ξ since $\xi < cf(\kappa)$. Therefore, applying the lemma with $\tau = \tau_\xi$, it follows that there are disjoint subsets J^ξ, $\tilde{J}_\nu^\xi$ $(\nu < \kappa)$ of I^ξ, and disjoint subsets A^ξ and R^ξ of S^ξ such that the corresponding conditions $(i)_\xi$–$(iv)_\xi$ are satisfied. This completes the inductive construction. Note that, by the construction, the sets J^ξ, $\tilde{J}_\nu^\xi$ are pairwise disjoint for all $\xi, \nu < \kappa$, $S^\xi = A^\xi \cup R^\xi$, $R^\xi \subseteq \bigcap_{\nu<\kappa} \mathcal{F}(\tilde{J}_\nu^\xi)$, J^ξ is a τ_ξ-fold cover of A^ξ, and $S = \bigcup_{\xi<cf(\kappa)} R^\xi \cup \bigcap_{\xi<cf(\kappa)} A^\xi$.

It should be remarked that in the inductive step above we applied the lemma for the family $\mathcal{F}^\xi = \langle F_i \cap S^\xi : i \in I^\xi \rangle$. But this is permissible since the $F_i \cap S^\xi$ are subintervals of the ordered set S^ξ.

If κ is regular, put

$$I_\nu = \bigcup_{\xi<\kappa} \tilde{J}^\xi_\nu \cup J^\nu \qquad (\nu < \kappa).$$

Then the I_ν are pairwise disjoint and $\mathcal{F}(I_\nu) \supseteq \bigcup_{\xi<\kappa} R_\xi \cup A^\nu = S$, as required.

Suppose κ is singular and that the theorem is true for smaller values of κ. Then, for each $\xi < cf(\kappa)$ there is a partition of J^ξ into τ_ξ pairwise disjoint sets J^ξ_ρ $(\rho < \tau_\xi)$ such that

$$A^\xi \subseteq \mathcal{F}(J^\xi_\rho) \qquad (\forall \rho < \tau_\xi).$$

For $\nu < \kappa$ let $\xi(\nu) < cf(\kappa)$ be minimal such that $\tau_{\xi(\nu)} > \nu$. Then the sets

$$I_\nu = \bigcup_{\xi<cf(\kappa)} \tilde{J}^\xi_\nu \cup J^{\xi(\nu)}_\nu \qquad (\nu < \kappa)$$

are pairwise disjoint and $\mathcal{F}(I_\nu) \supseteq \bigcup_{\xi<cf(\kappa)} R^\xi \cup A^{\xi(\nu)} = S$. This completes the proof of the theorem.

5. Proof of the main lemma

We now assume that τ is a fixed regular cardinal number less than κ (thus $\tau = 1$ if $\kappa = \omega$). For $x, y \in S$ write $x \sim y$ if there is a τ-fold cover $J \subseteq I$ of the closed interval

$$[x, y] \stackrel{\text{def}}{=} \{z \in S : x \leq z \leq y\} \cup \{z \in S : y \leq z \leq x\}$$

such that

$$\theta(J, S) \leq \begin{cases} \kappa & \text{if } \tau = 1, \\ \tau^+ & \text{if } \tau \text{ is infinite.} \end{cases}$$

Clearly $\sim$ is an equivalence relation, and the equivalence classes are subintervals of S. Let $\{E_\gamma : \gamma \in \Gamma\}$ be the set of all equivalence classes. For each $\gamma \in \Gamma$, let $I_\gamma = \{i \in I : F_i \cap E_\gamma \neq \emptyset\}$.

In the case when $\tau = 1$, it is obvious that, if E_γ is an equivalence class and $i \in I_\gamma$, then $F_i \subseteq E_\gamma$. The following lemma records this fact and says something similar for the case when τ is infinite.

LEMMA 5.1. *Let $\gamma \in \Gamma$. If $\tau = 1$, then $F_i \subseteq E_\gamma$ for all $i \in I_\gamma$. If τ is an infinite cardinal, then $|\{i \in I(\gamma) : F_i \not\subseteq E_\gamma\}| \leq \tau$.*

PROOF. We can assume that τ is infinite. Suppose the lemma is false. Then without loss of generality we may assume that $I' = \{i \in I(\gamma) : E_\gamma <_R F_i\}$ has cardinality $|I'| > \tau$. Then there is a subset $J = \{i_\rho : \rho < \tau^+\} \subseteq I'$ of cardinality τ^+ such that either (i) $\bigcap_{\rho<\tau^+} F_{i_\rho} \cap E_\gamma \neq \emptyset$, or (ii)

$$F_{i_0} <_L F_{i_1} <_L \dots .$$

There is some $b \in S \setminus E_\gamma$ such that the set $J(b)$ has cardinality $\geq \tau$. Fix $J' \subseteq J(b)$ such that $|J'| = \tau$. If (i) holds, choose any $a \in \bigcap_{\rho<\tau^+} F_{i_\rho} \cap E_\gamma$, and if (ii) holds choose $a \in \bigcap_{i \in J'} F_i \cap E_\gamma$. Then J' witnesses $a \sim b$, and this is a contradiction. □

In the case when $\tau = 1$, it is easy to see that to prove the main lemma, it is enough to prove it for the case when there is just one equivalence class. Unfortunately, Lemma 5.1 does not allow us to make the same assumption in the case when τ is infinite. The trouble is that, if E is an equivalence class, and we only consider those intervals $F_i \subseteq E$, then we may have omitted τ intervals F_i which overlap E but are not subsets of E, and so there may very well be elements $x, y \in E$ which are not equivalent for the reduced family. However, it will be enough to prove the following analogue of the main lemma for the equivalence classes E_γ.

LEMMA 5.2. *For each $\gamma \in \Gamma$ there are pairwise disjoint subsets J_γ and $\tilde{J}_{\gamma,\nu}$ $(\nu < \kappa)$ of I_γ and disjoint subsets $A_\gamma, R_\gamma \subseteq E_\gamma$ such that*

(i) $E_\gamma = A_\gamma \cup R_\gamma$,

(ii) $\theta(J_\gamma \cup \bigcup_{\nu<\kappa} \tilde{J}_{\gamma,\nu}, A_\gamma) \leq \kappa$,

(iii) *J_γ is a τ-fold cover of A_γ,*

(iv) $R_\gamma \subseteq \bigcap_{\nu<\kappa} \mathcal{F}(\tilde{J}_{\gamma,\nu})$, *and moreover such that (v)* $\mathcal{F}(\bigcup_{\nu<\kappa} \tilde{J}_{\gamma,\nu}) \subseteq E_\gamma$.

PROOF. We first show that the main lemma follows from Lemma 5.2 with $A = \bigcup_\gamma A_\gamma$, $R = \bigcup_\gamma R_\gamma$, $J = \bigcup_\gamma J_\gamma$ and $\tilde{J}_\nu = \bigcup_\gamma \tilde{J}_{\gamma,\nu}$ $(\nu < \kappa)$. Note that condition (v) ensures that the sets J and $\tilde{J}_\nu$ $(\nu < \kappa)$ are pairwise disjoint as required. Also, conditions (i), (iii) and (iv) of the main lemma follow immediately from the corresponding statements of Lemma 5.2. It only remains to check that

$$\theta(J \cup \bigcup_{\nu<\kappa} \tilde{J}_\nu, A) \leq \kappa.$$

Let $x \in A$. Then $x \in A_\gamma$ for some $\gamma \in \Gamma$. Consider the the indices $i \in J \cup \bigcup_{\nu<\kappa} \tilde{J}_\nu$ such that $x \in F_i$. By (ii) of Lemma 5.2 we know that x belongs to fewer than κ of the sets F_i with

$$i \in J_\gamma \cup \bigcup_{\nu<\kappa} \tilde{J}_{\gamma,\nu}.$$

Also, by Lemma 5.1, there are at most τ intervals F_i containing x with

$$i \in (J \cup \bigcup_{\nu<\kappa} \tilde{J}_\nu) \setminus (J_\gamma \cup \bigcup_{\nu<\kappa} \tilde{J}_{\gamma,\nu}).$$

Therefore, all the conditions of the main lemma are satisfied. □

5.1. Proof of Lemma 5.2. Before giving a proof of the lemma, we recall two well-known facts, and we give the simple proofs. The first is about linearly ordered sets.

LEMMA 5.3. *Let $(U,<)$ be a linearly ordered set, $|U| = \mu = cf(\mu)$. Then either (a) U contains a well-ordered or reverse well-ordered subset of cardinality μ, or (b) there is $u \in U$ such that $|\{v \in U : v < u\}| = |\{v \in U : u < v\}| = \mu$.*

PROOF. Let $A = \{u \in U : |\{v \in U : v < u\}| < \mu\}$, $B = \{u \in U : |\{v \in U : u < v\}| < \mu\}$. If either $|A| = \mu$ or $|B| = \mu$ then (a) holds; otherwise (b) holds. □

The next fact is the following Bernstein-type lemma.

LEMMA 5.4. *Let Y, K be sets, $|K| = \kappa$, $|Y| \leq \kappa$, and for each $y \in Y$, let $K_y \subseteq K$, $|K_y| = \kappa$. Then there are sets $K_{y,\nu} \subseteq K_y$ which are pairwise disjoint for all $y \in Y$ and $\nu < \kappa$, and $|K_{y,\nu} \cap K_y| = \kappa$.*

PROOF. Let $\langle (y_\rho, \nu_\rho) : \rho < \kappa \rangle$ be a sequence in $Y \times \kappa$ such that, for every pair $(y,\nu) \in Y \times \kappa$, $|\{\rho : (y_\rho, \nu_\rho) = (y,\nu)\}| = \kappa$. Now inductively choose distinct elements $x_\rho \in K_{y_\rho}$ $(\rho < \kappa)$, and put $K_{y,\nu} = \{x_\rho : (y_\rho, \nu_\rho) = (y,\nu)\}$. □

From now on let $E = E_\gamma$ be a fixed equivalence class, and let $\mu = cf(E)$ be the cofinality of E. Also, let $J_{\gamma,R} = \{i \in I_\gamma : E <_R F_i\}$, and define $I'_{\gamma,R} = I_\gamma \setminus J_{\gamma,R}$; by Lemma 5.1, $|J_{\gamma,R}| \leq \tau$ (and $J_{\gamma,R} = \emptyset$ if $\tau = 1$) so that $I'_{\gamma,R}$ is a κ-fold cover of E.

The next lemma shows that, under certain conditions, Lemma 5.1 can be improved. Of course, there is a similar result for the dual ordering of E. Call a subset $A \subseteq E$ *satisfactory* if it has a τ-fold cover J such that $\theta(J,S) \leq \kappa$ and call it *very satisfactory* if either it is satisfactory and $\tau = 1$, or there is a τ-fold cover J such that $\theta(J,S) \leq \tau^+$. In particular, $[x,y]$ is very satisfactory for all $x, y \in E$. Since the union of two (very) satisfactory sets is also (very) satisfactory it follows that if $E(\geq x)$ is very satisfactory for some $x \in E$, then this is true for every $x \in E$.

LEMMA 5.5. *Either (i) $E(\geq x)$ is very satisfactory for some $x \in E$, or (ii) $|i \in I_\gamma : E <_R F_i\}| < \tau$.*

PROOF. If $\tau = 1$ then (ii) holds by Lemma 5.1. Therefore, we may assume that τ is infinite and μ, the cofinality of the linearly ordered set $(E, \leq)$, is also infinite. Since the union of fewer than τ^+ very satisfactory sets is also very satisfactory, we may assume that $\mu = cf(\mu) > \tau$. If (ii) is false there are τ intervals F_i such that $E <_R F_i$ and $F_i \cap E \neq \emptyset$, and since $cf(E) > \tau$, there is $x \in E$ which belongs to all these τ intervals F_i and (i) holds. □

LEMMA 5.6. *Either (i) there is $x \in E$ such that $E(\geq x)$ is very satisfactory, or (ii) there is $x \in E$ such that $E(\geq x) \subseteq \mathcal{F}(I'_{\gamma,R}(x))\}$.*

PROOF. Suppose (ii) is false. Then $\mu = cf(E) \geq \omega$ and for any $x \in E$, there is an $f(x) \in E$ such that

$$\mathcal{F}(I'_{\gamma,R}(x)) \cap E(\geq f(x)) = \emptyset.$$

We can assume that $f(x) \leq f(y)$ holds for $x \leq y$ in E. We will show that $E(\geq x_0)$ is very satisfactory for any $x_0 \in E$.

Let x_ξ $(\xi < \mu)$ be a strictly increasing cofinal sequence in E, and for $\eta < \mu$, define

$$g(\eta) = \min\{\xi : f(x_\eta) < x_\xi\}.$$

Case 1. $\mu = \omega$. Define $\eta_0 = 0$, $\eta_{n+1} = g(\eta_n)$, and put $y_n = x_{\eta_n}$ $(n < \omega)$. Since $y_n \sim y_{n+1}$, there is a τ-fold cover J_n of $Y_n = [y_n, y_{n+1}]$ such that either $\tau = 1$ and $\theta(J_n, S) \leq \kappa$, or τ is infinite and $\theta(J_n, S) \leq \tau^+$. We can assume that $F_i \cap Y_n \neq \emptyset$ for all $i \in J_n$. Then, from the choice of the y_n, we see that $\mathcal{F}(J_n) \cap \mathcal{F}(J_m) = \emptyset$ if $|m - n| > 2$. It follows from this that $J = \bigcup_{n<\omega} J_n$ is a τ-fold cover of $E(\geq x_0)$ and $\theta(J, S) \leq \kappa$ if $\tau = 1$, and $\theta(J, S) \leq \tau^+$ if τ is infinite. Hence $E(\geq x_0)$ is very satisfactory.

Case 2. $\mu > \omega$. In this case there is a club $C = \{\xi_\rho : \rho < \mu\} \subseteq \mu$ such that $0 = \xi_0 < \xi_1 < \xi_2 < \dots$, each ξ_ρ is a limit for $1 \leq \rho < \mu$, and for any $\xi \in C$, $\eta < \xi$ implies that $g(\eta) < \xi$. Put $D_\rho = \{x : x_{\xi_\rho} \leq x < x_{\xi_{\rho+1}}\}$ $(\rho < \mu)$. Since D_ρ is very satisfactory it has a τ-fold cover $J_\rho \subseteq I$ such that $\theta(J_\rho, S) \leq \kappa$ if $\tau = 1$ and $\theta(J_\rho, S) \leq \tau^+$ if τ is infinite. We can assume that $F_i \cap D_\rho \neq \emptyset$ for all $i \in J_\rho$. Note that, if $x \in D_\rho$, then because $\xi_{\rho+1}$ is a limit, there is $\eta < \xi_{\rho+1}$ such that $x < x_\eta$, and hence $f(x) \leq f(x_\eta) < x_{g(\eta)} < \xi_{\rho+1}$. It follows from this that $\mathcal{F}(J_\rho) \cap \mathcal{F}(J_\sigma) = \emptyset$ for $\rho < \sigma < \mu$. It follows that $J = \bigcup_{\rho<\mu} J_\rho$ is a τ-fold cover of $E(\geq x_0)$ and $E(\geq x_0)$ is very satisfactory. □

LEMMA 5.7. *Let $J \subseteq I_\gamma$, and suppose that $x \in E$ is such that $|J(x)| < \kappa$ and $E(\geq x) \subseteq \mathcal{F}(I'_{\gamma,R}(x))$. Then either (i) $E(\geq x)$ is satisfactory, or (ii) there are an end section R of E, and κ pairwise disjoint subsets $\tilde{J}_\nu \subseteq I'_{\gamma,R}(x) \setminus J$ $(\nu < \kappa)$ such that $\theta(\bigcup_{\nu<\kappa} \tilde{J}_\nu, E \setminus R) \leq \kappa$, and*

$$E(\geq x) \subseteq R \subseteq \mathcal{F}(\tilde{J}_\nu) \subseteq E \qquad (\nu < \kappa).$$

PROOF. Suppose (i) is false. Since the union of fewer than κ (very) satisfactory sets is satisfactory it follows that $cf(E) = \mu \geq \kappa$. Since each of the sets $F_i \cap E(\geq x)$ $(i \in I'_{\gamma,R}(x))$ is a proper initial segment of $E(\geq x)$ and $E(\geq x) \subseteq \mathcal{F}(I'_{\gamma,R}(x))$, there is $\tilde{J} = \{i_\alpha : \alpha < \mu\} \subseteq I'_{\gamma,R}(x)$, such that $F_{i_\alpha} <_R F_{i_\beta}$ for $\alpha < \beta < \mu$ and $E(\geq x) \subseteq \mathcal{F}(\tilde{J})$. Since $|J(x)| < \kappa$, and since $|J_{\gamma.L} = \{i \in I_\gamma : F_i <_L E\}| \leq \tau < \kappa$ by Lemma 5.1, we may also assume that

$\tilde{J}$ is disjoint from $J \cup I_{\gamma,L}$. Thus $\tilde{J} \cap J = \emptyset$ and $E(\geq x) \subseteq \mathcal{F}(\tilde{J}') \subseteq E$ holds whenever $\tilde{J}'$ is a subset of $\tilde{J}$ of cardinality μ.

Now consider the left ends of the sets F_i for $i \in \tilde{J}$. If there is $\tilde{J}' \subseteq \tilde{J}$ of cardinality μ such that all the intervals F_i $(i \in \tilde{J}')$ have a common left-end, then (ii) holds with $R = \mathcal{F}(\tilde{J}')$ and $\tilde{J}_\nu$ $(\nu < \kappa)$ any κ pairwise disjoint subsets of $\tilde{J}'$ each having cardinality μ. Therefore, since μ is regular, we may assume that the left ends of the F_i $(i \in \tilde{J})$ are all distinct.

By Lemma 5.3, since $\mu \geq \kappa$, either (a) there is $M = \{\rho(\sigma) : \sigma < \mu\} \subseteq \mu$ such that either (*) $F_{i_{\rho(0)}} <_L F_{i_{\rho(1)}} <_L \dots$, or (**) $F_{i_{\rho(0)}} >_L F_{i_{\rho(1)}} >_L \dots$, or (b) there are $K', \tilde{J}' \subseteq \tilde{J}$ such that $|K'| = \kappa$, $|\tilde{J}'| = \mu$, and $F_i <_L F_j$ holds for all $i \in K'$ and $j \in \tilde{J}'$. If (a)(*) holds, put $R = \bigcap_{\sigma<\kappa} F_{i_{\rho(\sigma)}} \cup E(\geq x)$, and let $\tilde{J}_\nu$ $(\nu < \kappa)$ be κ pairwise disjoint subsets of $\{i_{\rho(\sigma)} : \sigma < \mu\}$ chosen so that, for each $\nu < \kappa$, we have both $|\tilde{J}_\nu \cap \{i_{\rho(\sigma)} : \sigma < \kappa\}| = \kappa$ and $|\tilde{J}_\nu \cap \{i_{\rho(\sigma)} : \kappa \leq \sigma < \mu\}| = \mu$. Clearly R is an end section of E containing $E(\geq x)$, and for any $y \in E \setminus R$, $|\{\sigma : y \in F_{i_{\rho(\sigma)}}\}| < \kappa$. Therefore (ii) holds. If (a)(**) holds, then (ii) again follows if we put $R = \bigcup_{\sigma<\mu} F_{i_{\rho(\sigma)}}$ and let the $\tilde{J}_\nu$ $(\nu < \kappa)$ be arbitrary pairwise disjoint subsets of $\{i_\rho : \rho \in M\}$ each having cardinality μ. Finally, suppose that (b) holds. In this case we put

$$R = E(\geq x) \cup \{y \in E : |\{k \in K' : y \in F_k\}| = \kappa\}$$

Then R is a final segment of E and R has coinitiality $\rho \leq |K'| = \kappa$. If $y \in E \setminus R$, then $|\{i \in K' \cup \tilde{J}' : y \in F_i\}| < \kappa$. Let $\tilde{J}'_\nu$ $(\nu < \kappa)$ be κ pairwise disjoint subsets of $\tilde{J}'$ each having cardinality μ. Let y_σ $(\sigma < \rho)$ be a decreasing sequence coinitial in R. For each $\sigma < \rho$, let $K'_\sigma = \{i \in K' : y_\sigma \in F_i\}$. Then $|K'_\sigma| = \kappa$ for each $\sigma < \rho$. By Lemma 5.4, there are κ subsets $K'_{\sigma,\nu}$ of K'_σ which are pairwise disjoint for all $\sigma < \rho$ and $\nu < \kappa$. Then (ii) holds in this case with $\tilde{J}_\nu = \tilde{J}'_\nu \cup \bigcup_{\sigma<\rho} K'_{\sigma,\nu}$ $(\nu < \kappa)$. □

PROOF. We now conclude the proof of Lemma 5.2. There are four cases to consider depending upon whether the statements

$$\exists x \in S \ (S(\geq x) \text{ is satisfactory}), \tag{7}$$

and

$$\exists y \in S \ (S(\leq y) \text{ is satisfactory}), \tag{8}$$

are true or false.

If (7) and (8) are both true then there is a τ-fold cover, J, of E such that $\theta(J,S) \leq \kappa$, and the lemma holds with $A = E$, $R = \emptyset$ and $\tilde{J}_\nu = \emptyset$ $(\nu < \kappa)$.

Suppose that (8) is true and (7) is false. By Lemma 5.6 there is some $x \in S$ such that $E(\geq x) \subseteq \mathcal{F}(I'_{\gamma,R}(x))$ and since $x \sim y$ and (8) holds, there is a τ-fold cover, J, of $E(\leq x)$ such that $\theta(J,S) \leq \kappa$. By Lemma 5.7, there is an end section R of E and κ pairwise disjoint sets $\tilde{J}_\nu \subseteq I_\gamma \setminus J$ ($\nu < \kappa$) such that $\theta(\bigcup_{\nu<\kappa} \tilde{J}_\nu, E \setminus R) \leq \kappa$ and $E(\geq x) \subseteq R \subseteq \mathcal{F}(\tilde{J}_\nu) \subseteq E$ ($\nu < \kappa$). Then all the requirements of the lemma are satisfied if we put $A = E \setminus R$. Similarly if (7) is true and (8) is false.

It remains only to consider the case when (7) and (8) are both false. By Lemma 5.6 there are $x, y \in E$ such that $y \leq x$, $E(\geq x) \subseteq \mathcal{F}(I'_{\gamma,R}(x))$ and $E(\leq y) \subseteq \mathcal{F}(I'_{\gamma,L}(y))$ (where $I'_{\gamma,L} = I_\gamma \setminus \{i \in I_\gamma : F_i <_L E\}$). Let $J \subseteq I$ be a τ-fold cover of the closed interval $[y, x]$ such that $\theta(J, S) \leq \kappa$. It follows from Lemma 5.7 that there are an end section R_x and pairwise disjoint sets $\tilde{J}'_\nu \subseteq I \setminus J$ ($\nu < \kappa$) such that $\theta(\bigcup_{\nu<\kappa} \tilde{J}'_\nu, E\setminus R) \leq \kappa$, and

$$E(\geq x) \subseteq R_x \subseteq \mathcal{F}(\tilde{J}'_\nu) \subseteq E \qquad (\nu < \kappa).$$

If $R_x = E$ the lemma holds with $A = \emptyset$ and $R = R_x$ and $\tilde{J}_\nu = \tilde{J}'_\nu$. So we may assume that $y \notin R_x$. Let $I' = J \cup \bigcup_{\nu<\kappa} \tilde{J}'_\nu$. Since $\theta(I', E \setminus R_x) \leq \kappa$, it follows from the dual of Lemma 5.7, that there are an initial segment R_y of E and disjoint sets $\tilde{J}''_\nu \subseteq I \setminus I'$ ($\nu < \kappa$) such that $\theta(\bigcup_{\nu<\kappa} \tilde{J}''_\nu, E \setminus R_y) \leq \kappa$, and

$$E(\leq y) \subseteq R_y \subseteq \mathcal{F}(\tilde{J}''_\nu) \subseteq E \qquad (\nu < \kappa).$$

Then the lemma holds with $R = R_x \cup R_y$, $A = S \setminus R$ and $\tilde{J}_\nu = \tilde{J}'_\nu \cup \tilde{J}''_\nu$ ($\nu < \kappa$). □

6. Concluding remarks

It is natural to ask if there is any analogue of Theorem 1.1 for product spaces. The most obvious generalization is false. For example, let $S_1 = \{0, 1\}$, $S_2 = \{0, 1, 2\}$, $\mathcal{F} = \{I_i : 1 \leq i \leq 6\}$, where $I_1 = \{0\} \times S_2$, $I_2 = \{0\} \times \{1\}$, $I_3 = S_1 \times \{0\}$, $I_4 = S_1 \times \{2\}$, $I_5 = \{1\} \times \{0, 1\}$, $I_6 = \{1\} \times \{1, 2\}$. Then $\mathcal{F}$ is a 2-fold cover of $S_1 \times S_2$ by rectangles (products of intervals) which does not factorize into two disjoint covers. However, Pach claims that, for every finite k, there is an $r = r(k)$ such that any r-fold cover of the plane by rectangles contains k disjoint 1-fold covers. We do not know the proof of this, and so we have no idea if it generalizes to the product of arbitrary linearly ordered sets. However, this does suggest the following question. *Is there an ω-fold cover of the plane by rectangles which is not the union of ω disjoint covers?*

In connection with the above question, let us remark that there does exist a set S and an ω-cover $\mathcal{F}$ of subsets of S which does not even factorize into two disjoint covers. For example, let $S = \{f \in {}^\omega 2 : f(n) \neq 0$ for infinitely many $n\}$, and let $\mathcal{F} = \{A_n : n < \omega\}$, where $A_n = \{f : f(n) = 1\}$.

A family, $\mathcal{F}$, of subsets of a set S is *independent* if $\bigcap \mathcal{F}_1 \cap \bigcap \bar{\mathcal{F}}_2 \neq \emptyset$ whenever $\mathcal{F}_1$ and $\mathcal{F}_2$ are disjoint subsets of $\mathcal{F}$ and $\bar{\mathcal{F}}_2 = \{S \setminus X : X \in \mathcal{F}_2\}$. We ask the following question: *Suppose $\mathcal{F}$ is an ω-fold cover of S and there is some integer $k \geq 3$ such that $\mathcal{F}$ does not contain a subfamily of k independent sets, is then $\mathcal{F}$ the disjoint union of ω subfamilies $\mathcal{F}_i$ ($i < k$) such that $\bigcup \mathcal{F}_i = \bigcup \mathcal{F}$?* (Note that, if $\mathcal{F}$ does not contain 2 independent sets, then the members of $\mathcal{F}$ form a chain under inclusion, and so there is an ω partition of $\mathcal{F}$ in this case.)

References

1. B. Dushnik and E.W. Miller, *Partially ordered sets.* Amer. J. of Math. **63** (1941), 605.
2. János Pach, *Covering the plane with convex polygons.* Discrete Comput. Geom. **1** (1986), 73-81.

Department of Mathematics, Technion, Haifa, Israel.

Rutgers, The State University of New Jersey, P.O. Box 1179, Piscataway, NJ 08855-1179

Department of Mathematics and Statistics, University of Calgary, Calgary, Alberta, Canada.

Contemporary Mathematics
Volume **192**, 1996

Souslin Absoluteness, Uniformization and Regularity Properties of Projective Sets

Eyal Amir and Haim Judah

ABSTRACT. We show that Souslin Absoluteness and Projective Regularity holds iff Souslin Uniformization does. As a result, Souslin Absoluteness plus Σ^1_n Projective Regularity implies Δ^1_{n+1} Projective Regularity.

1. Introduction

We have several goals in our research in descriptive set theory. Two of them are:

1. To find a statement about the reals that explains completely the theory of the reals in Solovay models.
2. To find a combinatorial statement equivalent to "Projective measurability" (as well as the Baire Property).

In both of these goals, we introduce major advances. The following notion plays an important role (general references for this article might be [**Je**], [**Je2**], [**Ku**], [**Mo**], and [**Ox**]).

NOTATION 1.1. In this article we deal only with the boldface classes in the projective hierarchy, so the lightface symbols (Σ^1_n) represent the boldface ($\mathbf{\Sigma^1_n}$).

DEFINITION 1.2 (cf [**JS**] §0). We say that a forcing notion $\mathbb{P}$ **is Souslin** iff the set $\mathbb{P}$ and the relations $\leq_{\mathbb{P}}$ and $\perp_{\mathbb{P}}$ are all Σ^1_1 sets and $\mathbb{P}$ is ccc.

As examples of Souslin Forcing notions, one has

$\mathbb{A}$: The Amoeba forcing notion.
$\mathbb{B}$: The Random forcing notion.
$\mathbb{C}$: The Cohen forcing notion.
$\mathbb{D}$: The Dominating forcing notion.
$\mathbb{E}$: The Amoeba meager (Universal Meager) forcing notion.

In the following, μ is the Lebesgue measure (on the appropriate field).

1991 *Mathematics Subject Classification.* Primary 03C55; Secondary 04A15.

We wish to thank the "Emmy Nother Institute in the Mathematics Department of Bar-Ilan University" for its financial support.

1.1. Latest Results. In a joint work of Judah and Bagaria ([**BJ**]), it was proved that in the Solovay model ([**So**]) Souslin Absoluteness holds.

DEFINITION 1.3 (cf [**BJ**], [**Ju**] §2). Let V be a universe of set theory. Given a forcing notion $\mathbb{P} \in V$, we say that V is **$\Sigma^1_n(\mathbb{P})$-absolute** iff for every Σ^1_n-sentence φ with parameters in V we have

$$V \models \varphi \text{ iff } V^{\mathbb{P}} \models \varphi.$$

The corresponding definition for Π^1_n is similar.

We say that V is **Souslin Absolute**, if it is $\Sigma^1_n(\mathbb{P})$-absolute for every Souslin Forcing notion $\mathbb{P}$ for every $n \in \mathbb{N}$.

In the second direction of interest, we already have the following results:

THEOREM 1.4 ([**Ju2**]). 1. *Δ^1_2-measurability iff Σ^1_3 (Random)-Absolute*
2. *Δ^1_2-categoricity iff Σ^1_3 (Cohen)-Absolute*
3. *Σ^1_2-measurability iff Σ^1_3 (Amoeba)-Absolute*
4. *Σ^1_2-categoricity iff Σ^1_3 (Hechler)-Absolute*

THEOREM 1.5 ([**Ju2**]). 1. *$\Sigma^1_4(\mathbb{B})$-Absolute + $\Sigma^1_3(\mathbb{A})$-Absolute $\rightarrow \Delta^1_3(L)$.*
2. *$\Sigma^1_4(\mathbb{C})$-Absolute + $\Sigma^1_3(\mathbb{D})$-Absolute $\rightarrow \Delta^1_3(B)$.*

Shelah proved the following :

THEOREM 1.6 (cf [**Ju2**] p.8). $\Sigma^1_3(L) \Rightarrow (\forall r \in \mathbb{R})(\omega_1^{\mathbb{L}[r]} < \omega_1)$.

Recently J. Brendle, using the ideas of [**Ju2**], proved the following

THEOREM 1.7 (cf [**Ju2**] p.14). *Σ^1_4 (Amoeba)-Absolute $\rightarrow \Sigma^1_3$-measurability.*

COROLLARY 1.8 (cf [**Ju2**] p.14). *Σ^1_4 (Amoeba)-Absolute $\rightarrow \Sigma^1_3$-categoricity.*

Another recent result by L. Halbeisen and the second author in [**HJ**] is

THEOREM 1.9 ([**HJ**]). *Σ^1_6-Mathias-absoluteness implies Σ^1_4-Ramsey-regularity ($\Sigma^1_4(\mathfrak{R})$).*

Looking at these results, the question arises whether there is a general connection between Souslin Absoluteness and Regularity (Measurability, Categoricity etc.). This will be the main motivation of the rest of this article.

2. Souslin Ideals - Foundations

DEFINITION 2.1. We say that a σ-ideal $\mathcal{I}$ is a **Souslin Ideal** if $\mathcal{I}$ is a nontrivial (i.e. $\mathbb{R} \notin \mathcal{I}$) Borel ccc absolute σ-ideal.

Notice that this context is quite similar to the one used in [**Ku2**] with the difference that Kunen requires that his ideal (which he calls "reasonable ideal") will have a form of the Fubini property. We will abuse notations by referring to $\mathcal{I}$, which is defined on the **Borel** σ-algebra, as its expansion to the real line.

In [**JuRo**] H. Judah and A. Rosłanowski tried to ensure the existence of "nice" Ideals, for some Forcing Notions.

DEFINITION 2.2. A forcing notion $\mathbb{P}$ is **countably-1-generated** if there are conditions $p_n \in \mathbb{P}$ (for $n \in \omega$) such that

$$(\forall p \in \mathbb{P})(\forall q \in \mathbb{P}, q \perp p)(\exists n \in \omega)(p_n \perp p \ \&\ p_n \not\perp q).$$

In this situation the conditions p_n $(n \in \omega)$ are called σ-1-generators of the forcing notion $\mathbb{P}$.

They showed that for each forcing notion $\mathbb{P}$ which is countably-1-generated, there is a Souslin Ideal $\mathcal{I}_{\mathbb{P}}$ on ω^ω such that:

COROLLARY 2.3. *The algebra* $\mathbf{Borel}(\omega^\omega)/\mathcal{I}_{\mathbb{P}}$ *is a ccc complete Boolean algebra, and the mapping*

$$\mathbb{P} \longrightarrow \mathbf{Borel}(\omega^\omega)/\mathcal{I}_{\mathbb{P}} : p \mapsto [\phi(p)]_{\mathcal{I}_{\mathbb{P}}}$$

is a dense embedding (so $RO(\mathbb{P}) \cong \mathbf{Borel}(\omega^\omega)/\mathcal{I}_{\mathbb{P}}$*). For each Borel code c:* $[\![\dot{r} \in A_c]\!]_{\mathbb{P}} = [A_c]_{\mathcal{I}_{\mathbb{P}}}$ □

For the rest of our work we will use the following context as our main point of view: Let $\mathbb{P} = \mathbf{Borel}(\omega^\omega)/\mathcal{I}$ be a Souslin forcing notion ($\mathcal{I}$ is a Souslin-ideal on ω^ω). All of the previous examples of Souslin Forcing notions are also examples of this situation. For Random and Cohen, the ideals are, correspondingly, the Null and the Meager sets.

DEFINITION 2.4. We say that a set A is **$\mathcal{I}$-regular** if there is a **Borel** set B such that $B\triangle A \in \mathcal{I}$. Let $n \geq 1$. $\Sigma^1_n(\mathcal{I})$ is the following statement:
Every Σ^1_n subset of the real line A is $\mathcal{I}$-regular.

Our general $\mathbb{P}$ satisfies many nice properties. We give two examples.

LEMMA 2.5. *Let* $\mathcal{M}$ *be a transitive model of ZFC. If* G *is an* $\mathcal{M}$*-generic filter on* $\mathbb{P}$*, then there is a unique real number* x_G *such that for all* $B \in \mathbb{P}$

$$x_G \in B^* \Leftrightarrow [B]_{\mathbb{P}} \in G \tag{1}$$

The formula (1) determines G *and hence* $\mathcal{M}[G] = \mathcal{M}[x_G]$.

PROOF. To start, we claim that there is at most one real number x that satisfies

$$\forall B \in \mathbf{Borel}(x \in B^* \Leftrightarrow [B] \in G). \tag{2}$$

If x satisfies (2), then x belongs to all B^* such that $[B] \in G$. If $x < y$ are two real numbers, let r be a rational number such that $x < r < y$, and let A be the interval $(r, \infty) \subseteq \mathbb{R}$. Either $[A]$ or $[\mathbb{R} \setminus A]$ belong to G, but $x \notin A^*$ and $y \notin (\mathbb{R} \setminus A)^*$.

In order to show that there exists a real number x that satisfies (2), let

$$x = sup\{r : r \text{ is a rational number and } [(r, \infty)] \in G\}. \tag{3}$$

By the genericity of G, there exists r such that $[(r, \infty)] \notin G$, and hence the supremum (3) exists. Note also that $x \notin \mathcal{M}$ (by the genericity of G). We shall show that x satisfies (2). We shall show, by induction on **Borel** codes in $\mathcal{M}$, that for every $c \in BC^{\mathcal{M}}$,

$$x \in A^*_c \iff [A_c] \in G. \tag{4}$$

First we consider the Σ^0_1-codes (in $\mathcal{M}$), and let us start with those $c \in \Sigma^0_1 \cap \mathcal{M}$ that code a rational interval, i.e., such that $c(n) = 1$ for exactly one n; then c codes the interval I_n. Let $I_n = (p, q)$. We have

$$\begin{aligned} x \in A^*_c &\iff p < x < q \\ &\iff p < sup\{r : [(r, \infty)] \in G\} < q \\ &\iff [(p, \infty)] \in G \wedge [(q, \infty)] \notin G \\ &\iff [(p, q)] \in G \iff [A_c] \in G. \end{aligned}$$

Now, if $c \in \Sigma_1^0$, then $A_c = \bigcup_{n=0}^{\infty} I_{k_n}$, where $\{k_n : n = 0, 1, \ldots\}$ is the set $\{k : c(k) = 1\}$, and we have

$$\begin{aligned} x \in A_c^* &\iff x \in \textstyle\bigcup_{n=0}^{\infty} I_{k_n}^* \\ &\iff \exists n(x \in I_{k_n}^*) \\ &\iff \exists n([I_{k_n}] \in G) \\ &\iff \textstyle\sum_{n=0}^{\infty} [I_{k_n}] \in G \\ &\iff [\textstyle\bigcup_{n=0}^{\infty} I_{k_n}] \in G \iff [A_c] \in G. \end{aligned}$$

Next let $\alpha < \omega_1^{\mathcal{M}}$, and let $c \in \Pi_\alpha^0 \cap \mathcal{M}$, and let us assume that (4) holds for all $c \in \Sigma_\alpha^0 \cap \mathcal{M}$. We may assume that $c(0) = 0$; then $u(c) \in \Sigma_\alpha^0 \cap \mathcal{M}$ and $A_{u(c)} = \mathbb{R} \setminus A_c$, and we have

$$x \in A_c^* \iff x \notin A_{u(c)}^* \iff [A_{u(c)}] \notin G \iff [A_c] \in G.$$

Finally, the induction step for Σ_α^0 is handled in a way similar to the case for $c \in \Sigma_1^0$. Thus (4) holds for every $c \in BC^{\mathcal{M}}$, and thus x is the unique real number that satisfies (1). □

One should notice that in fact we did not use the absoluteness of the ideal, but just the structure of the partial order (being of the form **Borel**/$\mathcal{I}$). The following lemma provides a characterization of $\mathbb{P}$-reals.

LEMMA 2.6. *A real number is a $\mathbb{P}$-real over $\mathcal{M}$ if and only if it does not belong to any Borel set $I \in \mathcal{I}$ with a code in $\mathcal{M}$.*

PROOF. On the one hand, if x is a $\mathbb{P}$-real over $\mathcal{M}$, let G be an $\mathcal{M}$-generic filter on $\mathbb{P}$ such that $x = x_G$. Then if $A_c \in \mathcal{I}$, then $[A_c] \notin G$, and by 2.5, $x \notin A_c^*$.

On the other hand, let x be such that $x \notin A_c^*$ whenever $A_c \in \mathcal{I}$ (and $c \in \mathcal{M}$). First we observe that if $[A_c] = [A_d]$ then $A_c \triangle A_d \in \mathcal{I}$, hence $A_c^* \triangle A_d^* \in \mathcal{I}^*$ (by the absoluteness of $\mathcal{I}$). It follows that x belongs to A_c^* if and only if x belongs to A_d^*. Let

$$G = \{[A_c] : x \in A_c^*\}. \tag{5}$$

It is easy to see that G is a filter on $\mathbb{P}$: If $[A_c] \in G$ and $[A_d] \in G$, then $x \in A_c^* \cap A_d^*$ and hence $[A_c \cap A_d] \in G$; similarly, if $[A_c] \geq [A_d]$ and $[A_c] \in G$, then $[A_d] \in G$ (recall that we use "$p \geq q$" to denote "p is stronger than q"). We shall show that G is $\mathcal{M}$-generic. Since $\mathbb{P}$ satisfies the ccc, it suffices to show that if $\{A_{c_n} : n \in \omega\} \in \mathcal{M}$ is such that $\sum_{n=0}^{\infty} [A_{c_n}] \in G$, then some $[A_{c_n}]$ is in G. But this is true because

$$\sum_{n=0}^{\infty} [A_{c_n}] = [\bigcup_{n=0}^{\infty} A_{c_n}] \text{ and } (\bigcup_{n=0}^{\infty} A_{c_n})^* = \bigcup_{n=0}^{\infty} A_{c_n}^*.$$

Finally, we claim that $x = x_G$. But this follows from (5), by the genericity of G. Thus a real number x is a $\mathbb{P}$-real over $\mathcal{M}$ if and only if $x \notin A_c^*$ for any **Borel** set $A_c \in \mathcal{I}^{\mathcal{M}}$. □

DEFINITION 2.7. We say that σ is a **$\mathbb{P}$-name for a $\mathbb{P}$-real** over a model V, if G is a $\mathbb{P}$-generic filter over V and a is the intersection of G, and σ is the $\mathbb{P}$-name of a. a is called a **$\mathbb{P}$-real**. We will denote the set of all $\mathbb{P}$-reals over $\mathcal{M}$ by $Pr(\mathcal{M})$.

COROLLARY 2.8.

$$Pr(\mathcal{M}) = \mathbb{R}^* \setminus \bigcup \{A_c^* : c \in BC^{\mathcal{M}} \wedge A_c \in \mathcal{I}\}.$$

PROOF. Notice that by the last lemma we get

$$Pr(\mathcal{M}) = \mathbb{R}^* \setminus \bigcup \{A_c^* : c \in BC^{\mathcal{M}} \wedge A_c^* \in \mathcal{I}\}.$$

Thus by the absoluteness of $\mathcal{I}$ we get

$$Pr(\mathcal{M}) = \mathbb{R}^* \setminus \bigcup \{A_c^* : c \in BC^{\mathcal{M}} \wedge A_c \in \mathcal{I}\}.$$

□

In the rest of our work we will abuse notations and use the notations of subsets of the plane, also for their class in $\mathbb{P}$ (modulo the ideal $\mathcal{I}$). For example B will also denote $[B]_{\mathbb{P}}$.

3. Souslin Uniformization

3.1. General Facts.

LEMMA 3.1. *Let σ be a $\mathbb{P}$-name for a real number. Then there is a* **Borel** *function f such that for a $\mathbb{P}$-real a over V,*

$$V[a] \models \sigma[a] = f(a)$$

PROOF. We define f by approximating it using simple functions. We work in $[0,1]$. Let $A_{i,n} = [\![\sigma \in (\frac{i}{2^n}, \frac{i+1}{2^n})]\!], i < 2^n$. Let

$$f_n(x) = \sum_{i<2^n} \frac{i}{2^n} \chi_{A_{i,n}}(x)$$

where $\chi_{A_{i,n}}$ is the characteristic function on $A_{i,n}$. So, each f_n is a simple **Borel** function. Let

$$f(x) = \lim_{n\to\infty} f_n(x)$$

Since $f(x) = y \Leftrightarrow \forall n \exists m \forall k \geq m(|\, y - f_k(x) \,| < \frac{1}{n})$, f is **Borel**. Now, let a be a $\mathbb{P}$-real over V. Pick $\varepsilon > 0$. For every n, there is a unique $i < 2^n$ such that $a \in A_{i,n}$. But if $a \in A_{i,n}$, $\sigma[a] \in (\frac{i}{2^n}, \frac{i+1}{2^n})$. Also $f_n(a) = \frac{i}{2^n}$. Hence, $|\, \sigma[a] - f_n(a) \,| < \frac{1}{2^n}$. Thus, we can find n such that $|\, \sigma[a] - f_n(a) \,| < \varepsilon$. □

LEMMA 3.2. *Let $n \geq 2$. Assume $\varphi(x)$ is a Π^1_n-formula and f is a* **Borel** *function ($Graph(f)$ is* **Borel***). Then $\varphi(f(x))$ is also a Π^1_n-formula in the additional parameter, the Borel code of f.*

PROOF. Saying that for x, $V \models \varphi(f(x))$ holds, is equivalent to saying

$$V \models (\forall x \exists y(\langle x, y\rangle \in Graph(f))) \wedge (\forall y(\langle x, y\rangle \in Graph(f) \Rightarrow \varphi(y))).$$

□

3.2. Souslin Uniformization. We will show that there is a strong relationship between Uniformization and Souslin absoluteness.

DEFINITION 3.3. Let $n \geq 1$. $\Pi^1_n(\mathbb{P})$-uniformization (or as we will mention it **Souslin Uniformization**) is the following statement:
For every A a Π^1_n subset of the plane, if $\{x : A_x = \emptyset\} \in \mathcal{I}$, then there is a **Borel** function $f : \mathbb{R} \to \mathbb{R}$ such that $\{x : f(x) \in A_x\}^c \in \mathcal{I}$.
$\Sigma^1_n(\mathbb{P})$-uniformization is defined similarly. We will sometimes use $\Sigma^1_n(\mathcal{I})$ instead of $\Sigma^1_n(\mathbb{P})$ since this definition depends only on the ideal $\mathcal{I}$.

COROLLARY 3.4. *$\Sigma^1_n(\mathbb{P})$-uniformization iff $\Pi^1_{n-1}(\mathbb{P})$-uniformization.*

PROOF. The forward direction is obvious. The backward direction is as follows: Take A a Σ^1_n subset of the plane, and assume that $\{x : A_x = \emptyset\} \in \mathcal{I}$. $A = \{(x, y) : \varphi(x, y)\}$ where φ is a Σ^1_n-formula. So $\varphi(x, y) = \exists z \psi(x, y, z)$, where ψ is a Π^1_{n-1}-formula. The idea is to use the function guaranteed from the $\Pi^1_{n-1}(\mathbb{P})$-uniformization to replace the "$\exists$" quantifier in φ. Take

$$\phi : \mathbb{R}^2 \overset{1:1 \text{ onto}}{\longrightarrow} \mathbb{R}$$

to be a **Borel** function mapping $\mathbb{R}^2$ to $\mathbb{R}$. By our assumption on A, $E = \{x : A_x = \emptyset\} \in \mathcal{I}$. So,

$$(\forall x \notin E)(\exists y, z\ \psi(x, y, z)).$$

Therefore, we have the needed assumption for the set $B = \{\langle x, \phi(y, z)\rangle : \psi(x, y, z)\}$. But this is a Π^1_{n-1}-set, so by $\Pi^1_{n-1}(\mathbb{P})$-uniformization there is a function f such that

$$I = \{x : f(x) \notin B_x\} \in \mathcal{I} \tag{6}$$

Take $g(x, y) = x$. Then $F(x) = g(\phi^{-1}(f(x)))$ is the needed function $((\forall x \notin I)(F(x) \in A_x))$. □

The following lemma is due to H. Woodin ([**Wo**] 1,2):

LEMMA 3.5 (Woodin). *Let L, B denote the Null and Meager Ideals. Then*
1. *$\Pi^1_n(L)$-uniformization implies Σ^1_{n+2}-absoluteness for Random.*
2. *$\Pi^1_n(B)$-uniformization implies Σ^1_{n+2}-absoluteness for Cohen.*

We will now rephrase and prove that lemma for our more general case:

LEMMA 3.6. *Let $n \geq 1$. $\Pi^1_n(\mathbb{P})$-uniformization implies Σ^1_{n+2}-absoluteness for $\mathbb{P}$.*

PROOF. Let us prove the case $n = 1$. The general case follows by induction on the complexity of the formula.

Let $\exists x \forall y \varphi(x, y, z)$ be a Σ^1_3-formula with parameters in V, where φ is Σ^1_1. Suppose that v is a $\mathbb{P}$-real over V and $V[v] \models \exists x \forall y \varphi(x, y, a)$, for some $a \in \mathbb{R} \cap V$.

Let b be a witness so that $V[v] \models \forall y \varphi(b, y, a)$. Choose in V a term τ for b. τ may be chosen as a **Borel** function g such that

$$V[v] \models \forall y \varphi(g(v), y, a) \tag{7}$$

(see 3.1).

Suppose $V \models \forall x \exists y \neg\varphi(x, y, a)$. Then, $V \models \forall x \exists y \neg\varphi(g(x), y, a)$. Let $A = \{(x, y) : \neg\varphi(g(x), y, a)\}$. By $\Pi^1_1(\mathbb{P})$uniformization, there is a **Borel** function f such that $\{x : (g(x), f(x)) \in A\}^c \in \mathcal{I}$. Choose a **Borel** set $B \subseteq \{x : (g(x), f(x)) \in A\}$ such that $B^c \in \mathcal{I}$. Hence,

$$V \models \forall x (x \in B \Rightarrow \neg\varphi(g(x), f(x), a)).$$

Since $\neg\varphi$ is Π^1_1, $\forall x (x \in B \Rightarrow \neg\varphi(g(x), f(x), a))$ is Π^1_2 with the **Borel** codes for B, f, g as additional parameters (see 3.2). So,

$$V[v] \models \forall x (x \in B \Rightarrow \neg\varphi(g(x), f(x), a)).$$

But since v is a $\mathbb{P}$-real over V, and since the complement of B is a **Borel** set in the ideal $\mathcal{I}$ in V, $v \in B$ (see 2.6). Therefore, $V[v] \models \neg\varphi(g(v), f(v), a)$, which contradicts (7) above.

The other direction (in proving absoluteness) is simply by Shoenfields theorem (cf [**Je**] Ch. 41) which gives us Σ^1_2-absoluteness, and in the induction step - by the induction hypothesis. □

LEMMA 3.7. *Fix $n > 0$.*
Assume Π^1_{n+1}-absoluteness for $\mathbb{P}$. Take $\varphi \in (\Sigma^1_n \cup \Pi^1_n)$, and let τ be the canonical $\mathbb{P}$-name for a $\mathbb{P}$-real. Assign $p = \{x : \varphi(x)\}$. Then

1. *if p is $\mathcal{I}$-regular, then*
$$[\![\varphi(\tau)]\!] = p/\mathcal{I}.$$
2. *if p contains an $\mathcal{I}$-regular subset q, then*
$$[\![\varphi(\tau)]\!] \geq q/\mathcal{I}.$$

PROOF. Let us prove *2* first, and then *1* will follow easily. Take p and q as mentioned above. Take a **Borel** set $F \subseteq q$ such that $F/\mathcal{I} = q/\mathcal{I}$ (By the assumption on q, where $\mathcal{I}$ is the σ-ideal mentioned above). Take $r = F/\mathcal{I}$. We claim that $r \Vdash \varphi(\tau)$. Take $a \in F^*$ $\mathbb{P}$-real over V (if there is none, then $q \in \mathcal{I}$ and we are done). Then
$$V[a] \models \varphi(a),$$
since by $\Pi^1_{n+1}[\Pi^1_n]$-absoluteness for $\mathbb{P}$,
$$V \models \forall x \in F(\varphi(x)) \Rightarrow V^{\mathbb{P}} \models \forall x \in F(\varphi(x)).$$
But $V[a] \models \tau[a] = a$ (τ is the canonical $\mathbb{P}$-name for a $\mathbb{P}$-real). So $V[a] \models \varphi(\tau)$ (and this is for each $a \in F^*$ $\mathbb{P}$-real over V), thus $r \Vdash \varphi(\tau)$ and
$$[\![\varphi(\tau)]\!] \geq q/\mathcal{I}$$
Now, for the first part of the lemma, notice that p satisfies the assumptions given in the second part, of q. Thus it is obvious that
$$[\![\varphi(\tau)]\!] \geq p/\mathcal{I}$$
For equality, just observe that for '$\leq$' we have
$$[\![\neg\varphi(\tau)]\!] \geq \{x : \neg\varphi(x)\}/\mathcal{I}$$
(by the assumption of Π^1_{n+1}-absoluteness), which implies
$$[\![\varphi(\tau)]\!] \leq p/\mathcal{I}$$
Thus
$$[\![\varphi(\tau)]\!] = p/\mathcal{I}$$
□

FACT 3.8. Let $A \subseteq \mathbb{R}$. $(A \cap [\![\varphi(\tau)]\!]) \notin \mathcal{I} \Rightarrow [\![\varphi(\tau)]\!] \notin \mathcal{I}$.

PROOF. $[\![\varphi(\tau)]\!] \in \mathcal{I} \Rightarrow A \cap [\![\varphi(\tau)]\!] \in \mathcal{I}$ ($\mathcal{I}$ is an ideal). □

The following corollary is an application of the previous lemma.

COROLLARY 3.9. *Fix $n > 0$.*

1. *Assume $\Sigma^1_n(L)$. Assume also that Π^1_{n+1}-absoluteness for Random holds. Take $\varphi \in (\Sigma^1_n \cup \Pi^1_n)$ and let τ be the canonical random name for a random real. Then*
$$\mu([\![\varphi(\tau)]\!]) = \mu(\{x : \varphi(x)\})$$

2. *Assume $\Sigma^1_n(B)$. Assume further that Π^1_{n+1}-absoluteness for Cohen holds. Take $\varphi \in (\Sigma^1_n \cup \Pi^1_n)$ and let τ be the canonical Cohen name for a Cohen real. Then*

$$\{x : \varphi(x)\} \text{ is not meager} \iff [\![\varphi(\tau)]\!] \text{ is not meager}$$

□

We will now use the notion defined in 2.4 to prove a converse of 3.6.

LEMMA 3.10. *$\Sigma^1_n(\mathcal{I})$ + $\Sigma^1_{n+2}(\mathbb{P})$-absoluteness implies $\Pi^1_n(\mathbb{P})$-uniformization.*

PROOF. Let $A = \{(x,y) : \varphi(x,y)\}$ be a Π^1_n subset of the plane. Suppose that $\{x : A_x = \emptyset\} \in \mathcal{I}$. Let $C \in \mathcal{I}$ be a **Borel** set such that $\{x : A_x = \emptyset\} \subseteq C$. Let $B = \{(x,y) : x \in C\}$. Thus B is a **Borel** set in $\mathcal{I} \times \mathcal{P}(\mathbb{R})$. Let $\psi(x,y)$ be an arithmetical formula that defines B. Then

$$V \models \forall x(\exists y\varphi(x,y) \vee \exists y\psi(x,y))$$

By Σ^1_{n+2}-absoluteness for $\mathbb{P}$,

$$V^{\mathbb{P}} \models \forall x(\exists y\varphi(x,y) \vee \exists y\psi(x,y))$$

Let τ be the canonical name for a $\mathbb{P}$-real in V.

$$V^{\mathbb{P}} \models \exists y\varphi(\tau,y) \vee \exists y\psi(\tau,y)$$

Moreover, if a is a $\mathbb{P}$-real over V, then $V[a] \models \tau[a] = a$. But since $\{x : B_x \neq \emptyset\}$ is a **Borel** set contained in $\mathcal{I}$ in V, $a \notin \{x : B_x \neq \emptyset\}^*$. Hence

$$V^{\mathbb{P}} \models \exists y\varphi(\tau,y)$$

Let σ be a $\mathbb{P}$-name for a real such that

$$V^{\mathbb{P}} \models \varphi(\tau,\sigma)$$

Then we can find a Borel function f such that for each $\mathbb{P}$-real a, $V[a] \models \sigma[a] = f(a)$. So

$$V^{\mathbb{P}} \models \varphi(\tau, f(\tau)) \tag{8}$$

Now assume $\{x : \neg\varphi(x, f(x))\} \notin \mathcal{I}$. Take $p = [\![\neg\varphi(\tau, f(\tau))]\!]$. Take $a \in p^*$ a $\mathbb{P}$-real over V (one may do so, since by corollary 3.7, $p \notin \mathcal{I}$). Then, $V[a] \models \neg\varphi(a, f(a))$ (p forces that), but that contradicts (8) above.

Therefore $\{x : \neg\varphi(x, f(x))\} \in \mathcal{I}$. □

LEMMA 3.11 ([**JuRo**]). *Every analytic set is $\mathcal{I}$-regular.*

In order for our theorem to be complete, it remains to show that Uniformization implies Regularity.

LEMMA 3.12. *$\Pi^1_n(\mathbb{P})$-uniformization implies $\Sigma^1_n(\mathcal{I})$.*

PROOF. Take a set $C \in \Sigma^1_n$. Take $A = (C^c \times \{0\}) \cup (C \times \{1\})$. By uniformization, we have a Borel function f that almost everywhere uniformizes A. So the preimage of 1 by f is **Borel** and is almost everywhere like C. □

THEOREM 3.13. *$\Sigma^1_{n+2}(\mathbb{P})$-absoluteness + $\Sigma^1_n(\mathcal{I})$ $\iff$ $\Pi^1_n(\mathbb{P})$-uniformization.*

PROOF. Obvious from lemmas 3.10, 3.6 and 3.12. □

COROLLARY 3.14. *Using the previous theorem:*

1. $\Sigma^1_{n+2}(\mathbb{B})$*-absoluteness* + $\Sigma^1_n(L)$ *iff* $\Pi^1_n(L)$*-uniformization.*
2. $\Sigma^1_{n+2}(\mathbb{C})$*-absoluteness* + $\Sigma^1_n(B)$ *iff* $\Pi^1_n(B)$*-uniformization.*

□

Using these results we can now establish a new link between Souslin Absoluteness and Regularity.

LEMMA 3.15. $\Sigma^1_n(\mathbb{P}$*-uniformization) implies* $\Delta^1_n(\mathcal{I})$.

PROOF. Take a set $C \in \Delta^1_n$. Take $A = (C^c \times \{0\}) \cup (C \times \{1\})$. A is a Σ^1_n-set (even Δ^1_n-set). By uniformization, we have a Borel function f that almost everywhere uniformizes A. So the preimage of 1 by f is **Borel** and is almost everywhere like C. □

COROLLARY 3.16. $\Sigma^1_{n+2}(\mathbb{P})$*-absoluteness* + $\Sigma^1_n(\mathcal{I}) \Rightarrow \Delta^1_{n+1}(\mathcal{I})$.

PROOF. By Theorem 3.13 we get $\Pi^1_n(\mathbb{P})$-uniformization. Using 3.15 and 3.4 we get $\Delta^1_{n+1}(\mathcal{I})$. □

We tried to show the full Induction Step (i.e $\Sigma^1_n \Rightarrow \Sigma^1_{n+1}$) using Partial Functions Uniformization. But J. Brendle showed that there are cases where Uniformization for sets which are not "Full" in the meaning of the Souslin Ideal $\mathcal{I}$ (i.e the guarantee to find a function for each set A in the proper Step in the Projective Hierarchy) implies Σ^1_{n+1} Regularity, while the Uniformization itself is (in these cases) equivalent to Δ^1_n Regularity (take for instance the case of $n = 1$ and $\mathbb{P} = \mathbb{C}$). One might have suggested the use of Partial Functions, but we proved that the existence of a partial function (under an assumption of our basic Uniformization scheme) implies the existence of a Complete Function for that same set. Thus this way of inquiry will not bear more fruits.

COROLLARY 3.17. *Souslin-Absoluteness implies* $\Delta^1_4(\mathbb{B})$ *and* $\Delta^1_4(\mathbb{C})$.

PROOF. Using 1.7 and 1.8 and then applying 3.16 leads to the corollary. □

PROBLEM 3.18. It is still open whether Souslin Absoluteness implies $\Sigma^1_n(\mathcal{I})$.

4. Acknowledgements

We would like to thank all of the people who read and verified the preprint version of this article including Saharon Shelah, Andrzej Roslanowski, Miroslav Repicky, Otmar Spinas, Tzvi Scarr, the referee and Tomek Bartoszyński. We would especially want to thank Joerg Brendle for his important corrections and remarks.

References

[BJ] J. Bagaria and H. Judah, *Amoeba forcing, Souslin forcing and additivity of measure*, Set Theory of the continuum (H. Judah, W. Just, H .Woodin, Eds.), Springer, 1992, 155-173.

[HJ] L. Halbeisen, H. Judah, *Mathias Absoluteness and the Ramsey Property* submitted to the Journal of Symbolic Logic.

[Je] T. Jech, *Set Theory* Academic Press, New York, 1978.

[Je2] ———, *Multiple Forcing* Cambridge Tracts in Mathematics **88**, New York, 1985.

[Ju] H. Judah, *Souslin Absoluteness*, submitted to the proceedings of the conference in Honour of Pf. Azriel Levy IMCP.

[Ju2] ———, *Absoluteness for projective sets*, Logic Colloquium 1990 (J. Oikkonen, J. Väänänen, Eds.), Lecture Notes in Logic **2**, Springer 1993, 145-154.

[JuBa] H. Judah, T. Bartoszyński, *Measure and Category in Set Theory — the Asymmetry*, Preprint.

[JuRo] H. Judah & A. Roslanowski, *Ideals determined by some Souslin forcing notions*, preprint.

[JS] H. Judah, S. Shelah, *Souslin Forcing*, The Journal of Symbolic Logic **53** (1988), 1182–1207.

[Ku] K. Kunen, *Set Theory*, studies in Logic and the Foundations of Mathematics **102**, North-Holland (1980).

[Ku2] ———, *Random and Cohen reals*, in: *Handbook of Set-Theoretic Topology* (K .Kunen, J.E. Vaughan eds), North-Holland 1984.

[Mo] Y.N.Moschovakis, *Descriptive Set Theory*, Studies in logic and the foundations of mathematics, **100**, North-Holland 1980.

[Ox] J.C.Oxtoby *Measure and Category*, Graduate texts in Mathematics, Springer-Verlag (1971).

[So] R.Solovay, *A model of set-theory in which every set of reals is Lebesgue measurable*, Ann.Math. **94** (1971), 201-245.

[Wo] W.H. Woodin *On the Consistency Strength of Projective Uniformization*, Proceedings of the Herbrand Symposium. Logic Colloquium '81. J.Stern (editor). 365-384. North-Holland Publishing Company, 1982.

Dept. of Mathematics and Computer Science, Bar-Ilan University, 52900 Ramat-Gan, Israel,

Current address: Computer Science Department, Stanford University, Stanford, California 94305, U.S.A.

E-mail address: amire@bimacs.cs.biu.ac.il

Dept. of Mathematics and Computer Science, Bar-Ilan University, 52900 Ramat-Gan, Israel

Contemporary Mathematics
Volume **192**, 1996

NOT EVERY γ-SET IS STRONGLY MEAGER

TOMEK BARTOSZYŃSKI AND IRENEUSZ RECŁAW

ABSTRACT. We present two constructions of γ-sets which are large in sense of category.

1. INTRODUCTION

Most of the notation used in this paper is standard. By reals we mean the space 2^ω with the operation $+\mod 2$. By rationals we mean the canonical dense subset of 2^ω,

$$\mathbb{Q} = \{x \in 2^\omega : \forall^\infty n \; x(n) = 0\}.$$

Finally, let $\mathcal{N}$ denote the ideal of measure zero subsets of 2^ω, with respect to the standard product measure in this space. **0** and **1** denote constant functions equal to 0 and 1 respectively.

Let us recall the following definitions:

Definition 1.1. (1) *A family $\mathcal{J} \subseteq P(X)$ is an ω-cover of X if for every finite set $F \subseteq X$ there exists $B \in \mathcal{J}$ such that $F \subseteq B$,*

(2) *A topological space X is a γ-set if for every $\mathcal{J}$, open ω-cover of X, there exists a family $\{D_n : n \in \omega\} \subseteq \mathcal{J}$ such that $X \subseteq \bigcup_m \bigcap_{n>m} D_n$,*

(3) *A set $X \subseteq \mathbb{R}$ is strongly meager if for every null set $G \subset \mathbb{R}$, $X + G \neq \mathbb{R}$.*

Galvin and Miller in [1] constructed a γ-set of reals of size continuum under Martin's Axiom. They showed that for every γ-set $X \subseteq \mathbb{R}$ and every meager set F, $X + F$ is meager. They asked whether the same is true for null sets. We give negative answer to this question.

Note that γ-sets are always meager, that is, if X is a γ-set then $X \cap P$ is meager in P for every perfect set $P \subseteq 2^\omega$ ([2]).

Since every γ-set is a strong measure zero set (see [2]), by a result of Laver it is consistent that every γ-set is countable so a sum with a null set is null.

On the other hand, for an ideal $\mathcal{I}$, Pawlikowski defined the cardinal coefficients $\mathsf{add}_t(\mathcal{I}) = \min\{|X| : \forall F \in \mathcal{I} \; X + F \notin \mathcal{I}\}$. He showed in [4], that $\mathsf{add}(\mathcal{N}) = \min\{\mathsf{add}_t(\mathcal{N}), \mathfrak{b}\}$. It is consistent that $\mathsf{add}(\mathcal{N}) < \mathfrak{p}$. Then $\mathsf{add}_t(\mathcal{N}) = \mathsf{add}(\mathcal{N}) < \mathfrak{p}$. So every subset of size $\mathsf{add}_t(\mathcal{N})$ is a γ-set, since all sets of size $< \mathfrak{p}$ are γ-sets ([3]), but there is a set X of size $\mathsf{add}_t(\mathcal{N})$ and a null set G with $X + G \notin \mathcal{N}$.

1991 *Mathematics Subject Classification.* 04A15 03E50.

First author partially supported by SBOE grant #95–041 and the second author supported by the research grant BW UG 5100-5-0148-4.

2. Main results

In this section we present two constructions of γ-sets.

Theorem 2.1. *Assume $\mathfrak{p} = 2^{\aleph_0}$. Then there is a γ-set $X \subseteq 2^\omega$ which is not strongly meager.*

Proof For $n \in \omega$ let $k_n = \sum_{i=0}^{n} 2^i$. Let $A_n = \{x \in 2^\omega : x[k_n, k_{n+1}) = \mathbf{0}\}$ for $n \in \omega$. Note that each set A_n is clopen and has measure measure $2^{-(n+1)}$. So $G = \bigcap_m \bigcup_{n>m} A_n$ is null.

We will construct a set $X' = \{x_\alpha : \alpha < 2^\omega\}$ such that

(1) $\forall \alpha\ x_\alpha \in 2^\omega$,
(2) $\forall \alpha\ \forall^\infty n\ x_{\alpha+1}(n) \leq x_\alpha(n)$,
(3) $X = X' \cup \mathbb{Q}$ is a γ-set,
(4) $\forall z \in 2^\omega\ \exists \alpha\ \exists^\infty n\ z[k_n, k_{n+1}) = x_\alpha[k_n, k_{n+1})$,
(5) $\forall \alpha\ \exists^\infty n\ x_\alpha[k_n, k_{n+1}) = \mathbf{1}$.

Note that if z and x_α are like in (4) then $z + x_\alpha \in G$. Therefore, condition (4) implies that $X + G = 2^\omega$.

Let $\mathcal{S} = \{\ddagger \in \in^\omega : \exists^\infty \backslash\ \ddagger[\|\backslash, \|\backslash_{+\infty}) = \mathbf{1}\}$. For $x \in 2^\omega$ let $[x]^\star = \{z \in \mathcal{S} : \forall^\infty n\ x(n) \geq z(n)\}$.

Lemma 2.2. *Let $\mathcal{J}$ be open ω-cover of $\mathbb{Q}$ and $x \in \mathcal{S}$. Then there is a sequence $D_n \in \mathcal{J}$ and $y \in [x]^\star$ such that $\mathbb{Q} \cup [y]^\star \subseteq \bigcup_m \bigcap_{n>m} D_n$.*

Proof Using the fact that $\mathcal{J}$ is open ω-cover of $\mathbb{Q}$ we can find a sequence $\langle l_n : n \in \omega \rangle$ of natural numbers and a sequence $\langle D'_n : n \in \omega \rangle$ of elements of $\mathcal{J}$ such that for every n,

$$\forall z \in 2^\omega \left(z[n, l_n) = \mathbf{0} \rightarrow z \in D'_n \right).$$

Without loss of generality we can assume that there exists a set $Z \subseteq \omega$ such that

$$x[k_n, k_{n+1}) = \begin{cases} \mathbf{1} & \text{if } n \in Z \\ \mathbf{0} & \text{if } n \notin Z \end{cases}.$$

Choose $Y \subseteq \omega$ and $Z' \subseteq Z$ such that

$$\bigcup_{n \in Z'} [k_n, k_{n+1}) \cap \bigcup_{n \in Y} [n, l_n) = \emptyset.$$

Define

$$y[k_n, k_{n+1}) = \begin{cases} \mathbf{1} & \text{if } n \in Z' \\ \mathbf{0} & \text{if } n \notin Z' \end{cases}.$$

It is clear that $y \in [x]^\star$. Suppose that $z \in [y]^\star$. Note that

$$\forall^\infty n \in Y\ z[n, l_n) = \mathbf{0},$$

which means that $z \in D'_n$ for all except finitely many $n \in Y$. Thus, in order to finish the proof it is enough to define $D_n = D'_{y(n)}$, where $y(n)$ is the n-th element of Y. □

Lemma 2.3. *Suppose that $\{x_\alpha : \alpha < \kappa < \mathfrak{p}\} \subseteq 2^\omega$ is a sequence such that $x_\alpha \in [x_\beta]^\star$ for $\alpha > \beta$. Then $\bigcap_{\alpha<\kappa} [x_\alpha]^\star \neq \emptyset$.*

PROOF Define $Y_\alpha = \{n : x_\alpha[k_n, k_{n+1}) = \mathbf{1}\}$. Then $Y_\alpha \subseteq^\star Y_\beta$ if $\alpha > \beta$. Since $\kappa < \mathfrak{p}$, there is Y such that $Y \subseteq^\star Y_\alpha$ for each $\alpha < \kappa$. Define

$$x[k_n, k_{n+1}) = \begin{cases} \mathbf{1} & \text{if } n \in Y \\ \mathbf{0} & \text{if } n \notin Y \end{cases}$$

It is clear that $x \in \bigcap_{\alpha<\kappa}[x_\alpha]^\star$. □

Let $\{\mathcal{J}_\alpha : \alpha < 2^\omega\}$ be an enumeration of all ω-covers of $\mathbb{Q}$, and let $\{z_\alpha : \alpha < 2^\omega\}$ be enumeration of all elements of 2^ω.

Assume that the set $X_\alpha = \{x_\beta : \beta < \alpha\}$ has been already constructed. Assume that $\mathcal{J}_\alpha$ is ω-cover of $X_\alpha \cup \mathbb{Q}$. Since $|\alpha| < \mathfrak{p}$ and all sets of size $< \mathfrak{p}$ are γ-sets, we can choose $\mathcal{J}'_\alpha = \{U_n : n \in \omega\}$, ω-subcover of this such that

$$X_\alpha \cup \mathbb{Q} \subseteq \bigcup_m \bigcap_{n>m} U_n.$$

If α is limit then apply 2.3 to get a real $x'_\alpha \in \bigcap_{\beta<\alpha}[x_\beta]^\star$. If α is not limit let $x'_\alpha = x_{\alpha-1}$.

Next apply 2.2 to x'_α and $\mathcal{J}'_\alpha$ and x'_α to get a real y_α. Finally let x_α be such that
(1) $\exists^\infty n\ z_\alpha[k_n, k_{n+1}) = x_\alpha[k_n, k_{n+1})$,
(2) $\exists^\infty n\ x_\alpha[k_n, k_{n+1}) = \mathbf{1}$,
(3) $\forall^\infty n\ x_\alpha(n) \leq y_\alpha(n)$.
This finishes the construction and the proof of the theorem. □

The set constructed above is a γ-set but it contains a subset which is not a γ-set. In the next theorem we will show how to build a set which is a hereditarily γ-set and is not strongly meager. The construction is a slight modification of Todorcevic's construction of a hereditarily γ-set from [1].

Theorem 2.4. *Assume $\diamondsuit$. Then there exists a hereditarily γ-set which is not strongly meager.*

PROOF For a tree $p \subseteq 2^{<\omega}$ let $[p]$ be the set of branches of p. Similarly, for a finite set $U \subseteq 2^{<\omega}$ let

$$[U] = \{x \in 2^\omega : \exists s \in U\ x\mathsf{dom}(s) = s\}.$$

Let $\{k_n : n \in \omega\}$ and G be the sequence and the set defined at the beginning of the proof of 2.1. Let $\mathcal{P}$ be the collection of all perfect trees p such that there exists a sequence $\{U_n : n \in \omega\}$ such that
(1) $U_n \subseteq 2^{[k_n, k_{n+1})}$ for all n,
(2) $\exists^\infty n\ U_n = 2^{[k_n, k_{n+1})}$,
(3) $[p] = \bigcap_{n\in\omega}[U_n]$.
By induction on levels we will build an Aronszajn tree consisting of elements of $\mathcal{P}$ ordered by inclusion. The γ-set we are looking for will be a selector from the elements of this tree.

For perfect trees p, q and a set $R \subseteq 2^n \cap q$ define

$$p \leq_R q \Rightarrow p \cap 2^n = R \ \&\ p \subseteq q.$$

If $R = q \cap 2^n$ we write $p \leq_n q$ instead of $p \leq_{q\cap 2^n} q$.

Let $\{z_\alpha : \alpha < \omega_1\}$ be enumeration of 2^ω.

We will build by induction a partial ordering $\prec$ on ω_1, $\{p_\alpha : \alpha \in \omega_1\}$ and $\{x_\alpha : \alpha \in \omega_1\}$ such that

(1) $T = (\omega_1, \prec)$ is an Aronszajn tree and for limit α, $\bigcup_{\beta<\alpha} T_\beta = \alpha$,
(2) $p_\alpha \in \mathcal{P}$ for all α,
(3) $\forall \alpha, \beta \left(\alpha \prec \beta \iff p_\alpha \subseteq p_\beta\right)$,
(4) $x_\alpha \in [p_\alpha]$ for all α,
(5) $\exists^\infty n\ x_\alpha[k_n, k_{n+1}) = z_\alpha[k_n, k_{n+1})$,
(6) if $\mathcal{J}$ is an "appropriate" ω-cover then there exists α and a sequence $\langle D_n : n \in \omega\rangle$ of elements of $\mathcal{J}$ such that for all $p \in T_\alpha$ $[p] \subseteq \bigcup_m \bigcap_{n>m} D_n$,
(7) if $\beta > \alpha$ and $q \in T_\alpha$, $R \subseteq q \cap 2^m$ then there exists $p \in T_\beta$ such that $p \leq_R q$.

Note that the condition (7) guarantees that the construction will not terminate after countably many steps. Condition (6) is rather vague, but with the right interpretation of the word "appropriate", together with the condition (4) it will guarantee that the set $X = \{x_\alpha : \alpha < \omega_1\}$ is a hereditary γ-set. Finally (5) yields that $X + G = 2^\omega$.

For a set $Y \subseteq 2^\omega$ and a perfect tree p let $p(Y)$ be the tree representing the closure of $[p] \cap Y$.

We use $\diamondsuit$ to construct oracle sequences $\{\mathcal{J}_\alpha, X_\alpha, \langle p^\beta : \beta < \alpha\rangle : \alpha < \omega_1\}$ such that for any $Y \subseteq X$ and an open ω-cover of Y, $\mathcal{J}$, there are stationary many α's such that
(1) $\mathcal{J}_\alpha = \mathcal{J}$,
(2) $X_\alpha = \{x_\beta : \beta < \alpha\} \cap Y$,
(3) $p^\beta = p_\beta(Y)$ for $\beta < \alpha$.

Note that these sequences are easy to obtain from an ordinary $\diamondsuit$ sequence by coding $\mathcal{J}$'s and Y's by subsets of ω_1 (even ω in case of $\mathcal{J}$).

We will build the tree T (or $\prec$) by induction on levels. If $\alpha = \beta + 1$ and T_β is already constructed then T_α is any extension of T_β satisfying the requirements.

Suppose that α is a limit ordinal. We look for a sequence $\{D_n : n \in \omega\} \subseteq \mathcal{J}_\alpha$ such that
(i) $\forall y \in X_\alpha\ \forall^\infty n\ y \in D_n$,
(ii) $\forall \beta < \alpha\ \forall m\ \forall R \subseteq p_\beta \cap 2^m\ \exists \delta \left(p_\delta \leq_R p_\beta\ \&\ \forall^\infty n\ [p^\delta] \subseteq D_n\right)$.

If such a sequence $\{D_n : n \in \omega\}$ does not exist then T_α is an arbitrary extension of $\bigcup_{\beta<\alpha} T_\beta$.

Otherwise we fix a sequence $\{D_n : n \in \omega\}$ satisfying the above conditions then for every $\beta < \alpha$ and every $R \subseteq p_\beta \cap 2^m$ we build a chain $\{\delta_n : n \in \omega\} \subseteq \alpha$ and $\{l_n : n \in \omega\}$ such that
(1) $\delta_n \prec \delta_{n+1}$ for all n,
(2) $p_{\delta_{n+1}} \leq_{l_n} p_{\delta_n}$ for all n,
(3) $\bigcap_{n\in\omega} p_{\delta_n} \in \mathcal{P}$,
(4) $p_{\delta_0} \leq_R p_\beta$,
(5) $\forall^\infty n\ [p^{\delta_0}] \subseteq D_n$.

The branch $\{\delta_n : n \in \omega\}$ will be extended on level α by, say, ρ and the corresponding set is $p_\rho = \bigcap_{n\in\omega} p_{\delta_n}$. Note that in this way we extend only countably many branches.

Observe that condition (3) will follow from (2) if only the sequence l_n is increasing fast enough. Conditions (4) and (5) follow from the condition (ii) above. This concludes the construction. It remains to show that X is a hereditary γ-set.

Suppose that $Y \subseteq X$ and let $\mathcal{J}$ be an ω-cover of Y. Let α be a limit ordinal such that

(1) $\mathcal{J}_\alpha = \mathcal{J}$,
(2) $X_\alpha = \{x_\beta : \beta < \alpha\} \cap Y$,
(3) $p^\beta = p_\beta(Y)$ for $\beta < \alpha$.

To finish the proof it is enough to check that there exists a sequence $\{D_n : n \in \omega\} \subseteq \mathcal{J}_\alpha$ satisfying conditions (i) and (ii). In this case, according to the construction, $Y \subseteq \bigcup_m \bigcap_{n>m} D_n$.

Let $\{\langle \beta_n, R_n\rangle : n \in \omega\}$ be enumeration of the set

$$\{\langle \beta, R\rangle : \beta < \alpha \ \& \ \exists m\ R \subseteq p_\beta \cap 2^m\}.$$

It is enough to construct by induction a sequence $\{D_n : n \in \omega\} \subseteq \mathcal{J}_\alpha$ such that
(iii) $\forall i < n\ x_{\beta_i} \in D_n$,
(iv) $\forall i < n\ \exists \delta\ p_\delta \leq_{R_i} p_{\beta_i}\ \&\ [p^\delta] \subseteq D_n$.

We describe how to construct set D_n. For each $i < n$ and $s \in R_i$ choose $x^i_s \in Y$ (if possible) such that $s \subseteq x^i_s$. Let D_n be an element of $\mathcal{J}_\alpha$ such that

$$\{x_{\beta_i} : i < n\} \cup \{x^i_s : i < n, s \in R_i\} \subseteq D_n.$$

Such D_n exists since $\mathcal{J}_\alpha$ is an ω-cover of Y.

We need to verify condition (iv). Fix $i < n$ and choose m so large that $[x^i_s m] \subseteq D_n$ for all $s \in R_i$. If x^i_s does not exist let x_s be any element of $[p_{\beta_i}]$ extending s. Let $R = \{x^i_s m : s \in R_i\} \cup \{x_s m : s \in R_i\}$. By inductive hypothesis there exists δ such that $p_\delta \leq_R p_{\beta_i}$. Note that $[p^\delta] \subseteq \bigcup_{s \in R_i} [x^i_s m]$. Thus $[p^\delta] \subseteq D_n$, which finishes the construction and the proof. □

We can show the same theorems for the algebraic structure of the real line. Let us take a standard function $f : 2^\omega \to [0,1]$, $f(x) = \sum_{i\in\omega} x(i)/2^i$. f is continuous so $f(X)$ is a γ-set, where X is as in 2.1 or 2.4. Let $H = \bigcap_m \bigcup_{n>m} f(A_n)$. It is easy to see that $[0,1] \subseteq f(X) + H$. Let $G = H + \mathbb{Q}$ then $f(X) + G = \mathbb{R}$. □

References

[1] Fred Galvin and Arnold W. Miller, *γ-sets and other singular sets of real numbers*, Topology and its Applications **17** (1984), no. 2, 145–155.
[2] J. Gerlits and Zs. Nagy, *Some properties of $C(X)$, I*, Topology Appl. **14** (1982), 151–161.
[3] Arnold W. Miller, *Special subsets of the real line*, Handbook of Set Theoretic Topology (Amsterdam) (K. Kunen and J. E. Vaughan, eds.), North-Holland, Amsterdam, 1984, pp. 201–235.
[4] Janusz Pawlikowski, *Powers of transitive bases of measure and category*, Proceedings of the American Mathematical Society **93**, 1985, pp. 719–729

Department of Mathematics, Boise State University, Boise, Idaho 83725
E-mail address: tomek@math.idbsu.edu

Institute of Mathematics, University Gdansk, Wita Stwosza 57, 80-952 Gdansk, Poland
E-mail address: matir@halina.univ.gda.pl

Contemporary Mathematics
Volume **192**, 1996

Reductions Between Cardinal Characteristics of the Continuum

ANDREAS BLASS

ABSTRACT. We discuss two general aspects of the theory of cardinal characteristics of the continuum, especially of proofs of inequalities between such characteristics. The first aspect is to express the essential content of these proofs in a way that makes sense even in models where the inequalities hold trivially (e.g., because the continuum hypothesis holds). For this purpose, we use a Borel version of Vojtáš's theory of generalized Galois-Tukey connections. The second aspect is to analyze a sequential structure often found in proofs of inequalities relating one characteristic to the minimum (or maximum) of two others. Vojtáš's max-min diagram, abstracted from such situations, can be described in terms of a new, higher-type object in the category of generalized Galois-Tukey connections. It turns out to occur also in other proofs of inequalities where no minimum (or maximum) is mentioned.

1. Introduction

Cardinal characteristics of the continuum are certain cardinal numbers describing combinatorial, topological, or analytic properties of the real line $\mathbb{R}$ and related spaces like ${}^{\omega}\omega$ and $\mathcal{P}(\omega)$. Several examples are described below, and many more can be found in [**4, 14**]. Most such characteristics, and all those under consideration in this paper, lie between $\aleph_1$ and the cardinality $\mathfrak{c} = 2^{\aleph_0}$ of the continuum, inclusive. So, if the continuum hypothesis (CH) holds, they are equal to $\aleph_1$. The theory of such characteristics is therefore of interest only when CH fails.

That theory consists mainly of two sorts of results. First, there are equations and (non-strict) inequalities between pairs of characteristics or sometimes between one characteristic and the maximum or minimum of two others. Second,

1991 *Mathematics Subject Classification.* 03E05, 03E40, 03E75.

Partially supported by NSF grant DMS-9204276 and NATO grant LG 921395.

The content of my talks at BEST had already been submitted for publication elsewhere. So this paper corresponds rather to the talk I gave at Oberwolfach in October, 1993.

This paper is in final form, and no version of it will be submitted for publication elsewhere.

there are independence results showing that other equations and inequalities are not provable in Zermelo-Fraenkel set theory (ZFC). As examples of results of the first sort, we mention in particular the work of Rothberger **[10]** and Bartoszyński **[1]** relating the characteristics associated to Lebesgue measure and Baire category; as examples of the second sort we mention **[2]** and the earlier work cited there. Many more examples can be found in **[4, 14]** and the references there.

A curious aspect of the proofs of inequalities (and of equations, which we regard as pairs of inequalities) in this theory is that they contain significant information whether or not CH holds, even though CH makes the inequalities themselves trivial. In other words, the proofs establish additional information beyond the inequalities. Vojtáš **[15]** introduced a framework in which one can attempt to formulate such additional information. He associates cardinal characteristics to binary relations on the reals and shows that proofs of inequalities between characteristics usually amount to the construction of a suitable pair of functions between the domains and the ranges of the corresponding relations. He calls these pairs of functions generalized Galois-Tukey connections, but for brevity we shall call them morphisms. The existence of such morphisms seems a plausible candidate for the "additional information" established by typical proofs of inequalities between cardinal characteristics.

It turns out, however, by a result of Yiparaki **[16]**, that this additional information is still trivial in the presence of CH. We shall show in Section 2 how to modify Vojtáš's framework so as to produce non-trivial results even in the presence of CH. The key idea is to add a definability (or absoluteness) requirement on the functions that constitute a morphism.

Section 3 contains some information about Baire category that will be used in subsequent examples.

Section 4 is concerned with a structure often found in proofs of inequalities of the form $x \geq \min(y, z)$, a structure that Vojtáš described with his max-min diagram. We show that this diagram can be neatly interpreted in terms of a construction, which we call *sequential composition*, on the relations associated to y and z. This construction allows us to analyze the "flow of control" in proofs of such three-cardinal inequalities.

Section 5 is devoted to showing that sequential composition is necessary for the results in Section 4. In particular, certain simpler compositions proposed by Vojtáš are not adequate for these results.

Finally, in Section 6, we discuss a situation where sequential composition arises in the natural proof of an inequality involving just two cardinal characteristics, not a maximum or minimum.

I thank Peter Vojtáš and Janusz Pawlikowski for useful discussions of the topic of this paper.

2. Borel Galois-Tukey Connections

To motivate Vojtáš's framework, we define a few cardinal characteristics of the continuum and discuss the proofs of some inequalities relating them. The characteristics to be used in this example are the *bounding number* $\mathfrak{b}$, the *dominating number* $\mathfrak{d}$, the *unsplitting number* $\mathfrak{r}$, and the *splitting number* $\mathfrak{s}$, defined as follows. (See [**4, 14**] for more information about these cardinals.) For functions f and g from ω to ω, we write $f \leq^* g$ to mean that $f(n) \leq g(n)$ for all but finitely many n. A family $\mathcal{B} \subseteq {}^\omega\omega$ of such functions is *unbounded* if there is no $g \in {}^\omega\omega$ such that all elements of $\mathcal{B}$ are $\leq^* g$. The smallest possible cardinality for an unbounded family is $\mathfrak{b}$. A family $\mathcal{D} \subseteq {}^\omega\omega$ is *dominating* if every $g \in {}^\omega\omega$ is $\leq^*$ some $f \in \mathcal{D}$. The smallest possible cardinality for a dominating family is $\mathfrak{d}$. A subset X of ω *splits* another such set Y if both $Y \cap X$ and $Y - X$ are infinite. A family $\mathcal{S} \subseteq \mathcal{P}(\omega)$ is a *splitting family* if every infinite $Y \subseteq \omega$ is split by some member of $\mathcal{S}$. The smallest possible cardinality for a splitting family is $\mathfrak{s}$. A family $\mathcal{R}$ of infinite subsets of ω is an *unsplit family* (sometimes called *refining* or *reaping*) if no single X splits all the members of $\mathcal{R}$. The smallest possible cardinality for an unsplit family is $\mathfrak{r}$.

It is easy to see that all four of the cardinals just defined are between $\aleph_1$ and $\mathfrak{c}$ inclusive and that $\mathfrak{b} \leq \mathfrak{d}$. As an example for future analysis, we give the proof of the less trivial yet well-known inequality $\mathfrak{s} \leq \mathfrak{d}$. Given a dominating family $\mathcal{D}$, we shall assign to each $f \in \mathcal{D}$ a set $X_f \subseteq \omega$ in such a way that all these X_f's constitute a splitting family. This will clearly suffice to prove $\mathfrak{s} \leq \mathfrak{d}$. The construction is as follows. Given f, partition ω into finite intervals $[a_0, a_1)$, $[a_1, a_2)$, ... each satisfying $f(a_n) < a_{n+1}$. To be specific, let $a_0 = 0$ and $a_{n+1} = 1 + \max\{a_n, f(a_n)\}$. Then let X_f be the union of the even-numbered intervals $[a_{2n}, a_{2n+1})$. To see that the X_f's constitute a splitting family, let an arbitrary infinite $Y \subseteq \omega$ be given, and let $g : \omega \to \omega$ be the function sending each natural number n to the next larger element of Y. As $\mathcal{D}$ is dominating, it contains an $f \geq^* g$. In the construction of X_f we have, for all sufficiently large n, that the next element of Y after a_n is $g(a_n) \leq f(a_n) < a_{n+1}$ and therefore lies in the interval $[a_n, a_{n+1})$. So Y meets all but finitely many of these intervals and therefore contains infinitely many members of X_f and infinitely many members of $\omega - X_f$. That is, X_f splits Y, as required.

We now present Vojtáš's framework for describing cardinal characteristics and proofs of inequalities, using the preceding proof as an example. First, each characteristic was defined as the smallest possible cardinality for a set $\mathcal{Z}$ of reals such that every real is related in a certain way to one in $\mathcal{Z}$. More precisely, in each case we had a triple $\mathbf{A} = (A_-, A_+, A)$ of two sets $A_\pm$ and a binary relation A between them such that the characteristic is

$$\|(A_-, A_+, A)\| = \min\{|\mathcal{Z}| : \mathcal{Z} \subseteq A_+ \text{ and } \forall x \in A_- \exists z \in \mathcal{Z}\, A(x, z)\},$$

which we call the *norm* of $\mathbf{A}$. For example, $\mathfrak{d}$ is the norm of $({}^\omega\omega, {}^\omega\omega, \leq^*)$, and $\mathfrak{s}$ is the norm of $(\mathcal{P}_\infty(\omega), \mathcal{P}(\omega), \text{is split by})$. (Here $\mathcal{P}_\infty$ means the family of infinite subsets.) To decribe $\mathfrak{b}$ and $\mathfrak{r}$ as norms, it suffices to take the descriptions for $\mathfrak{d}$ and $\mathfrak{s}$ and *dualize* them in the following sense: interchange A_- with A_+, and replace A with the complement of the converse relation. In general, we write

$$(A_-, A_+, A)^\perp = (A_+, A_-, \{(z,x) : \text{not } A(x,z)\}.$$

(Vojtáš calls the norms of $\mathbf{A}$ and of its dual the dominating and bounding numbers of $\mathbf{A}$, respectively, by analogy with the example of $\mathfrak{d}$ and $\mathfrak{b}$ above.) We sometimes call triples (A_-, A_+, A) relations, although strictly speaking it is only the third component A that is a relation. To avoid trivialities, we shall tacitly assume that our relations have $A_\pm \neq \emptyset$, that each element of A_- is A-related to some element of A_+ (so $\|\mathbf{A}\|$ is defined), and that not all elements of A_- are A-related to any single element of A_+ (so $\|\mathbf{A}^\perp\|$ is defined). These assumptions amount to requiring the norms of both $\mathbf{A}$ and $\mathbf{A}^\perp$ to be at least 2.

The proof of $\mathfrak{s} \leq \mathfrak{d}$ presented above consists of three parts. First, there was a construction $\xi_+ : {}^\omega\omega \to \mathcal{P}(\omega)$ sending each f to X_f. Second, there was a (simpler) construction $\xi_- : \mathcal{P}_\infty(\omega) \to {}^\omega\omega$ sending each Y to the function $\xi_-(Y) = g$ that maps n to the next larger element of Y. Finally, there was the verification that if $g \leq^* f$ then Y is split by X_f. Abstracting this structure, we obtain Vojtáš's notion of a generalized Galois-Tukey connection; we call it simply a morphism and write it in the opposite direction to Vojtáš's.

DEFINITION. A *morphism* ξ from (A_-, A_+, A) to (B_-, B_+, B) is a pair of functions $\xi_- : B_- \to A_-$ and $\xi_+ : A_+ \to B_+$ such that, for all $b \in B_-$ and all $a \in A_+$,

$$A(\xi_-(b), a) \implies B(b, \xi_+(a)).$$

Our convention for the direction of morphisms was chosen partly to work well with other uses of the same category [**8, 3**] and partly so that the direction of a morphism agrees with the direction of the implication displayed in the definition.

The existence of a morphism ξ from (A_-, A_+, A) to (B_-, B_+, B) immediately implies the norm inequality $\|(A_-, A_+, A)\| \geq \|(B_-, B_+, B)\|$. Indeed, if $\mathcal{Z}$ is as in the definition of norm for (A_-, A_+, A), then $\xi_+(\mathcal{Z})$ has no greater cardinality and serves the same purpose in the definition of the norm of (B_-, B_+, B). Because of this, we write $\mathbf{A} \geq \mathbf{B}$ to indicate the existence of such a morphism. (Our inequalities, unlike our morphisms, go in the same direction as Vojtáš's.)

A morphism from $\mathbf{A}$ to $\mathbf{B}$ becomes, just by interchanging its two components, a morphism in the opposite direction between the dual objects. Thus, for example, the proof above of $\mathfrak{s} \leq \mathfrak{d}$, exhibiting a morphism from $({}^\omega\omega, {}^\omega\omega, \leq^*)$ to $(\mathcal{P}_\infty(\omega), \mathcal{P}(\omega), \text{is split by})$, also exhibits a morphism from $(\mathcal{P}(\omega), \mathcal{P}_\infty(\omega), \text{does not split})$ to $({}^\omega\omega, {}^\omega\omega, \not\geq^*)$ and thus proves that $\mathfrak{b} \leq \mathfrak{r}$. (In this form, the inequality is essentially due to Solomon [**11**].)

One can similarly exhibit morphisms that capture the combinatorial content of the proofs of a great many other inequalities between cardinal characteristics. This applies both to trivial inequalities like $\mathfrak{b} \leq \mathfrak{d}$ and deep results like Bartoszyński's theorem that the additivity of category is at least the additivity of measure. See [5] for a presentation of Bartoszyński's theorem that makes the morphism explicit, and see [15] for more examples.

For future reference, we mention that the converse, $\mathfrak{d} \leq \mathfrak{s}$, of the inequality proved above is known not to be provable in ZFC (though of course it holds in some models of ZFC, for example models of CH). It fails, for example, in the model obtained from a model of CH by adding a set $C \subseteq {}^\omega\omega$ of $\aleph_2$ Cohen reals. In this model, $\mathfrak{d} = \aleph_2$ because any $\aleph_1$ reals lie in a submodel generated by $\aleph_1$ members of C and therefore cannot dominate the other members of C. On the other hand, $\mathfrak{s} = \aleph_1$ because the set of ground model reals is non-meager (i.e., of second Baire category) in the extension and any non-meager subset of $\mathcal{P}(\omega)$ is a splitting family (because the reals that fail to split any particular Y form a meager set).

We shall call an equation or (non-strict) inequality between cardinal characteristics "correct" if it is provable in ZFC and "incorrect" if it is independent of ZFC. (It cannot be refutable in ZFC, because it holds in models of CH; recall that we deal only with characteristics that lie between $\aleph_1$ and $\mathfrak{c}$.) In models of CH, the difference between correct and incorrect inequalities is hidden, because all the inequalities are true there. Nevertheless, it seems reasonable to say, even in such models, that correct inequalities, like $\mathfrak{s} \leq \mathfrak{d}$, hold for understandable, combinatorial reasons while incorrect inequalities, like $\mathfrak{d} \leq \mathfrak{s}$, hold only because CH "happens" to be true. Can one make mathematical sense of such statements?

Vojtáš's theory, in particular the fact that proofs of inequalities between characteristics usually exhibit morphisms between the corresponding relations, suggests an affirmative answer to this question. The understandable, combinatorial reason for a correct inequality is given by the morphism. So one might hope that there are, even in the presence of CH, no morphisms corresponding to incorrect inequalities. Then, when incorrect inequalities hold in a model, this would not be because of good reasons (i.e., morphisms) but because of other properties of the model (like CH).

A theorem of Yiparaki [16], Chapter 5, dashes this hope. She shows that, if

$$||\mathbf{A}|| = |A_+| = ||\mathbf{B}^\perp|| = |B_-|,$$

then there is a morphism from $\mathbf{A}$ to $\mathbf{B}$. In particular, in all our examples there is such a morphism if CH holds, because all the cardinals in the displayed equation are then equal to $\aleph_1$. So models of CH not only satisfy all inequalities, correct or incorrect, they also contain morphisms to justify all these inequalities.

These morphisms, however, are highly non-constructive; their definition involves a multitude of arbitrary choices. We propose therefore to eliminate them by working not with arbitrary morphisms but with well-behaved ones. To be specific, we restrict our attention to objects (A_-, A_+, A) in which $A_\pm$ are Borel sets of reals and A is a Borel relation between them, and we consider only morphisms ξ both of whose components $\xi_\pm$ are Borel functions. The objects and morphisms considered above, in connection with the definitions of $\mathfrak{d}$ and $\mathfrak{s}$ (and their duals) and the proof of $\mathfrak{s} \leq \mathfrak{d}$ (and its dual) are all Borel in this sense. Yiparaki's proof, on the other hand, involves non-Borel morphisms.

As an example of what the restriction to Borel morphisms accomplishes, we show that there is no Borel morphism corresponding to the incorrect inequality $\mathfrak{d} \leq \mathfrak{s}$.

PROPOSITION 1. *There is no Borel morphism*

$$(\xi_-, \xi_+) : (\mathcal{P}_\infty(\omega), \mathcal{P}(\omega), \textit{is split by}) \to ({}^\omega\omega, {}^\omega\omega, \leq^*).$$

PROOF. Suppose we had such a morphism (ξ_-, ξ_+) consisting of two Borel maps. Let $c \in {}^\omega\omega$ be Cohen-generic over the universe V (in some Boolean extension), and by abuse of notation write (ξ_-, ξ_+) also for the pair of Borel maps in $V[c]$ having the same codes as the original (ξ_-, ξ_+) had in V. Since the ground model reals remain a non-meager and hence splitting set in the extension, $\xi_-(c)$ is split by some X in the ground model. Because ξ is a morphism, it follows that $c \leq^* \xi_+(X)$. But the ground model contains X and the code for ξ_+ and therefore also $\xi_+(X)$. This is absurd, as no real from the ground model can dominate a Cohen real. □

More generally, we can show that there are no Borel morphisms between Borel relations when the corresponding inequality of characteristics can be violated by forcing. To express this precisely, suppose we have Borel **A** and **B** such that some notion of forcing P forces $\|\mathbf{A}\| < \|\mathbf{B}\|$, where we have, as above, abused notation by writing **A** and **B** for the objects in the forcing extension having the same Borel codes as the original **A** and **B** in the ground model. Then there is no Borel morphism (in the ground model) $\xi : \mathbf{A} \to \mathbf{B}$. Indeed, for ξ to be such a morphism would be a Π^1_1 assertion about the Borel codes of ξ, **A**, and **B**. Such assertions are preserved by forcing extensions. But P forces that there is no such morphism, because it forces the opposite ordering of the norms. (Note that we proved Proposition 1 by adding a single Cohen real, whereas the general argument just given would involve adding at least $\aleph_2$ Cohen reals.)

We emphasize that, in the situation of the preceding paragraph, the inequality $\|\mathbf{A}\| \geq \|\mathbf{B}\|$ may well be true in the ground model, but there cannot be a Borel morphism causing it.

It should be noted that independence proofs for inequalities between cardinal characteristics (e.g., [**2, 14**]) typically take the form considered above, i.e., they produce a P forcing a strict inequality in the opposite direction. In many cases, the proofs in the literature involve not only a forcing construction but some requirements on the ground model over which the forcing is done. But these requirements, usually CH or Martin's axiom, can themselves be forced, so there is a P as in the discussion above.

Summarizing, we have that incorrect inequalities between cardinal characteristics, though they may be true in some models and may be given by morphisms in some models, are not given by Borel morphisms in any model, provided their incorrectness can be established by a forcing argument.

3. Baire Category

In Section 4, we shall use some cardinal characteristics related to Baire category as an example to motivate and illustrate the operation of sequential composition of relations. In preparation for this, we devote the present section to introducing the notation and preliminary results needed to make the later discussion proceed smoothly. Along the way, we shall see a few more examples of the structures discussed in Section 2.

In what follows, we shall use the Cantor space ${}^{\omega}2$, with its usual (product) topology, as the underlying space in all our discussions of Baire category. One of the characteristics that we shall need is the additivity of Baire category, $\mathbf{add}(B)$, the smallest number of meager (= first category) sets whose union is not meager. Like the characteristics discussed in the preceding section, $\mathbf{add}(B)$ lies between $\aleph_1$ and $\mathfrak{c}$ inclusive. It can be described as the norm of $(B, B, \not\supseteq)$, where B is the collection of meager subsets of ${}^{\omega}2$. This description is not amenable, as it stands, to the Borel considerations of the preceding section, because B is not a set of reals but a set of sets of reals. One can, however, replace B with the collection of meager F_σ sets, i.e., countable unions of nowhere-dense closed sets; this does not affect $\mathbf{add}(B)$, since every meager set is included in a meager F_σ set. It is easy to code meager F_σ sets by reals in such a way that the set of codes is Borel, and with some additional work one can arrange that the relations we need, like $\not\supseteq$, are also Borel in the codes. We omit the details of this since we shall soon introduce a different way of viewing $\mathbf{add}(B)$ for which these matters are easier to handle.

The other Baire category characteristic that we shall need is the covering number, $\mathbf{cov}(B)$, the smallest number of meager sets needed to cover ${}^{\omega}2$. This is the norm of $({}^{\omega}2, B, \in)$, and again we can replace B by the collection of meager F_σ sets or by the collection of their codes. It is obvious that $\mathbf{add}(B) \leq \mathbf{cov}(B)$. As expected, this trivial inequality corresponds to a trivial morphism, with $\xi_- : B \to {}^{\omega}2$ sending any meager set to some real not in it and with $\xi_+ : B \to B$ being the identity map. (The map ξ_- involves an arbitrary choice, but from a code for a meager F_σ set one can obtain, in a Borel fashion, a specific real not in that set, just by following the proof of the Baire category theorem.)

It will be convenient to use a particular, easily coded basis for the ideal of meager sets. To introduce this basis, we first define a *chopped real* to be a pair (x, Π), where $x \in {}^{\omega}2$ and where Π is a partition of ω into finite intervals $I_0 = [a_0, a_1)$, $I_1 = [a_1, a_2)$, ..., with $0 = a_0 < a_1 < a_2 < \dots$. The idea is that a real $x : \omega \to 2$ has been chopped into finite pieces by the partition Π of its domain. We say that a real $y \in {}^{\omega}2$ *matches* the chopped real (x, Π) if y agrees with x on infinitely many of the intervals of Π, i.e., if there are infinitely many $n \in \omega$ such that $y \restriction I_n = x \restriction I_n$. We write $\mathrm{Match}(x, \Pi)$ for the set of all y that match (x, Π). It is easy to verify that $\mathrm{Match}(x, \Pi)$ is a dense G_δ set. Talagrand [**12**], in obtaining a combinatorial description of meager filters, proved that every dense G_δ set in ${}^{\omega}2$ has a subset of the form $\mathrm{Match}(x, \Pi)$; equivalently, every meager set is included in the complement of Match(x,Π) for some chopped real.

Thus, we can use chopped reals and the corresponding matching sets instead of arbitrary meager sets or meager F_σ sets in describing the cardinal characteristics of Baire category. Specifically, let us define CR to be the set of chopped reals and define

$$\mathbf{U} = (CR, {}^\omega 2, \text{matches}^\smile),$$

where $\smile$ means to take the converse of a relation. Then the norm of the dual of $\mathbf{U}$ is the smallest size for a family of chopped reals such that no single real matches them all. Equivalently, by Talagrand's result, it is the smallest number of dense G_δ sets with empty intersection. Taking the complements of those sets, we find that this is precisely the covering number. So we have $\mathbf{cov}(B) = \|\mathbf{U}^\perp\|$. (The norm of $\mathbf{U}$ itself is the smallest cardinality of a non-meager set of reals, called the *uniformity* of category; it is important in its own right, but we shall not need it here. The reason for defining $\mathbf{U}$ as we did, rather than dually, is to avoid excessive negations and to avoid some dualizations in the next section.)

For a similar description of the additivity of category, we need to describe, in terms of chopped reals, the inclusion relation between the corresponding dense G_δ sets. We say that one chopped real (x,Π) *engulfs* another (x',Π') if all but finitely many intervals of Π include intervals of Π' on which x and x' agree.

LEMMA. *Match$(x,\Pi) \subseteq$ Match$(x',\Pi)'$ if and only if (x,Π) engulfs (x',Π').*

PROOF. First, suppose (x,Π) engulfs (x',Π') and y matches (x,Π). So there are infinitely many intervals I of Π on which y agrees with x. Except for finitely many, each such I includes an interval J of Π' on which x and x' agree. Thus we get infinitely many intervals J of Π' on which y and x' agree, so y matches (x',Π').

Conversely, suppose (x,Π) does not engulf (x',Π'), so there are infinitely many intervals I of Π that contain no interval of Π' on which x and x' agree. Discarding some of these intervals I, we can arrange that no interval of Π' meets more than one of them. Define $y \in {}^\omega 2$ by making it agree with x on the union of these intervals I and making it disagree with x' everywhere else. Then y matches (x,Π) but not (x',Π'). □

Define

$$\mathbf{V} = (CR, CR, \text{is engulfed by}).$$

Then the norm of the dual of V is the minimum number of chopped reals such that no single real engulfs them all. Equivalently, by Talagrand's result and the lemma, it is the minimum number of dense G_δ sets such that no single dense G_δ set is included in them all. Taking complements, we find that this is just the additivity of category. So $\mathbf{add}(B) = \|\mathbf{V}^\perp\|$. (The norm of $\mathbf{V}$ itself is the *cofinality* characteristic of Baire category.)

The trivial inequality $\mathbf{add}(B) \leq \mathbf{cov}(B)$ is given by a trivial morphism $\mathbf{U}^\perp \to \mathbf{V}^\perp$ or equivalently $\mathbf{V} \to \mathbf{U}$ in which one component is the identity map $CR \to CR$ while the other component $CR \to {}^{\omega}2$ sends a chopped real to its first component (forgetting the partition and keeping only the real).

A non-trivial inequality [7], namely $\mathbf{add}(B) \leq \mathfrak{b}$, is also easy to see from this point of view. Notice that $\mathfrak{b}$ is, as discussed in Section 2, the norm of $\mathbf{W}^{\perp}$, where

$$\mathbf{W} = ({}^{\omega}\omega, {}^{\omega}\omega, \leq^*).$$

So to prove the inequality in question, it suffices to exhibit a morphism $\mathbf{W}^{\perp} \to \mathbf{V}^{\perp}$ or equivalently $\xi : \mathbf{V} \to \mathbf{W}$. Define ξ_- to map any $f \in {}^{\omega}\omega$ to the chopped real $(0, \Pi)$ where 0 is the identically zero function $\omega \to 2$ and where Π is chosen so that, for any $n \in \omega$, $f(n)$ is no more than one interval past n (i.e., if $n \in I_k$ then $f(n) \in I_l$ for some $l \leq k+1$). (To get a Borel map ξ_-, the intervals should be chosen in a canonical manner, for example by choosing each endpoint a_n as small as possible subject to the constraints that $f(n)$ be at most one interval past n and that all the intervals be nonempty.) Define ξ_+ to map any chopped real (x, Π) to the function in ${}^{\omega}\omega$ sending each $n \in \omega$ to the right endpoint of the interval after the one that contains n. It is straightforward to verify that, if $\xi_-(f)$ is engulfed by (x, Π), then $f \leq^* \xi_+(x, \Pi)$, so ξ serves as the required morphism.

The inequalities above combine to give $\mathbf{add}(B) \leq \min\{\mathbf{cov}(B), \mathfrak{b}\}$. In fact, equality holds here [**13, 7**], but the converse inequality cannot be separated into two simpler inequalities each proved by exhibiting a morphism. On the contrary, this converse inequality, like any inequality of the form $x \geq \min\{y, z\}$ (or $x \leq \max\{y, z\}$) involves an interaction of all three cardinals. In the next section, we shall review the proof of $\mathbf{add}(B) \geq \min\{\mathbf{cov}(B), \mathfrak{b}\}$ and use it to motivate a construction that allows such proofs to be presented as morphisms between suitable relations.

4. Sequential Composition

To treat inequalities involving the maximum or minimum of two cardinal characteristics, it is natural to seek a construction which, given two relations $\mathbf{A}$ and $\mathbf{B}$, produces another relation $\mathbf{C}$ with $||C|| = \max\{||A||, ||B||\}$ and $||C^{\perp}|| = \min\{||A^{\perp}||, ||B^{\perp}||\}$. Such a construction is the product $\mathbf{A} \times \mathbf{B}$ given by letting C_- be the disjoint union $A_- \sqcup B_-$, letting C_+ be the product $A_+ \times B_+$, and defining $C(x, (a, b))$ to hold when either $x \in A_-$ and $A(x, a)$ or $x \in B_-$ and $B(x, b)$. This is the product in the category-theoretic sense with respect to our definition of morphisms. (With Vojtáš's convention it is, as he points out, the coproduct.) It would be pleasant if proofs of three-cardinal inequalities, like the $\mathbf{add}(B) \geq \min\{\mathbf{cov}(B), \mathfrak{b}\}$ mentioned at the end of the preceding section, could be presented as constructions of morphisms from a product, and Vojtáš asks ([**15**]) whether this can be done, after noting that the usual proof has a more complicated structure (described below).

An earlier (preprint) version of [**15**] contained a different sort of product, which we shall call the *old product* to distinguish it from the (categorical) product in the preceding paragraph. The old product $\mathbf{C}$ of $\mathbf{A}$ and $\mathbf{B}$ has $C_- = A_- \times B_-$, $C_+ = A_+ \times B_+$, and $C((x,y),(a,b))$ if and only if both $A(x,a)$ and $B(y,b)$. The norm of $\mathbf{C}$ and its dual are the maximum and minimum of the norms of the factors and their duals, respectively, as long as the norms are infinite. (More precisely, $\|\mathbf{C}^\perp\| = \min\{\|\mathbf{A}^\perp\|, \|\mathbf{B}^\perp\|\}$ and $\max\{\|\mathbf{A}\|, \|\mathbf{B}\|\} \leq \|\mathbf{C}\| \leq \|\mathbf{A}\| \cdot \|\mathbf{B}\|$. The last two inequalities can both be strict, but of course only when the norms involved are finite. For example, $(3, 3, \neq)$ has norm 2 but its old product with itself has norm 3.) In the preprint version of [**15**], Vojtáš asked whether proofs of certain three-cardinal inequalities could be presented as constructions of morphisms from an old product.

We shall show that the answer to his question is negative for both the product and the old product if we require morphisms to consist of Borel maps. (Without this requirement, Yiparaki's result applies and shows that an affirmative answer is consistent, being true in models of CH.) Thus the more complicated structure, described by Vojtáš in his max-min diagram, is essential. But we shall see that this more complicated structure can also be described in terms of a construction of a suitable (more complicated) $\mathbf{C}$ from $\mathbf{A}$ and $\mathbf{B}$.

To motivate this construction and set the stage for the proof of its necessity, we review the proof that $\mathbf{add}(B) \geq \min\{\mathbf{cov}(B), \mathfrak{b}\}$, using the notation and machinery of the preceding section.

Let (x_α, Π_α) for $\alpha < \kappa$ be $\kappa < \min\{\mathbf{cov}(B), \mathfrak{b}\}$ chopped reals. We must produce a single chopped real (y, Θ) engulfing them all. Since $\kappa < \mathbf{cov}(B)$, fix a real y matching all the (x_α, Π_α). This will be the first component of the chopped real we seek; it remains to produce Θ such that, for each $\alpha < \kappa$, for all but finitely many blocks J of Θ, there is a block I of Π_α such that $I \subseteq J$ and $x_\alpha \restriction I = y \restriction I$.

For each α, define $f_\alpha : \omega \to \omega$ by letting $f_\alpha(n)$ be the right endpoint of the next interval of Π_α, after the one containing n, on which y agrees with x_α. (Such an interval exists because y matches (x_α, Π_α).) As $\kappa < \mathfrak{b}$, fix some $g : \omega \to \omega$ eventually majorizing every f_α. Then choose Θ so that, for each of its intervals, g of the left endpoint is smaller than the right endpoint. For each $\alpha < \kappa$, for all but finitely many intervals $J = [a, b]$ of Θ, we have $f_\alpha(a) \leq g(a) < b$ and therefore, by definition of f_α, y agrees with x_α on some interval of Π_α that starts after a and ends before b and thus is contained in J. This shows that (y, Θ) is as required.

Let us describe this proof in terms of the relations $\mathbf{U}$, $\mathbf{V}$, and $\mathbf{W}$ from the preceding section. Recall that

$$\mathbf{U} = (CR, {}^\omega 2, \text{matches}^\smile),$$
$$\mathbf{V} = (CR, CR, \text{is engulfed by}),$$
$$\mathbf{W} = ({}^\omega\omega, {}^\omega\omega, \leq^*),$$

so $\|\mathbf{U}^\perp\| = \mathbf{cov}(B)$, $\|\mathbf{V}^\perp\| = \mathbf{add}(B)$, and $|\mathbf{W}^\perp\| = \mathfrak{b}$. (Thus the inequality $\mathbf{add}(B) \geq \min\{\mathbf{cov}(B), \mathfrak{b}\}$ should correspond to a morphism from some sort of "product" of **U** and **W** to **V**; the problem is to define an appropriate sort of product.)

The proof of $\mathbf{add}(B) \geq \min\{\mathbf{cov}(B), \mathfrak{b}\}$ above began by regarding the given elements (x_α, Π_α) of V_- (for which we needed to find a V-related element of V_+) as elements of U_- and finding a $y \in U_+$ that is U-related to them all. Thus, the proof implicitly used the identity map to convert elements of V_- into elements of U_-. (In other applications, a non-trivial map will occur here.) The next step was to define, from y and (x_α, Π_α), the element $f_\alpha \in W_-$; so the proof uses a map $V_- \times U_+ \to W_-$. (Our definition of f_α presupposed that y matches (x_α, Π_α); to get a total map, let f_α be identically zero if y fails to match (x_α, Π_α).) Finally, from y and an element $g \in W_+$ that is W-related to each f_α, we produced the required $(y, \Theta) \in V_+$ that is V-related to each (x_α, Π_α). The constructions in the proof can be summarized as three maps (where we have omitted the subscripts α from x, Π, and f)

$$\begin{aligned} \alpha &: V_- \to U_- : (x, \Pi) \mapsto (x, \Pi), \\ \beta &: V_- \times U_+ \to W_- : ((x, \Pi), y) \mapsto f, \text{ and} \\ \gamma &: U_+ \times W_+ \to V_+ : (y, g) \mapsto (y, \Theta). \end{aligned}$$

Their key property is that from $U(\alpha(x, \Pi), y)$ and $W(\beta((x, \Pi), y), g)$ we were able to infer $V((x, \Pi), \gamma(y, g))$.

Notice that if β were a function of only (x, Π) rather than both (x, Π) and y, then the situation above would precisely describe a morphism ξ from the old product of **U** and **W** to **V**. Indeed, α and β could then serve as the two components of $\xi_- : V_- \to U_- \times W_-$ and γ could serve as ξ_+.

But β in the proof definitely needs y as an argument, so this proof is not described by a morphism from an old product. It is even less describable in terms of the (new, categorical) product; that would require each element of V_- to have an image in U_- *or* in W_-, not both.

The awkward U_+ in the domain of β can be moved to the codomain by considering what category theorists call the exponential adjoint and computer scientists call currying:

$$\hat{\beta} : V_- \to {}^{U_+}W_- : (x, \Pi) \mapsto (y \mapsto \beta((x, \Pi), y)).$$

The key property of α, β, and γ can, of course, be trivially rewritten in terms of α, $\hat{\beta}$, and γ. The result is that the first two of these are the components of ξ_- and the third is ξ_+ for a morphism ξ to **V** from the following compound of **U** and **W**, which we call their *sequential composition* (cf. also [**3**]):

$$\mathbf{U};\mathbf{W} = (U_- \times {}^{U_+}W_-, U_+ \times W_+, \{((a, \rho), (u, w)) \mid U(a, u) \text{ and } W(\rho(u), w)\}.$$

Notice that, quite generally, a morphism from a sequential composition to another relation can be regarded (via exponential adjointness) as consisting of three functions that enjoy the key property of α, β, and γ discussed above. Such a triple of functions is precisely what Vojtáš [15] describes with his max-min diagram.

The following proposition records the connection between sequential composition and maxima and minima of cardinal characteristics. Note that this proposition has a product of cardinals where the product of relations has a maximum of cardinals and the old product has (as discussed above) something between the maximum and the product of cardinals. Of course, for infinite cardinals (the case we're interested in) all these cardinals coincide.

PROPOSITION 2. *For any relations* $\mathbf{A}$ *and* $\mathbf{B}$, *we have*

$$\|\mathbf{A};\mathbf{B}\| = \|\mathbf{A}\| \cdot \|\mathbf{B}\| \quad \text{and} \quad \|(\mathbf{A};\mathbf{B})^{\perp}\| = \min\{\|\mathbf{A}^{\perp}\|, \|\mathbf{B}^{\perp}\|\}.$$

PROOF. We prove only the least trivial part, namely the $\geq$ half of the first equation. By definition, $\|\mathbf{A};\mathbf{B}\|$ is the smallest possible cardinality for a set $S \subseteq A_+ \times B_+$ such that, for every element $x \in A_-$ and every function $\rho : A_+ \to B_-$, some $(a, b) \in S$ satisfies $A(x, a)$ and $B(\rho(a), b)$. Fix such an S of minimum cardinality; we must show that this cardinality is at least the product of the norms of $\mathbf{A}$ and $\mathbf{B}$.

For each $a \in A_+$, let $S_a = \{b \in B_+ \mid (a, b) \in S\}$. Let $X = \{a \in A_+ \mid |S_a| \geq \|\mathbf{B}\|\}$. It will suffice to prove that X has cardinality at least $\|\mathbf{A}\|$, for then we have at least $\|\mathbf{A}\|$ elements a each S-related to at least $\|\mathbf{B}\|$ elements b and so we have at least $\|\mathbf{A}\| \cdot \|\mathbf{B}\|$ pairs $(a, b) \in S$.

So suppose toward a contradiction that $|X| < \|\mathbf{A}\|$. By definition of norm, we can fix some $x \in A_-$ that is not A-related to any element of X. Using the definition of norm again along with the definition of X, we can choose, for each $a \in A_+ - X$, an element $\rho(a) \in B_-$ that is not B-related to any $b \in S_a$. Extend ρ arbitrarily to a function $\rho : A_+ \to B_-$. By our choice of S, there is some $(a, b) \in S$ with $A(x, a)$ and $B(\rho(a), b)$. From $A(x, a)$ we infer, by our choice of x, that $a \notin X$. But then from $B(\rho(a), b)$ and our choice of ρ it follows that $b \notin S_a$. That contradicts $(a, b) \in S$. □

By virtue of this proposition and the general properties of morphisms, we see that the existence of a morphism $\mathbf{A};\mathbf{B} \to \mathbf{C}$ implies $\|\mathbf{C}\| \leq \|\mathbf{A}\| \cdot \|\mathbf{B}\|$ ($= \max\{\|\mathbf{A}\|, \|\mathbf{B}\|\}$ if the norms are infinite) and $\|\mathbf{C}^{\perp}\| \geq \min\{\|\mathbf{A}^{\perp}\|, \|\mathbf{B}^{\perp}\|\}$. In other words, morphisms from sequential compositions provide a way to present proofs of three-cardinal inequalities relating (in the non-trivial direstion) one cardinal to the maximum or minimum of two others. The example of $\mathbf{add}(B) \geq \min\{\mathbf{cov}(B), \mathfrak{b}\}$ shows that this sort of presentation occurs naturally in one example; other examples are given in Vojtáš's discussion of the max-min diagram [15], and another example (involving a sequential composition one of whose factors is an old product) is discussed in [3].

One unpleasant aspect of sequential composition needs to be discussed. Because of the function-space construction in the definition of sequential composition, even if $\mathbf{A}$ and $\mathbf{B}$ are, as we advocated in Section 2, Borel relations between Borel sets of reals, $\mathbf{A};\mathbf{B}$ will not be of this form, simply because its domain is not a set of reals at all but rather a set one type higher. So it does not yet make sense to talk about Borel morphisms involving sequential compositions.

Fortunately, when we deal (as we did above) with morphisms ξ *from* a sequential composition $\mathbf{A};\mathbf{B}$ to some $\mathbf{C}$, then the troublesome (higher type) part is $\xi_- : C_- \to A_- \times {}^{A_+}B_-$ or more precisely its second component $C_- \to {}^{A_+}B_-$. We can declare such a function to be Borel if and only if its exponential adjoint $C_- \times A_+ \to B_-$ is Borel. In this sense, the morphism involved in our proof of $\mathbf{add}(B) \geq \min\{\mathbf{cov}(B), \mathfrak{b}\}$ is Borel, and so are the morphisms involved in the other proofs to which we alluded two paragraphs ago.

The exponential adjoint trick does not work for morphisms *to* a sequential composition, for then the function space ${}^{A_+}B_-$ occurs in the domain rather than the codomain of a function. The only sensible way to talk about such functions being Borel seems to be to work not with the space of all functions $A_+ \to B_-$ but with the subspace of Borel functions, or rather with the set of Borel codes for such functions. We postpone further discussion of this until after we see, in the next section, an occurrence "in nature" of something that ought to be a Borel morphism to a sequential composition.

5. Non-existence of Borel morphisms

The purpose of this section is to show that, in the proof of the inequality $\mathbf{add}(B) \geq \min\{\mathbf{cov}(B), \mathfrak{b}\}$, one cannot replace the sequential composition $\mathbf{U};\mathbf{W}$ with the product or the old product. In fact, we shall see that $\mathbf{V}$ is in some sense equivalent to $\mathbf{U};\mathbf{W}$. We begin with a proposition showing that products (old or new) do not suffice. It also shows that the order of the components in a sequential composition is essential.

PROPOSITION 3. *Let* $\mathbf{U}$, $\mathbf{V}$, *and* $\mathbf{W}$ *be as in the preceding two sections. There is no Borel morphism to* $\mathbf{V}$ *from the product of* $\mathbf{U}$ *and* $\mathbf{W}$, *nor from their old product, nor from* $\mathbf{W};\mathbf{U}$.

PROOF. Notice first that there are morphisms from $\mathbf{W};\mathbf{U}$ to the old product and from the old product to the product. Both morphisms are the identity on the $+$-components $W_+ \times U_+$. On the $-$-components, we have $W_- \sqcup U_- \to W_- \times U_- \to W_- \times {}^{W_+}U_-$, where the first map is defined by fixing elements $w_0 \in W_-$ and $u_0 \in U_-$ and sending any $w \in W_-$ to (w, u_0) and any $u \in U_-$ to (w_0, u), and the second map is the identity on the first component W_- while on the second component it sends any $u \in U_-$ to the constant function in ${}^{W_+}U_-$ with value u. It is trivial to check that these functions define Borel morphisms. So it suffices to show that there is no Borel morphism $\xi : \mathbf{W};\mathbf{U} \to \mathbf{V}$.

Suppose there were such a ξ. Form a forcing extension of the universe M by first adding a Mathias real m and then, over the resulting model, adding a Cohen real c. Extend ξ to a Borel morphism in $M[m, c]$, still called ξ, with the same Borel code. More precisely, the two functions that constitute ξ correspond (via exponential adjointness in one component) to three Borel maps

$$\alpha : V_- \to W_-, \quad \beta : V_- \times W_+ \to U_-, \quad \gamma : W_+ \times U_+ \to V_+$$

with the key property that, for all $x \in V_-$, $w \in W_+$, and $u \in U_+$, if $W(\alpha(x), w)$ and $U(\beta(x, w), u)$, then $V(x, \gamma(w, u))$. We use the same symbols $U_\pm$, U, $V_\pm$, V, $W_\pm$, W, α, β, and γ for the objects in $M[m, c]$ having the same Borel codes (in M). The key property is a Π^1_1 sentence, so it remains true in the forcing extension.

We apply the key property with an arbitrary $x \in V_- \cap M$, with $w = m$ and with $u = c$. Then, since α is coded in M and since a Mathias real dominates all ground model reals, $\alpha(x) \leq^* m$, i.e., $W(\alpha(x), m)$. Similarly, since β is coded in the ground model M, the chopped real $\beta(x, m)$ is in $M[m]$ and is therefore matched by c (since matching $\beta(x, m)$ is a comeager requirement coded in $M[m]$ and c is a Cohen real over $M[m]$). That is, $U(\beta(x, m), c)$.

By the key property, it follows that $V(x, \gamma(m, c))$. In other words, we have a chopped real $\gamma(m, c) \in M[l, c]$ that engulfs every chopped real x from the ground model M. Recalling the connection between chopped reals (related by engulfing) and meager sets (related by inclusion) from Section 3, we find that all the meager Borel sets coded in M are subsets of a single meager set in $M[m, c]$. But Pawlikowski has shown that $M[m, c]$ has no meager set that includes all the meager Borel sets coded in M; see [**9**], Proposition 1.2 and the discussion following Corollary 1.3. This contradiction shows that no such morphism ξ can exist. □

The preceding proposition tells us that, of the various combinations of **U** and **W** considered so far, **U**; **W** is the only one admitting a Borel morphism to **V** and therefore the only one that can be used to prove $\mathbf{add}(B) \geq \min\{\mathbf{cov}(B), \mathfrak{b}\}$. In fact, Corollary 1.3 of [**9**] strongly suggests that **U**; **W** is in some sense equivalent to **V**. We state this corollary as the next proposition and give a proof somewhat different from that in [**9**] (in that we use more forcing but do not use, e.g., the Kuratowski-Ulam theorem) in order to suggest what the proper sense of equivalence between **U**; **W** and **V** should be.

PROPOSITION 4 (PAWLIKOWSKI). *Let M be any inner model. The union of all the meager Borel sets coded in M is meager if and only if there is a Cohen real c over M and there is a real d dominating all the reals of $M[c]$.*

PROOF. The "if" half is immediate from the existence of a Borel morphism ξ : **U**; **W** $\to$ **V** with code in M. Indeed, if c and d are as in the statement of the proposition, if x is any chopped real in M, and if $\alpha : V_- \to U_-$, $\beta : V_- \times U_+ \to W_-$, and $\gamma : U_+ \times W_+ \to V_+$ are the parts of ξ (as above), then we have $U(\alpha(x), c)$ (because c is a Cohen real over M which contains $\alpha(x)$), and $W(\beta(x, c), d)$ (because $\beta(x, c)$ is in $M[c]$ which d dominates), and therefore $V(x, \gamma(c, d))$ (because ξ is a morphism). Thus, $\gamma(c, d)$ engulfs all chopped reals $x \in M$, and therefore it codes a meager set that includes all meager Borel sets coded in M.

For the "only if" direction, suppose we have a meager set that includes all the meager Borel sets coded in M. Then, by Section 3, we have a chopped real (y, Θ) that engulfs all the chopped reals of M. Then clearly y matches every chopped real from M and is therefore Cohen generic over M. It remains to produce a real g dominating all the reals of $M[y]$. We claim that such a g is obtained by letting $g(n)$ be the right endpoint of the next interval in Θ after the one that contains n.

To verify that this g works, we consider any $f : \omega \to \omega$ in $M[y]$ and show that $f \leq^* g$. We may suppose without loss of generality that f is non-decreasing. Since f belongs to the Cohen extension $M[y]$, it is the denotation with respect to y of some name $\dot{f} \in M$. (Here and in the rest of this proof, names and forcing are with respect to the usual notion of forcing for adding a Cohen real, ${}^{<\omega}2$ ordered by reverse inclusion.) We assume without loss of generality that all conditions force "$\dot{f}$ is a function from ω to ω."

We define a chopped real $(x, \Pi) \in M$ with the property that, if $[a, b]$ is any interval in Π and if p is any Cohen condition that (has length at least b and) agrees with x on $[a, b]$, then p forces "$\dot{f}(a) \leq b$." We proceed by induction. After n intervals of Π and the restrictions of x to those intervals have been defined, we produce the next interval $[a, b]$ and the restriction of x to it as follows. Of course a is the first number not in the intervals already defined. Fix a list $u_0, u_1, \ldots, u_{r-1}$ of all the functions $a \to 2$. We inductively define functions $x_i : [a, l_i) \to 2$ with $a = l_0 \leq l_1 \leq \cdots \leq l_r$ by starting with the empty function as x_0 and obtaining x_{i+1} as an extension of x_i such that $u_i \cup x_{i+1}$ forces a particular value v_i for $\dot{f}(a)$. Such an extension exists because $\dot{f}$ is forced to be a total function on ω. After x_r and l_r have been reached, let b be the largest of l_r and all the v_i, and extend x_r arbitrarily to $[a, b]$; this defines the restriction of the desired x to the next interval $[a, b]$ of Π. If a condition p agrees with x on $[a, b]$ then it agrees with some u_i on a and therefore extends $u_i \cup x_{i+1}$ and forces "$\dot{f}(a) = v_i \leq b$" as required.

As (x, Π) is in M, it is engulfed by (y, Θ). That is, each interval $[m, n]$ of Θ, with only finitely many exceptions, includes an interval $[a, b]$ of Π on which x and y agree. Then the initial segment $y \restriction b$ of y is a condition of the sort considered in defining x, so it forces "$\dot{f}(a) \leq b$." Since y is Cohen generic, this forced statement is true in $M[y]$, i.e., $f(a) \leq b$. But then, as f is non-decreasing, we have, for all elements k of the interval of Θ immediately preceding $[m, n]$, that

$$f(k) \leq f(a) \leq b \leq n = g(k).$$

This applies to all sufficiently large $k \in \omega$ because $[m, n]$ can be any interval of Θ with only finitely many exceptions. Therefore, $f \leq^* g$, as required. □

We would like to regard the "only if" half of the preceding proof as presenting a morphism $\eta : \mathbf{V} \to \mathbf{U}; \mathbf{W}$ in the direction opposite to the ξ involved in the "if" half. This would mean that the proof involves (1) a contruction η_- whose input consists of a chopped real (z, Ψ) (from U_-) and a function $\Phi : U_+ \to W_-$ and whose output is a chopped real (x, Π) and (2) a construction η_+ whose input consists of a chopped real (y, Θ) and whose output consists of a real c and an element g of W_+. The required key property for a morphism is that (with the notation of (1) and (2)), if (y, Θ) engulfs (x, Π), then c matches (z, Ψ) and g dominates $\Phi(c)$.

Our proof looks vaguely but not exactly like this. Notice that a Cohen-forcing name $\dot{f}$ of a real, as in our proof of the proposition, defines a function Φ into ${}^{\omega}\omega$ from a comeager subset D of ${}^{\omega}2$, namely, $\Phi(u)(n)$ is the unique number forced to be the value of $\dot{f}(n)$ by some initial segment of u. (D consists of those u whose initial segments force values for all $\dot{f}(n)$; this D is comeager because $\dot{f}$ is forced to name a real.) For a Φ of this special sort (or, more precisely, for any extension of such a Φ to all of ${}^{\omega}2 = U_+$), our proof produced, from $\dot{f}$, a certain (x, Π). For the output of η_-, when the input consists of (z, Ψ) and such a special Φ, we take a chopped real that engulfs (z, Ψ), the (x, Π) constructed from $\dot{f}$ in the proof, and an (x', Π') such that $\text{Match}(x', \Pi')$ is included in the proper domain D of Φ. (It is trivial to produce a chopped real engulfing any finitely (or even countably) many given chopped reals.) For the output of η_+ on input (y, Θ), we take the pair whose first component is y and whose second component is the g constructed from Θ during the proof (the "end of the next interval" function). Then the key property that η needs is just what is established by the proof.

All this, however, was done only for the very special Φ's that correspond to names $\dot{f}$. The proof gives no hint what η_- should do if its second argument Φ is not of this form. This difficulty is, of course, connected with the difficulty mentioned earlier that sets like ${}^{U_+}W_-$ are one type higher than the (Borel) sets of reals that our theory is equipped to handle.

Both difficulties can be attacked by working not with all functions $U_+ \to W_-$ but with some subfamily that can be coded by reals (or with the codes rather than the functions). In the case at hand, it is tempting to take this family to be just those Φ's that are given by Cohen names $\dot{f}$; these names, which are essentially reals, can then serve as the codes. But this choice of a subfamily is obviously tailored to just this one example. We can do much better by noticing that any Borel function $\Phi : U_+ \to W_-$ agrees on some comeager set A with the function given by some $\dot{f}$. Thus, we can allow Borel functions as inputs to η_- at the cost of redefining the output of η_- so that it also engulfs a chopped real whose matching set is included in A.

The restriction to Borel functions is one that can be sensibly imposed in general. That is, we can modify the general definition of sequential composition $\mathbf{A}; \mathbf{B}$, when all the components of $\mathbf{A}$ and $\mathbf{B}$ are Borel sets and relations, by replacing the set ${}^{A_+}B_-$ of functions with the subset of Borel functions or, better, with the set of Borel codes for such functions.

To be specific about the coding, we note that Borel functions are precisely the functions recursive in a real and the type 2 object ${}^2E$ (see for example [**6**], Theorem VI.1.8). So as a code for such a function we can use the pair of the Kleene index (a natural number) for this recursive function and the real parameter.

This framework covers the standard examples, including the particular ones we have discussed. It should, however, probably be extended a bit because the set of codes of Borel maps from a Borel set to a Borel set is in general not itself a Borel set; it is only $\mathbf{\Pi}^1_1$. So it seems reasonable to allow the domains and codomains of relations to be $\mathbf{\Pi}^1_1$ (i.e., semi-recursive in ${}^2E$ and a real) sets of reals; the relations should probably still be required to be Borel in the weak sense that they and their complements relative to the domain and codomain are $\mathbf{\Pi}^1_1$ (so that dualization works). Morphisms should be functions with $\mathbf{\Pi}^1_1$ graphs, i.e., partial recursive in ${}^2E$ and reals but total on the $\mathbf{\Pi}^1_1$ sets in question.

Rather than speculate further on the basis of very limited examples, let me just list what I would like to see in an ideal framework. The sets, relations, and morphisms should come from a class that is broad enough to encompass the known examples but narrow enough to have absoluteness properties so that non-existence of incorrect morphisms can be proved. Furthermore, the framework should be closed under the naturally occurring constructions, including a suitable version of sequential composition. I believe that the framework described in the preceding paragraph may do all this, but this has not yet been fully checked.

I should perhaps mention that Pawlikowski told me that, with suitable coding, the components of morphisms that occur in classical proofs of cardinal characteristic inequalities can be taken to be not merely Borel but continuous. It is not clear, however, that this helps in the present situation, since the (natural) codes for continuous functions form only a $\mathbf{\Pi}^1_1$ set.

6. Unsplitting

In this section, we briefly discuss a situation where sequential composition arises naturally in connection with an inequality involving just two cardinal characteristics, and a rather trivial inequality at that. Recall that the unsplitting number $\mathfrak{r}$ is the smallest number of infinite subsets of ω such that no single set splits them all into two infinite pieces. Equivalently, it is the norm of $\mathbf{R} = ({}^\omega 2, \mathcal{P}_\infty(\omega), R)$ where $R(f, Y)$ means that f is almost constant on Y, i.e., is constant on $Y - F$ for some finite F.

We define a similar characteristic $\mathfrak{r}_3$ using splittings into three pieces rather than two. That is, $\mathfrak{r}_3$ is the norm of $\mathbf{R}_3 = ({}^\omega 3, \mathcal{P}_\infty(\omega), R_3)$ where again $R_3(f, Y)$ means that f is almost constant on Y.

It is easy to see that $\mathfrak{r}_3 = \mathfrak{r}$. In fact, there are morphisms $\xi : \mathbf{R}_3 \to \mathbf{R}$ and $\eta : \mathbf{R}; \mathbf{R} \to \mathbf{R}_3$ defined as follows. ξ_- is the inclusion map ${}^\omega 2 \to {}^\omega 3$ and ξ_+ is the identity map of $\mathcal{P}_\infty(\omega)$. To define η, we first introduce, for $f : \omega \to 3$, the notations f' and f'' for the functions $\omega \to 2$ obtained by identifying 2, whenever it occurs as a value of f, with 0 or 1 respectively. That is,

$$f' : \omega \to 2 : n \mapsto \begin{cases} 0, & \text{if } f(n) = 0 \\ 1, & \text{otherwise,} \end{cases}$$

and f'' is defined similarly except that the first case is "if $f(n) = 0$ or 2." Also fix, for each infinite $Y \subseteq \omega$ a bijection $e_Y : \omega \to Y$, for example the unique increasing bijection. Now we define η as follows. η_- sends $f : \omega \to 3$ to the pair consisting of f' and the map $G : \mathcal{P}_\infty(\omega) \to {}^\omega 2$ defined by $G(Y) = f'' \circ e_Y$. (In other words, $G(Y)$ is essentially $f'' \restriction Y$ but with its domain shifted from Y to ω by e_Y.) η_+ sends a pair of infinite sets (Y, Y') to $e_Y(Y')$, the set that "occupies in Y the locations that Y' occupies in ω." It is easy to check that ξ and η are morphisms, and therefore $\mathfrak{r} \leq \mathfrak{r}_3 \leq \mathfrak{r} \cdot \mathfrak{r} = \mathfrak{r}$.

The idea behind the morphism η is that to get a 3-unsplit family (in the obvious sense), it suffices to start with a 2-unsplit family $\mathcal{R}$ and then, within each of its members Y, form a new 2-unsplit family by transferring $\mathcal{R}$ from ω to Y via e_Y. The union of these new, transferred families is 3-unsplit because, if $f : \omega \to 3$, then there is some $Y \in \mathcal{R}$ on which f takes at most two values infinitely often and there is some $Y' \in e_Y[\mathcal{R}]$ on which those two values are reduced to one.

Notice that the "second order" families $e_Y[\mathcal{R}]$ depend in an essential way on the Y's. In other words, the domain of η in this proof apparently needs to be a sequential composition. There is no evident way to use a product or even an old product instead.

We therefore conjecture that there is no Borel morphism from the old product of two copies of $\mathbf{R}$ to $\mathbf{R}_3$. Intuitively, this means that, in the two uses of a 2-unsplit family $\mathcal{R}$ to produce a 3-unsplit family, the second use cannot be made independent of the first.

Although the conjecture seems highly plausible, we do not know how to prove even the weaker conjecture that there is no Borel morphism from $\mathbf{R}$ to $\mathbf{R}_3$. The intuitive meaning of this is merely that, to produce a 3-unsplit family, we must use the 2-unsplit family $\mathcal{R}$ twice, not just once.

To see the difficulty in proving these conjectures, recall that our previous proofs of non-existence of Borel morphisms involved the construction of suitable forcing models. To apply this method directly to prove the conjecture, we would want to find a forcing extension that contains an infinite subset Y of ω unsplit by all ground model functions $\omega \to 2$ but split by some ground model function $f : \omega \to 3$. This clearly cannot be achieved, since a set unsplit by f' and f'' (in the notation of the definition of η above) is also unsplit by f.

An alternative approach would be concentrate on the dual morphism $\mathbf{R}_3{}^\perp \to \mathbf{R}^\perp$. Now the forcing method suggests building a forcing extension in which every function $\omega \to 2$ is almost constant on an infinite Y from the ground model but some $f : \omega \to 3$ is not. But this cannot be achieved either. If $f : \omega \to 3$, then there is an infinite Y in the ground model on which f' is almost constant. Furthermore, $f'' \circ e_Y$ is almost constant on some Y' in the ground model. If e_Y was defined reasonably (e.g., as the increasing bijection), then it too is in the ground model, and so is $e_Y(Y')$ on which f is almost constant.

So the forcing method seems to be of no use in proving even the weak form of the conjecture. Some direct analysis of Borel morphisms seems to be needed.

References

1. T. Bartoszyński, *Additivity of measure implies additivity of category*, Trans. Amer. Math. Soc. **281** (1984), 209–213.
2. T. Bartoszyński, H. Judah, and S. Shelah, *The Cichoń diagram*, J. Symbolic Logic **58** (1993), 401–423.
3. A. Blass, *Questions and answers — a category arising in linear logic, complexity theory, and set theory*, Proceedings of a linear logic workshop held at MSI, Cornell, in June, 1993 (J.Y. Girard, Y. Lafont, and L. Regnier, eds.).
4. E. van Douwen, *The integers and topology*, Handbook of Set Theoretic Topology (K. Kunen and J. Vaughan, eds.), North-Holland, 1984, pp. 111-168.
5. D. Fremlin, *Cichoń's diagram*, Séminaire Initiation à l'Analyse (G. Choquet, M. Rogalski, and J. Saint-Raymond, eds.), Univ. Pierre et Marie Curie, 1983/84, pp. (5-01)–(5-13).
6. P. Hinman, *Recursion Theoretic Hierarchies*, Perspectives in Mathematical Logic, Springer-Verlag, 1977.
7. A. Miller, *Some properties of measure and category*, Trans. Amer. Math. Soc. **266** (1981), 93-114.
8. V.C.V. de Paiva, *A Dialectica-like model of linear logic*, Category Theory and Computer Science (D.H. Pitt, D.E. Rydeheard, P. Dybjer, A. Pitts, and A. Poigné, eds.), Lecture Notes in Computer Science 389, Springer-Verlag, 1989, pp. 341–356.
9. J. Pawlikowski, *Why Solovay real produces Cohen real*, J. Symbolic Logic **51** (1986), 957–968.
10. F. Rothberger, *Eine Äquivalenz zwischen der Kontinuumhypothese und der Existenz der Lusinschen und Sierpińskischen Mengen*, Fund. Math. **30** (1938), 215–217.
11. R.C. Solomon, *Families of sets and functions*, Czechoslovak Math. J. **27** (1977), 556–559.
12. M. Talagrand, *Compacts de fonctions mesurables et filtres non mesurables*, Studia Math. **67** (1980), 13–43.
13. J. Truss, *Sets having calibre* $\aleph_1$, Logic Colloquium 76 (R.O. Gandy and J.M.E. Hyland, eds.), Studies in Logic and Foundations of Math. 87, North-Holland, 1977, pp. 595–612.
14. J. Vaughan, *Small uncountable cardinals and topology*, Open Problems in Topology, ed. by J. van Mill and G.M. Reed, North-Holland, 1990, pp. 195-218.
15. P. Vojtáš, *Generalized Galois-Tukey connections between explicit relations on classical objects of real analysis*, Set Theory of the Reals (H. Judah, ed.), Israel Mathematical Conference Proceedings 6, American Mathematical Society, 1993, pp. 619–643.
16. O. Yiparaki, *On Some Tree Partitions*, Ph.D. thesis, University of Michigan, 1994.

MATHEMATICS DEPT., UN VERSITY OF MICHIGAN, ANN ARBOR, MI 48109, U.S.A.

E-mail address: ablass@umich.edu

Contemporary Mathematics
Volume **192**, 1996

Filter Games and Combinatorial Properties of Strategies

CLAUDE LAFLAMME

ABSTRACT. We characterize winning strategies in various infinite games involving filters on the natural numbers in terms of combinatorics or structural properties of the given filter. These generalize several ultrafilter games of Galvin.

1. Introduction

We look at various infinite games between two players **I** and **II** involving filters on the natural numbers in which **I** either plays cofinite sets, members of $\mathcal{F}$ or $\mathcal{F}^+$ and player **II** responds with an element or a finite subset of **I** 's move, depending on the game. In each version, the outcome depends on the set produced by player **II** , whether it belongs to the given filter $\mathcal{F}$, $\mathcal{F}^+$ or even $\mathcal{F}^*$, the dual ideal.

In each game considered, we will characterize winning strategies of either player in terms of combinatorics of the given filter $\mathcal{F}$; these combinatorics turn out to be generalizations of the classical notions of P-points, Q-points and selectivity for ultrafilters. In the case of ultrafilters, $\mathcal{F} = \mathcal{F}^+$ and most of our games to ones already studied by Galvin in unpublished manuscripts [**5**]; the various generalized combinatorics enjoyed by the filters become equivalent.

Several characterizations of Ramsey ultrafilters and P-points were known from works of Booth ([**4**]) and Kunen ([**8**]), and some generalizations of these combinatorics to filters were already made by Grigorieff in [**6**] where for example the notion of P-filter is characterized in terms of branches through certain trees; we shall see that this is very much in the spirit of winning strategies for certain games.

1991 *Mathematics Subject Classification.* Primary 04A20; Secondary 03E05,03E15,03E35.
This research was partially supported by NSERC of Canada.

Other variations of these games for ultrafilters can be found in Chapter VI of Shelah's book [**11**], and two of the games below have been analyzed by Bartoszynski and Scheepers [**2**].

We wish to thank Chris Leary and the referree for helpful suggestions and corrections regarding the present paper.

Our terminology is standard but we review the main concepts and notation. The natural numbers will be denoted by ω, $\wp(\omega)$ denotes the collection of all its subsets. Given $X \in \wp(\omega)$, we write $[X]^\omega$ and $[X]^{<\omega}$ to denote the infinite or finite subsets of X respectively. We use the well known 'almost inclusion' ordering between members of $[\omega]^\omega$, i.e. $X \subseteq^* Y$ if $X \setminus Y$ is finite. We identify $\wp(\omega)$ with ${}^{\omega}2$ via characteristic functions. The space ${}^{\omega}2$ is further equipped with the product topology of the discrete space $\{0,1\}$. A basic neighbourhood is then given by sets of the form

$$\mathcal{O}_s = \{f \in {}^{\omega}2 : s \subseteq f\}$$

where $s \in {}^{<\omega}2$, the collection of finite binary sequences. The terms "nowhere dense", "meager", "Baire property" all refer to this topology. Concatenation of elements $\overline{s}, \overline{t} \in {}^{<\omega}\omega$ will be written $\overline{s}\,{}^\frown\overline{t}$.

A filter is a collection of subsets of ω closed under finite intersections, supersets and containing all cofinite sets; it is called proper if it contains only infinite sets. For a filter $\mathcal{F}$, $\mathcal{F}^+$ denotes the collection of all sets X such that $\langle \mathcal{F}, X\rangle$ is a proper filter; it is useful to notice that $X \in \mathcal{F}^+$ if and only if $X^c \notin \mathcal{F}$. $(\mathcal{F}^+)^c = \wp(\omega) \setminus \mathcal{F}^+$, the collection of sets incompatible with $\mathcal{F}$ is the dual ideal and is usually denoted by $\mathcal{F}^*$. The *Fréchet* filter is the collection of cofinite sets, denoted by $\mathfrak{Fr}$.

The families $\mathcal{F}$ and $\mathcal{F}^+$ are dual in a different sense; this means that a set X containing an element of each member of $\mathcal{F}$ (resp. $\mathcal{F}^+$) must belong to $\mathcal{F}^+$ (resp. $\mathcal{F}$). In particular $\mathfrak{Fr}$ and $[\omega]^\omega$ are dual. From more general work of Aczel ([**1**]) and Blass ([**3**]), there is a duality between games in which a player chooses $X_k \in \mathcal{F}$ while the other player responds with $n_k \in X_k$, and games in which a player chooses $Y_k \in \mathcal{F}^+$ while the other player responds with $n_k \in Y_k$. The point is that the statements

$$(\forall X \in \mathcal{F})(\exists n \in X)\phi(n) \text{ and } (\exists Y \in \mathcal{F}^+)(\forall n \in Y)\phi(n)$$

are equivalent.

The following important result characterizes meager filters in terms of combinatorial properties.

PROPOSITION 1.1. *(Talagrand ([**12**])) The following are equivalent for a filter $\mathcal{F}$:*

(i) *$\mathcal{F}$ has the Baire property.*
(ii) *$\mathcal{F}$ is meager.*

(iii) *There is a sequence* $n_0 < n_1 < \cdots$ *such that*

$$(\forall X \in \mathcal{F})(\forall^\infty k)\ X \cap [n_k, n_{k+1}) \neq \emptyset.$$

Combinatorial properties of filters have played an important role in applications of Set Theory, and the classical notions of a filter being meager, a P-filter or selective have been around a long time. These concepts will be generalized below in terms of trees and other structural properties; these combinatorial ideas have their roots in Ramsey theory where P-points and selective ultrafilters (sometimes called 'Ramsey') have characterizations in term of these trees; this can be found in the papers by Booth [**4**] and Grigorieff [**6**].

We call a tree $\mathcal{T} \subseteq {}^{<\omega}\omega$ an $\mathcal{X}$-tree for some $\mathcal{X} \subseteq [\omega]^\omega$ ($\mathcal{X}$ will usually be a filter $\mathcal{F}$ or $\mathcal{F}^+$), if for each $\overline{s} \in \mathcal{T}$, there is an $X_{\overline{s}} \in \mathcal{X}$ such that $\overline{s}\,{}^\frown n \in \mathcal{T}$ for all $n \in X_{\overline{s}}$. Similarly we call a tree $\mathcal{T} \subseteq {}^{<\omega}([\omega]^{<\omega})$ an $\mathcal{X}$-tree of finite sets for some $\mathcal{X} \subseteq [\omega]^\omega$, if for each $\overline{s} \in \mathcal{T}$, there is an $X_{\overline{s}} \in \mathcal{X}$ such that $\overline{s}\,{}^\frown a \in \mathcal{T}$ for each $a \in [X_{\overline{s}}]^{<\omega}$. A branch of such a tree is thus an infinite sequence of finite sets and we will be interested in the union of such a branch, an infinite subset of ω.

Here are a few more combinatorial properties of filters that we will consider.

DEFINITION 1.2. *Let* $\mathcal{F}$ *be a filter on* ω.

(i) $\mathcal{F}$ *is called a Q-filter if for any partition of* ω *into finite sets* $\langle s_k : k \in \omega\rangle$, *there is an* $X \in \mathcal{F}$ *such that* $|\ X \cap s_k\ | \leq 1$ *for all* k.

(ii) $\mathcal{F}$ *is called a weak Q-filter if for any partition of* ω *into finite sets* $\langle s_k : k \in \omega\rangle$, *there is an* $X \in \mathcal{F}^+$ *such that* $|\ X \cap s_k\ | \leq 1$ *for all* k.

(iii) $\mathcal{F}$ *is called diagonalizable if there is an* $X \in [\omega]^\omega$ *such that* $X \subseteq^* Y$ *for all* $Y \in \mathcal{F}$.

(iv) $\mathcal{F}$ *is called* ω-+-*diagonalizable if there are* $\langle X_n \in \mathcal{F}^+ : n \in \omega\rangle$ *such that for each* $Y \in \mathcal{F}$, *there is an* n *such that* $X_n \subseteq^* Y$.

(v) $\mathcal{F}$ *is called* ω-*diagonalizable if there are* $\langle X_n \in [\omega]^\omega : n \in \omega\rangle$ *such that for each* $Y \in \mathcal{F}$, *there is an* n *such that* $X_n \subseteq^* Y$.

(vi) *A set* $X \subseteq [\omega]^{<\omega}$ *is called* $\mathcal{Z}$-*universal* ($\mathcal{Z}$ *will be* $\mathcal{F}$ *or* $\mathcal{F}^+$) *if for each* $Y \in \mathcal{Z}$, *there is an* $x \in X \cap [Y]^{<\omega}$. $\mathcal{F}$ *is called* ω-diagonalizable by $\mathcal{Z}$-universal sets *if there are* $\mathcal{Z}$-*universal sets* $\langle X_n : n \in \omega\rangle$ *such that for all* $Y \in \mathcal{F}$, *there is an* n *such that* $x \cap Y \neq \emptyset$ *for all but finitely many* $x \in X_n$.

(vii) $\mathcal{F}$ *is a* P-filter *if given any sequence* $\langle X_n : n \in \omega\rangle \subseteq \mathcal{F}$,*there is an* $X \in \mathcal{F}$ *such that* $X \subseteq^* X_n$ *for each* n.

(viii) $\mathcal{F}$ *is a* P$^+$-filter *if every* $\mathcal{F}^+$-*tree of finite sets has a branch whose union is if* $\mathcal{F}^+$.

(ix) $\mathcal{F}$ *is a* weak P-filter *if given any sequence* $\langle X_n : n \in \omega\rangle \subseteq \mathcal{F}$, *there is an* $X \in \mathcal{F}^+$ *such that* $X \subseteq^* X_n$ *for each* n. *Equivalently, every* $\mathcal{F}$-*tree of finite sets has a branch whose union is in* $\mathcal{F}^+$.

(x) $\mathcal{F}$ *is* Ramsey *if any* $\mathcal{F}$-*tree has a branch in* $\mathcal{F}$; *equivalently,* $\mathcal{F}$ *is both a Q-filter and a P-filter.*

(xi) $\mathcal{F}$ *is* +-Ramsey *if every* $\mathcal{F}^+$*-tree has a branch in* $\mathcal{F}^+$.

(xii) $\mathcal{F}$ *is* weakly Ramsey *if any* $\mathcal{F}$*-tree has a branch in* $\mathcal{F}^+$.

(xiii) $\mathcal{F}$ *is a P-point if it is an ultrafilter that is also a P-filter.*

The following diagram shows the provable implications among these filter properties.

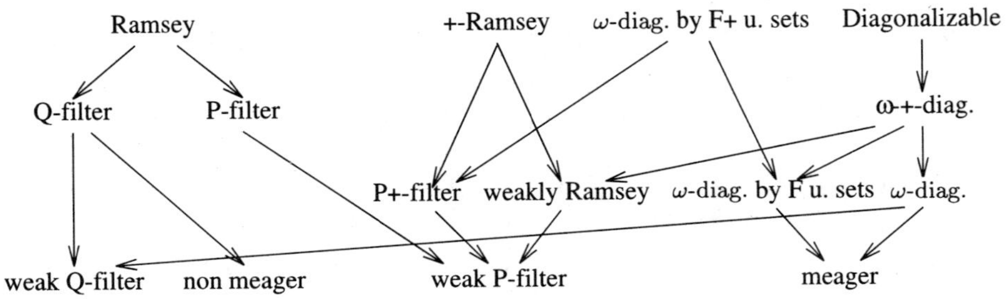

These implications can be proved directly, but will all follow from the results of this paper except for a few trivial ones. Also interesting, if not more, are the *non* implications; most are routine but we present here a few interesting cases.

EXAMPLE 1.1. *A filter* $\mathcal{F}$ ω*-diagonalizable by* $\mathcal{F}^+$*-universal sets but not a weak Q-filter.*

Partition ω into disjoint intervals $\mathcal{I}_n$ such that $|\mathcal{I}_n| \geq n$. Define

$$\mathcal{F} = \{X \subseteq \omega : (\exists m)(\forall n)\ |X^c \cap \mathcal{I}_n| \leq m\}.$$

This is certainly a filter and is ω-diagonalized by the $\mathcal{F}^+$-universal sets $\mathcal{S}_k = \bigcup_n [\mathcal{I}_n]^k$, $k \in \omega$. However, $\mathcal{F}$ contains the complement of each set which meets each $\mathcal{I}_n$ in at most one point and therefore is not a weak Q-filter. □

EXAMPLE 1.2. *An* ω*-diagonalized filter* $\mathcal{F}$ *which is not a weak P-filter and neither* ω*-diagonalizable by* $\mathcal{F}$*-universal sets.*

Let

$$\mathcal{F} = \mathfrak{Fr} \otimes \mathfrak{Fr} = \{X \subseteq \omega \times \omega : \{n : \{m :< n, m >\in X\} \text{ is cofinite }\} \text{ is cofinite }\}.$$

Then $\mathcal{F}$ is ω-diagonalized by the sets $\{< n, m >: m \in \omega\}$, for $n \in \omega$.
$\mathcal{F}$ is not a weak P-filter as the sets $X_n = \{< k, m >: k \geq n, m \in \omega\} \in \mathcal{F}$ show. Also given $\mathcal{F}$-universal sets $\mathcal{S}_n$, $n \in \omega$, we can easily find an $X \in \mathcal{F}$ such that X has empty intersection with infinitely many members of each $\mathcal{S}_n$, and therefore they do not ω-diagonalize $\mathcal{F}$. □

EXAMPLE 1.3. *Diagonalizable,* ω*-diagonalizable and* ω*-+-diagonalizable filters.*

If $\mathcal{F}$ is a P-filter, then $\mathcal{F}$ is diagonalizable if and only if it is ω-diagonalizable if and only if it is $\omega - +$-diagonalizable. On the other hand, these notions are distinct.
The filter

$$\mathcal{F} = \mathfrak{Fr} \otimes \mathfrak{Fr} = \{X \subseteq \omega \times \omega : \{n : \{m : (n,m) \in X\} \text{ is cofinite } \} \text{ is cofinite } \}$$

is ω-diagonalizable, but not ω-$+$-diagonalizable. Ssimilarly, if $\mathcal{G}$ is any non-diagonalizable filter (any non-meager filter will do), then the filter

$$\mathcal{F} = \{X \subseteq \omega \times \omega : \text{ for each } n, \{m : (n,m) \in X\} \in \mathcal{G}, \text{ and } \{n : \{m : (n,m) \in X\} \text{ is cofinite } \} \text{ is cofinite } \}$$

is ω-$+$-diagonalizable, but not diagonalizable. □

If $\mathcal{F}$ is an ultrafilter, then $\mathcal{F}$ is a P-filter if and only if it is a weak P-filter if and only if it is a P$^+$-filter; it is a Q-filter if and only if it is a weak Q-filter, and $\mathcal{F}$ is Ramsey if and only if it is weakly Ramsey if and only if it is $+$-Ramsey. Therefore these notions generalize the classical combinatorial properties of ultrafilters, but again, these notions can be seen to be different for filters in general.

The notions of ω-diagonalizability by $\mathcal{F}$-(resp $\mathcal{F}^+$)-universal sets are generalizations of the regular ω-(resp. $+$)-diagonalizability. Observe that a filter $\mathcal{F}$ is ω-diagonalizable by $\mathfrak{Fr}$-universal sets if and only if diagonalizable by a single $\mathfrak{Fr}$-universal set if and only if it is meager, and ω-$+$-diagonalizability implies ω-diagonalizability by $\mathcal{F}$-universal sets. This notion appears to be new.

Tree combinatorics is what most interest us in this paper as they naturally occur in terms of winning strategies; our main effort is then to express these combinatorics in terms of more familiar concepts. The following lemma shows the spirit of the paper.

LEMMA 1.3. *$\mathcal{F}$ is a non-meager P-filter if and only if every $\mathcal{F}$-tree of finite sets has a branch whose union is in $\mathcal{F}$.*

Proof: Assume first that every $\mathcal{F}$-tree of finite sets has a branch whose union is in $\mathcal{F}$. Given a descending sequence $\langle A_n; n \in \omega\rangle \subseteq \mathcal{F}$, define an $\mathcal{F}$-tree $\mathcal{T} \subseteq {}^{<\omega}([\omega]^{<\omega})$ such that $X_{\overline{s}} = A_n$ for each $\overline{s} \in \mathcal{T} \cap {}^{n}([\omega]^{<\omega})$. Any branch through $\mathcal{T}$ whose union is in $\mathcal{F}$ shows that $\mathcal{F}$ is a P-filter. To verify that $\mathcal{F}$ is non-meager, consider an increasing sequence of natural numbers $n_0 < n_1 < \dots$ and build again an $\mathcal{F}$-tree of finite sets $\mathcal{T}$ as follows. Having already $\overline{s} \in \mathcal{T}$, choose k such that $\overline{s} \in {}^{<\omega}([n_k]^{<\omega})$ and let $X_{\overline{s}} = \omega \setminus n_{k+1}$. The union of any branch through $\mathcal{T}$ misses infinitely many intervals of the form $[n_k, n_{k+1})$ and therefore shows that $\mathcal{F}$ is also non-meager.

Now assume that $\mathcal{F}$ is a non-meager P-filter and let $\mathcal{T}$ be an $\mathcal{F}$-tree of finite sets. Define $n_0 = 0$ and $A_0 = X_\emptyset$. More generally, given $A_0 \supseteq A_1 \supseteq \dots \supseteq A_k$ and $n_0 < n_1 < \dots < n_k$, let $n_{k+1} > n_k$ in A_k and put

$$A_{k+1} = \bigcap \{X_{\langle s_0, s_1, \cdots, s_i\rangle} : i \leq k+1, s_j \subseteq n_{k+1}\},$$

Now as $\mathcal{F}$ is a P-filter, there is a $Y \in \mathcal{F}$ such that $Y \subseteq^* A_n$ for each n and we may as well assume by a reindexing that

$$Y \setminus n_{k+1} \subseteq A_k \text{ for each } k.$$

Now as $\mathcal{F}$ is also non-meager, we can find an infinite set $K = \{k_\ell : \ell \in \omega\}$ and we might as well assume that for each ℓ,

$$Y \cap [n_{k_\ell}, n_{k_\ell+1}) = \emptyset.$$

Define $s_\ell = Y \cap [n_{k_\ell}, n_{k_{\ell+1}}) = Y \cap [n_{k_\ell+1}, n_{k_{\ell+1}})$ for each ℓ. We claim that $\langle s_k : k \in \omega \rangle$ is a branch through $\mathcal{T}$; indeed

$$\begin{aligned} s_\ell &= Y \cap [n_{k_\ell}, n_{k_{\ell+1}}) \\ &= Y \cap [n_{k_\ell+1}, n_{k_{\ell+1}}) \\ &\subseteq A_{k_\ell} \subseteq X_{\langle s_0, s_1, \cdots, s_{\ell-1} \rangle}. \end{aligned}$$

But this concludes the proof as its union Y is in $\mathcal{F}$. □

2. Filter Games

We will be interested in infinite games of the form $\mathfrak{G}(\mathcal{X}, \mathcal{Y}, \mathcal{Z})$ where $\mathcal{X}$ will usually be a filter $\mathcal{F}$ or $\mathcal{F}^+$, $\mathcal{Y}$ will be either ω or $[\omega]^{<\omega}$, and $\mathcal{Z}$ will be either $\mathcal{F}$, $\mathcal{F}^+$, $\mathcal{F}^c$, the complement of $\mathcal{F}$ in $\wp(\omega)$ or else $(\mathcal{F}^+)^c = \mathcal{F}^*$, the dual ideal.

The game $\mathfrak{G}(\mathcal{X}, \mathcal{Y}, \mathcal{Z})$ is played by two players **I** and **II** as follows: at stage $k < \omega$, **I** chooses $X_k \in \mathcal{X}$, then **II** responds with either $n_k \in X_k$ in the case that $\mathcal{Y}$ is ω, or else responds with a nonempty $s_k \in [X_k]^{<\omega}$ in the case that $\mathcal{Y}$ is $[\omega]^{<\omega}$. At the end of the game, **II** is declared the winner if $\{n_k : k \in \omega\}$(resp. $\bigcup_{k\in\omega} s_k) \in \mathcal{Z}$.

A few variations of these games have been considered in the literature, see for example in [**2**] and [**11**] for some special cases. In particular, the game $\mathfrak{G}(\mathfrak{Fr}, \omega, \mathcal{Z})$ is equivalent to the game in which at stage k player **I** chooses $m_k \in \omega$ and **II** responds with $n_k > m_k$, the outcome being that **II** wins the play if $\{n_k : k \in \omega\} \in \mathcal{Z}$ as before. Therefore we start with a result of [**2**].

THEOREM 2.1. *([**2**]) Fix a filter $\mathcal{F}$ and consider the game $\mathfrak{G}(\mathfrak{Fr}, \omega, \mathcal{F})$. Then*

(i) ***I** has a winning strategy if and only if $\mathcal{F}$ is not a Q-filter.*
(ii) ***II** has no winning strategy.*

But this game generalizes in many ways and we have the following results.

THEOREM 2.2. *Fix a filter $\mathcal{F}$ and consider the game $\mathfrak{G}(\mathfrak{Fr}, \omega, \mathcal{F}^+)$. Then*

(i) ***I** has a winning strategy if and only if $\mathcal{F}$ is not a weak Q-filter .*
(ii) ***II** has a winning strategy if and only if $\mathcal{F}$ is ω-diagonalized.*

A much less interesting game is the following:

THEOREM 2.3. *Fix a filter $\mathcal{F}$ and consider the game $\mathfrak{G}(\mathfrak{Fr}, \omega, \mathcal{F}^c)$. Then the game is determined and*

(i) ***II** always has a winning strategy.*

THEOREM 2.4. *Fix a filter $\mathcal{F}$ and consider the game $\mathfrak{G}(\mathfrak{Fr}, \omega, \mathcal{F}^*)$. Then the game is determined and*

(i) ***I** has a winning strategy if and only if $\mathcal{F} = \mathfrak{Fr}$ if and only if **II** has no winning strategy.*

Because of the duality between $\mathfrak{Fr}$ and $\mathfrak{Fr}^+ = [\omega]^\omega$, we will get

THEOREM 2.5. *Fix a filter $\mathcal{F}$, then the following games are dual of each other; that is a player has a winning strategy in one game if and only if the other player has a winning strategy in the other game.*

(i) $\mathfrak{G}([\omega]^\omega, \omega, \mathcal{F})$ *and* $\mathfrak{G}(\mathfrak{Fr}, \omega, \mathcal{F}^c)$.
(ii) $\mathfrak{G}([\omega]^\omega, \omega, \mathcal{F}^+)$ *and* $\mathfrak{G}(\mathfrak{Fr}, \omega, \mathcal{F}^*)$.
(iii) $\mathfrak{G}([\omega]^\omega, \omega, \mathcal{F}^c)$ *and* $\mathfrak{G}(\mathfrak{Fr}, \omega, \mathcal{F})$.
(iv) $\mathfrak{G}([\omega]^\omega, \omega, \mathcal{F}^*)$ *and* $\mathfrak{G}(\mathfrak{Fr}, \omega, \mathcal{F}^+)$.

Now we consider the more interesting games where **I** plays members of the filter $\mathcal{F}$.

THEOREM 2.6. *Fix a filter $\mathcal{F}$ and consider the game $\mathfrak{G}(\mathcal{F}, \omega, \mathcal{F})$. Then*

(i) ***I** has a winning strategy if and only if $\mathcal{F}$ is not a Ramsey filter.*
(ii) ***II** never has a winning strategy.*

THEOREM 2.7. *Fix a filter $\mathcal{F}$ and consider the game $\mathfrak{G}(\mathcal{F}, \omega, \mathcal{F}^+)$. Then*

(i) ***I** has a winning strategy if and only if $\mathcal{F}$ is not weakly Ramsey.*
(ii) ***II** has a winning strategy if and only if $\mathcal{F}$ is ω-+-diagonalizable.*

THEOREM 2.8. *Fix a filter $\mathcal{F}$ and consider the game $\mathfrak{G}(\mathcal{F}, \omega, \mathcal{F}^c)$. Then*

(i) ***I** never has a winning strategy.*
(ii) ***II** has a winning strategy if and only if $\mathcal{F}$ is not a Ramsey ultrafilter.*

THEOREM 2.9. *Fix a filter $\mathcal{F}$ and consider the game $\mathfrak{G}(\mathcal{F}, \omega, \mathcal{F}^*)$. Then*

(i) ***I** has a winning strategy if and only if $\mathcal{F}$ is countably generated.*
(ii) ***II** has a winning strategy if and only if $\mathcal{F}$ is not a +-Ramsey filter.*

Again the duality of $\mathcal{F}$ and $\mathcal{F}^+$ will provide the following result.

THEOREM 2.10. *Fix a filter $\mathcal{F}$, then the following games are dual of each other; that is a player has a winning strategy in one game if and only if the other player has a winning strategy in the other game.*

(i) $\mathfrak{G}(\mathcal{F}^+, \omega, \mathcal{F})$ *and* $\mathfrak{G}(\mathcal{F}, \omega, \mathcal{F}^c)$.

(ii) $\mathfrak{G}(\mathcal{F}^+, \omega, \mathcal{F}^+)$ *and* $\mathfrak{G}(\mathcal{F}, \omega, \mathcal{F}^*)$.
(iii) $\mathfrak{G}(\mathcal{F}^+, \omega, \mathcal{F}^c)$ *and* $\mathfrak{G}(\mathcal{F}, \omega, \mathcal{F})$.
(iv) $\mathfrak{G}(\mathcal{F}^+, \omega, \mathcal{F}^*)$ *and* $\mathfrak{G}(\mathcal{F}, \omega, \mathcal{F}^+)$.

Before we turn to games in which player **II** chooses finite sets at each round, consider the following infinite game $\mathfrak{G}_1(\mathcal{F})$ defined in [**2**]: at stage k, player **I** chooses $m_k \in \omega$ and **II** responds with n_k. At the end, **II** is declared the winner if

(i) $n_1 < n_2 < \cdots < n_k < \cdots$,
(ii) $m_k < n_k$ for infinitely many k, and
(iii) $\{n_k : k \in \omega\} \in \mathcal{F}$.

It is proved in [**2**] that **II** does not have any winning strategy in $\mathfrak{G}_1(\mathcal{F})$ and that **I** has a winning strategy if and only if $\mathcal{F}$ is meager. We have the following.

THEOREM 2.11. *Fix a filter $\mathcal{F}$, the the games $\mathfrak{G}_1(\mathcal{F})$ and $\mathfrak{G}(\mathfrak{Fr}, [\omega]^{<\omega}, \mathcal{F})$ are equivalent; that is a player has a winning strategy in one game if and only the same player has a winning strategy in the other game. Therefore, by* [**2**], *we have for either game,*

(i) ***I*** *has a winning strategy if and only if $\mathcal{F}$ is meager.*
(ii) ***II*** *never has a winning strategy.*

THEOREM 2.12. *Fix a filter $\mathcal{F}$, then the game $\mathfrak{G}(\mathfrak{Fr}, [\omega]^{<\omega}, \mathcal{F}^+)$ is dual to the game $\mathfrak{G}(\mathfrak{Fr}, [\omega]^{<\omega}, \mathcal{F})$. Therefore*

(i) ***I*** *never has a winning strategy.*
(ii) ***II*** *has a winning strategy if and only if $\mathcal{F}$ is meager.*

As above, the following game is uninteresting.

THEOREM 2.13. *Fix a filter $\mathcal{F}$, then the game $\mathfrak{G}(\mathfrak{Fr}, [\omega]^{<\omega}, \mathcal{F}^c)$ is equivalent to $\mathfrak{G}(\mathfrak{Fr}, \omega, \mathcal{F}^c)$. Therefore*

(i) ***II*** *always has a winning strategy.*

THEOREM 2.14. *Fix a filter $\mathcal{F}$, the the game $\mathfrak{G}(\mathfrak{Fr}, [\omega]^{<\omega}, \mathcal{F}^*)$ is equivalent to $\mathfrak{G}(\mathfrak{Fr}, \omega, \mathcal{F}^*)$. Therefore*

(i) ***I*** *has a winning strategy if and only if $\mathcal{F} = \mathfrak{Fr}$ if and only if* ***II*** *has no winning strategy.*

We turn to those games where **I** plays members of $\mathcal{F}$ and **II** responds with finite subsets.

THEOREM 2.15. *Fix a filter $\mathcal{F}$ and consider the game $\mathfrak{G}(\mathcal{F}, [\omega]^{<\omega}, \mathcal{F})$. Then*

(i) ***I*** *has <u>no</u> winning strategy if and only if $\mathcal{F}$ is a non-meager P-filter.*
(ii) ***II*** *never has a winning strategy.*

THEOREM 2.16. *Fix a filter $\mathcal{F}$ and consider the game $\mathfrak{G}(\mathcal{F}, [\omega]^{<\omega}, \mathcal{F}^+)$. Then*

(i) ***I*** *has* <u>*no*</u> *winning strategy if and only if $\mathcal{F}$ is a weak P-filter.*
(ii) ***II*** *has a winning strategy if and only if $\mathcal{F}$ is ω-diagonalizable by $\mathcal{F}$-universal sets.*

THEOREM 2.17. *Fix a filter $\mathcal{F}$, then the game $\mathfrak{G}(\mathcal{F}, [\omega]^{<\omega}, \mathcal{F}^c)$ is equivalent to the game $\mathfrak{G}(\mathcal{F}, \omega, \mathcal{F}^c)$. Therefore*

(i) ***I*** *never has a winning strategy.*
(ii) ***II*** *has a winning strategy if and only if $\mathcal{F}$ is not a Ramsey ultrafilter.*

THEOREM 2.18. *Fix a filter $\mathcal{F}$, then the game $\mathfrak{G}(\mathcal{F}, [\omega]^{<\omega}, \mathcal{F}^*)$ is equivalent to the game $\mathfrak{G}(\mathcal{F}, \omega, \mathcal{F}^*)$. Thus*

(i) ***I*** *has a winning strategy if and only if $\mathcal{F}$ is countably generated.*
(ii) ***II*** *has a winning strategy if and only if $\mathcal{F}$ is not a Ramsey ultrafilter.*

Finally, we turn to games where **I** plays members of $\mathcal{F}^+$ while **II** replies with finite subsets. Note that we do not have here the same duality as when **II** responded with natural numbers.

THEOREM 2.19. *Fix a filter $\mathcal{F}$ and consider the game $\mathfrak{G}(\mathcal{F}^+, [\omega]^{<\omega}, \mathcal{F})$. Then*

(i) ***I*** *has a winning strategy if and only if $\mathcal{F}$ is not a P-point.*
(ii) ***II*** *never has a winning strategy.*

THEOREM 2.20. *Fix a filter $\mathcal{F}$ and consider the game $\mathfrak{G}(\mathcal{F}^+, [\omega]^{<\omega}, \mathcal{F}^+)$.*

(i) ***I*** *has a winning strategy if and only if $\mathcal{F}$ is not a P^+-filter.*
(ii) ***II*** *has a winning strategy if and only if $\mathcal{F}$ is ω-diagonalizable by $\mathcal{F}^+$-universal sets.*

THEOREM 2.21. *Fix a filter $\mathcal{F}$, then the game $\mathfrak{G}(\mathcal{F}^+, [\omega]^{<\omega}, \mathcal{F}^c)$ is dual to the game $\mathfrak{G}(\mathcal{F}, \omega, \mathcal{F})$. Therefore*

(i) ***I*** *never has a winning strategy.*
(ii) ***II*** *has a winning strategy if and only if $\mathcal{F}$ is not a Ramsey filter.*

THEOREM 2.22. *Fix a filter $\mathcal{F}$, then the game $\mathfrak{G}(\mathcal{F}^+, [\omega]^{<\omega}, \mathcal{F}^*)$ is dual to the game $\mathfrak{G}(\mathcal{F}, \omega, \mathcal{F}^+)$. Therefore*

(i) ***I*** *has a winning strategy if and only if $\mathcal{F}$ is ω-+-diagonalizable.*
(ii) ***II*** *has a winning strategy if and only if $\mathcal{F}$ is not weakly Ramsey.*

3. Proofs

In this section we verify the results of section 2. We start with two general results.

LEMMA 3.1. *If a family $\mathcal{Z} \subseteq \wp(\omega)$ is closed under supersets, then the two games $\mathfrak{G}(\mathcal{X}, \omega, \mathcal{Z}^c)$ and $\mathfrak{G}(\mathcal{X}, [\omega]^{<\omega}, \mathcal{Z}^c)$ are equivalent.*

Proof: Since player **II** is trying to get out of $\mathcal{Z}$ which is assumed to be closed under supersets (therefore $\mathcal{Z}^c$ is closed under subsets), the best strategy for **II** is to play finite sets as small as possible, namely singleton since a legal move must be nonempty. □

The next Lemma regards the duality mentioned in the introduction and is taken from the work of Aczel [**1**] and Blass (see [**3**], Theorem 1). We include a hint of the proof for completeness.

THEOREM 3.2. *([**1**], [**3**]) For a given filter $\mathcal{F}$, the game $\mathfrak{G}(\mathcal{F}, \omega, \mathcal{Z})$ and the game $\mathfrak{G}(\mathcal{F}^+, \omega, \mathcal{Z}^c)$ are dual; that is a player has a winning strategy in one of these games if and only if the other player has a winning strategy in the other game.*

Proof: Suppose **II** has a winning strategy \$ in the game $\mathfrak{G}(\mathcal{F}, \omega, \mathcal{Z})$, we define a strategy \$\$ for **I** in the game $\mathfrak{G}(\mathcal{F}^+, \omega, \mathcal{Z}^c)$ as follows:
I starts with $\$\$(\emptyset) = \{\$(X) : X \in \mathcal{F}\} \in \mathcal{F}^+$. When **II** responds with n_0, **I** remembers one set $X_0 \in \mathcal{F}$ such that $\$(X_0) = n_0$.
At stage k, **I** has remembered k sets $X_0, X_1, \cdots, X_{k-1}$ from $\mathcal{F}$ while **II** responded with $\langle n_0, n_1, \cdots, n_{k-1}\rangle$. **I** then plays

$$\$\$(\langle n_0, n_1, \cdots, n_{k-1}\rangle) = \{\$(\langle X_0, X_1, \cdots, X_{k-1}, X\rangle) : X \in \mathcal{F}\} \in \mathcal{F}^+;$$

II responds with n_k and **I** remembers one set $X_k \in \mathcal{F}$ such that

$$\$(\langle X_0, X_1, \cdots, X_{k-1}, X_k\rangle) = n_k.$$

Thus a play in the new game corresponds to a play in the former game and thus the outcome $\{n_k : k \in \omega\} \in \mathcal{Z}$ and **I** 's strategy is a winning strategy in $\mathfrak{G}(\mathcal{F}^+, \omega, \mathcal{Z}^c)$.

The other cases are quite similar and left to the reader. □

Now we are ready to attack the proofs of section 2.

PROOF OF THEOREM 2.2: We first deal with player **I** . So suppose that $\mathcal{F}$ is not a weak Q-filter and therefore there is a partition of ω into finite sets $\langle s_k : k \in \omega\rangle$ such that no $X \in \mathcal{F}^+$ meet each s_k in at most one point. Then **I** 's strategy at stage k, after **II** has played $\langle n_0, n_1, \cdots, n_{k-1}\rangle$, is to respond with $\bigcup\{s_i : s_i \cap \{n_0, n_1, \cdots, n_{k-1}\} = \emptyset\}$.
Now suppose that $\mathcal{F}$ is a weak Q-filter and we show that any strategy \$ for **I** is not a winning strategy. Define a sequence of integers $\langle \pi_k : k \in \omega\rangle$ such that $[\pi_0, \infty) \subseteq \$(\emptyset)$ and more generally

$$[\pi_{k+1}, \infty) \subseteq \bigcap\{\$(\langle n_0, n_1, \cdots, n_i\rangle) : n_0 < n_1 < \cdots < n_i < \pi_k\}.$$

By assumption there is an $X \in \mathcal{F}^+$ which meets each interval $[\pi_k, \pi_{k+1})$ in at most one point. But $X = X_0 \cup X_1$ where $X_i = X \cap \bigcup_k [\pi_{2k+i}, \pi_{2k+i+1})$ and therefore $X_i \in \mathcal{F}^+$ for some i. Write X_i in increasing order as $\langle n_k : k \in \omega\rangle$; but then X_i is the outcome of a legal play won by **II** , and thus \$ was not a winning strategy for **I** .

Now we deal with player **II** . Suppose that $\mathcal{F}$ is ω-diagonalized by $\langle X_n : n \in \omega\rangle \subseteq [\omega]^\omega$. Fix a surjective map $\sigma : \omega \to \omega$ such that the preimage of each n is infinite. Here is **II** 's strategy: at stage k, after **I** played $Y_k \in \mathfrak{Fr}$, **II** responds with an element of $Y_k \cap X_{\sigma(k)} \setminus k$. At the end of the play, **II** 's outcome is a set with infinite intersection with each X_n and therefore belongs to $\mathcal{F}^+$, thus this is a winning strategy for **II** .

Now let \$ be a winning strategy for **II** in the game, we show that $\mathcal{F}$ is ω-diagonalizable. We claim that

$$(\forall Y \in \mathcal{F})(\exists n = n(Y))(\exists s = s(Y) \in^n \omega)(\forall^\infty t \in^{n+1} \omega) s < t \implies \$(t) \in Y.$$

Indeed otherwise one quickly produces a winning play for **I** . But then the collection $\{\{\$(s^n); n \in \omega\}; s \in {}^{<\omega}\omega\}$ ω-diagonalize $\mathcal{F}$. This completes the proof. □

<u>PROOF OF THEOREM 2.3:</u> This is trivial; **II** chooses an infinite $X \notin \mathcal{F}$ and plays continually members of X. □

<u>PROOF OF THEOREM 2.4:</u> If $\mathcal{F} = \mathfrak{Fr}$, then $\mathcal{F}^+ = [\omega]^\omega$ and **I** 's strategy is to ensure that **II** 's outcome is infinite, and of course **II** has no winning strategy. If however $\mathcal{F} \neq \mathfrak{Fr}$, **II** chooses an infinite $X \notin \mathcal{F}^+$ and continually plays members of X; this constitutes a winning strategy for **II** . □

<u>PROOF OF THEOREM 2.5:</u> This follows from Theorem 3.2 as $\mathfrak{Fr}^+ = [\omega]^\omega$. □

<u>PROOF OF THEOREM 2.6:</u> If **II** had a winning strategy in this game, it would also be a winning strategy for $\mathfrak{G}(\mathfrak{Fr}, \omega, \mathcal{F})$, which is impossible.

Now for player **I** . If $\mathcal{F}$ is not a Q-filter, then **I** fixes a partition of ω into finite sets $\langle s_k : k \in \omega\rangle$ and plays to ensure that **II** 's outcome meets each s_k in at most one point; therefore **I** wins. If on the other hand $\mathcal{F}$ is not a P-filter, **I** then fixes a sequence $\langle X_n : n \in \omega\rangle \subseteq \mathcal{F}$ such that no $X \in \mathcal{F}$ is almost included in each X_n. It suffices for **I** to play $\cap_{i<k} X_i$ at stage k to produce a winning strategy. This leaves us with the more interesting situation in which $\mathcal{F}$ is a a Q-filter P-filter and we must show that any strategy \$ for **I** cannot be a winning one. Fixing such a strategy \$, **II** first chooses $Y \in \mathcal{F}$ such that $Y \subseteq^* \$(s)$ for all $s \in {}^{<\omega}\omega$. Now **II** defines a sequence $\langle X_n : n \in \omega\rangle \subseteq \mathcal{F}$ and a sequence $\langle n_k : k \in \omega\rangle$ as follows: $X_0 = \$(\emptyset)$ and n_0 is such that $Y \setminus n_0 \subseteq X_0$. Now given $n_0 < n_1 < \cdots < n_k$, put

$$X_{k+1} = \bigcap \{\$(\langle m_0, m_1, \cdots, m_i\rangle) : i \leq k, m_j \in [n_j, n_k] \cap Y\}.$$

Then **II** chooses n_{k+1} such that $Y \setminus n_{k+1} \subseteq X_{k+1}$. Now because $\mathcal{F}$ is a Q-filter, **II** knows very well that there is a set in $\mathcal{F}$ missing infinitely many intervals, say

$[n_{k_n}, n_{k_n+1})$, for an infinite set $\{k_n : n \in \omega\}$, and therefore by selectivity again **II** can find a set $Y' = \{y_n : n \in \omega\} \subseteq Y$ in $\mathcal{F}$ such that Y' misses all these intervals $[n_{k_n}, n_{k_n+1})$ and further intersects each other interval $[n_{k_n}, n_{k_{n+1}})$ in at most one point.
But we claim now that Y' is a legal play in the game! Indeed, for each k, say $y_k \in [n_{k_n}, n_{k_{n+1}}) \cap Y' = [n_{k_n+1}, n_{k_{n+1}}) \cap Y'$, then $y_k \in X_{k_n+1} \subseteq \$(\langle y_0, y_1, \cdots, y_{k-1}\rangle)$. Thus **I** 's strategy \$ was definitely not a winning one. □

PROOF OF THEOREM 2.7: We first look at player **I** . If $\mathcal{F}$ is not weakly Ramsey, there is an $\mathcal{F}$-tree $\mathcal{T}$ such that no branches belong to $\mathcal{F}^+$; this gives a winning strategy for **I** by playing along the tree.
Now suppose that $\mathcal{F}$ is weakly Ramsey and that \$ is a strategy for **I** , we will produce a winning play for **II** showing that **I** cannot have a winning strategy. Let $X_0 = \$(\emptyset)$ and choose $n_0 \in X_0$. Having produced $n_0 < n_1 < \cdots < n_k$, let

$$X_{k+1} = \bigcap\{\$(y_0, y_1, \cdots, y_i) : i \leq k, y_0 < y_1 < \cdots y_k \leq n_k\} \setminus n_k.$$

$X_{k+1} \in \mathcal{F}$ and choose $n_{k+1} \in X_{k+1}$. Now define a $\mathcal{F}$-tree $\mathcal{T}$ inductively by letting $X_\emptyset = X_0$, and given $s = \langle y_0, y_1, \cdots, y_i\rangle \in \mathcal{T}$, say $n_{k-1} \leq y_i < n_k$, then let $X_s = X_{k+1}$. This $\mathcal{F}$-tree $\mathcal{T}$ must contain a branch $\{y_k : k \in \omega\} \in \mathcal{F}^+$ by assumption, but this clearly is a legal play of the game in which **II** wins.

Now we deal with player **II** . Suppose that $\mathcal{F}$ is ω-+-diagonalized by $\langle X_n : n \in \omega\rangle \in \mathcal{F}^+$. Fix a surjective map $\sigma : \omega \to \omega$ such that the preimage of each n is infinite. Here is **II** 's strategy: at stage k, after **I** played $Y_k \in \mathcal{F}$, **II** responds with an element of $Y_k \cap X_{\sigma(k)} \setminus k$. Notice the importance of having each $X_n \in \mathcal{F}^+$. At the end of the play, **II** 's outcome is a set with infinite intersection with each X_n and therefore belongs to $\mathcal{F}^+$, thus it is a winning strategy for **II** .
Now let \$ be a winning strategy for **II** in the game, we show that $\mathcal{F}$ is ω-+-diagonalizable. We first define an $\mathcal{F}^+$-tree as follows. Let $X_\emptyset = \{\$(X) : X \in \mathcal{F}\} \in \mathcal{F}^+$, and for each $n \in X_\emptyset$, select an $X_\emptyset^n \in \mathcal{F}$ such that $\$(X_\emptyset^n) = n$. More generally, given $X_{\overline{s}}^n \in \mathcal{F}$, for $n \in X_{\overline{s}} \in \mathcal{F}^+$, say $\overline{s} = \langle s_0, s_1, \cdots, s_i\rangle$, let

$$X_{\overline{s}\,^\wedge n} = \{\$(X_\emptyset^{s_0}, X_{\langle s_0\rangle}^{s_1}, X_{\langle s_0,s_1\rangle}^{s_2}, \cdots, X_{\langle s_0,s_1,\cdots,s_{i-1}\rangle}^{s_i}, X) : X \in \mathcal{F}\} \in \mathcal{F}^+,$$

and for each $k \in X_{\overline{s}\,^\wedge n}$, choose $X_{\overline{s}\,^\wedge n}^k \in \mathcal{F}$ such that

$$\$(X_\emptyset^{s_0}, X_{\langle s_0\rangle}^{s_1}, X_{\langle s_0,s_1\rangle}^{s_2}, \cdots, X_{\langle s_0,s_1,\cdots,s_{i-1}\rangle}^{s_i}, X_{\overline{s}\,^\wedge n}^k) = k.$$

Therefore we obtain a $\mathcal{F}^+$-tree $\mathcal{T}$ each of whose branches is a legal play of the game, and therefore belongs to $\mathcal{F}^+$ as \$ was a winning strategy for **II** and this means that the sets $\{X_{\overline{s}} : \overline{s} \in \mathcal{T}\} \subseteq \mathcal{F}^+$ must ω-+-diagonalize $\mathcal{F}$. □

PROOF OF THEOREM 2.8: As in $\mathfrak{G}(\mathfrak{Fr}, \omega, \mathcal{F}^c)$, **I** has no winning strategy.

As for **II** , let us consider the case when $\mathcal{F}$ is not a Ramsey ultrafilter first. There are three possibilities which **II** figures out. If $\mathcal{F}$ is not an ultrafilter, **II** chooses $X \in \mathcal{F}^+ \setminus \mathcal{F}$ and constantly plays members of X, and therefore wins the

game. Otherwise **II** checks whether $\mathcal{F}$ is a Q-filter and if not chooses a partition $\langle s_k : k \in \omega \rangle$ of ω into finite sets for which $\mathcal{F}$ contains no selector. But then **II** 's strategy is to play a selector for the partition, therefore winning again. Finally **II** realizes that it must be that $\mathcal{F}$ is not a P-filter and thus selects $\langle X_n : n \in \omega \rangle \subseteq \mathcal{F}$ with no $Y \in \mathcal{F}$ almost included in each X_n. Then **II** 's strategy at the kth move is to play a member of $\cap_{i<k} X_i$ and again has a winning strategy.
Now we suppose that $\mathcal{F}$ is a Ramsey ultrafilter and must show that **II** cannot have a winning strategy. But as $\mathcal{F}$ is an ultrafilter, $\mathcal{F} = \mathcal{F}^+$ and by the duality theorem 3.2, the game $\mathfrak{G}(\mathcal{F}, \omega, \mathcal{F}^c)$ is dual to the game $\mathfrak{G}(\mathcal{F}, \omega, \mathcal{F})$, and as $\mathcal{F}$ is a Q-filter P-filter, **I** has no winning strategy in $\mathfrak{G}(\mathcal{F}, \omega, \mathcal{F})$ by theorem 2.6 and therefore **II** has no winning strategy in $\mathfrak{G}(\mathcal{F}, \omega, \mathcal{F}^c)$. □

PROOF OF THEOREM 2.9: As far as player **I** is concerned, if $\mathcal{F}$ is generated by countably many sets $\langle X_n : n \in \omega \rangle$, then it suffices for **I** to play $\cap_{i<k} X_i \setminus k$ at stage k, and thus the outcome of the play is a set $Y \subseteq^* X_n$ for each n, definitely in $\mathcal{F}^+$ and **I** wins.
If on the other hand **I** has a winning strategy \$ in the game, then the filter $\mathcal{F}$ must be generated by $\langle \$(s) : s \in {}^{<\omega}\omega \rangle$, as can easily be verified.

Now for player **II** . If $\mathcal{F}$ is not +-Ramsey, then there is an $\mathcal{F}^+$-tree $\mathcal{T}$ none of whose branches belong to $\mathcal{F}^+$; therefore **II** 's strategy is to play along the tree $\mathcal{T}$.
Suppose finally that $\mathcal{F}$ is +-Ramsey and we show that any strategy \$ for **II** is not a winning strategy.
We first define a $\mathcal{F}^+$-tree as follows. Let $X_\emptyset = \{\$(X) : X \in \mathcal{F}\} \in \mathcal{F}^+$, and for each $n \in X_\emptyset$, select an $X_\emptyset^n \in \mathcal{F}$ such that $\$(X_\emptyset^n) = n$. More generally, given $X_{\overline{s}}^n \in \mathcal{F}$, for $n \in X_{\overline{s}} \in \mathcal{F}^+$, say $\overline{s} = \langle s_0, s_1, \cdots, s_i \rangle$, let

$$X_{\overline{s}\,^\wedge n} = \{\$(X_\emptyset^{s_0}, X_{\langle s_0 \rangle}^{s_1}, X_{\langle s_0, s_1 \rangle}^{s_2}, \cdots, X_{\langle s_0, s_1, \cdots, s_{i-1} \rangle}^{s_i}, X) : X \in \mathcal{F}\} \in \mathcal{F}^+,$$

and for each $k \in X_{\overline{s}\,^\wedge n}$, choose $X_{\overline{s}\,^\wedge n}^k \in \mathcal{F}$ such that

$$\$(X_\emptyset^{\overline{s}_0}, X_{\langle s_0 \rangle}^{s_1}, X_{\langle s_0, s_1 \rangle}^{s_2}, \cdots, X_{\langle s_0, s_1, \cdots, s_{i-1} \rangle}^{s_i}, X_{\overline{s}\,^\wedge n}^k) = k.$$

Therefore we obtain a $\mathcal{F}^+$-tree $\mathcal{T}$ all of whose branches are a legal play of the game, but there is such a branch in $\mathcal{F}^+$ as $\mathcal{F}$ is +-Ramsey and therefore **II** 's strategy is not a winning strategy. □

PROOF OF THEOREM 2.10: Follows immediately from Theorem 3.2. □

PROOF OF THEOREM 2.11: We must show that the games $\mathfrak{G}_1(\mathcal{F})$ is equivalent to the game $\mathfrak{G}(\mathfrak{Fr}, [\omega]^{<\omega}, \mathcal{F})$. As **I** is trying to produce an outcome out of $\mathcal{F}$ in the game $\mathfrak{G}(\mathfrak{Fr}, [\omega]^{<\omega}, \mathcal{F})$, **I** plays without loss of generality cofinite sets of the form $[n, \infty)$ which we identify with n to simplify notation.
Suppose first that **I** has a winning strategy \$ in the game $\mathfrak{G}_1(\mathcal{F})$, we define a strategy $\overline{\$}$ for **I** in the game $\mathfrak{G}(\mathfrak{Fr}, [\omega]^{<\omega}, \mathcal{F})$ by

$$\overline{\$}(s_0, s_1, \cdots, s_i) = \$(\bigcup_{j \le i} s_j),$$

where $(\bigcup_{j\le i} s_j)$ is considered as an element of $[\omega]^{<\omega}$. This gives a winning strategy for **I** in $\mathfrak{G}(\mathfrak{Fr}, [\omega]^{<\omega}, \mathcal{F})$.
Now let $\$$ be a winning strategy for **I** in $\mathfrak{G}(\mathfrak{Fr}, [\omega]^{<\omega}, \mathcal{F})$, we define a strategy $\overline{\$}$ for **I** in $\mathfrak{G}_1(\mathcal{F})$ as follows:
I starts without loss of generality with $p_0 = m_0 = \overline{\$}(\emptyset) = \$(\emptyset) = 0$. After **II** has replied with n_k, then **I** replies with $m_{k+1} = \$(\{n_0, n_1, \dots, n_k\})$ until **I** notices that **II** actually replied with some $n_{k+1} > m_{k+1}$; at this point **I** settles for $s_0 = \{n_0, \dots, n_k\}$ and sets $p_1 = m_{k+1}$. Notice that if **II** never plays some $n_{k+1} > m_{k+1}$, **I** trivially wins the game. Now more generally, after **I** has collected $\langle s_0, s_1, \dots, s_\ell\rangle$ and defined $p_\ell = m_{q+1}$ say, the play goes on with **II** playing $n_{q+1}, n_{q+2}, \dots, n_r$ and **I** replying with

$$m_{r+1} = \$(s_0, \dots, s_\ell, \{n_{q+1}, n_{q+2}, \dots, n_r\})$$

until **I** notices again that **II** replied with some $n_{r+1} > m_{r+1}$; at this point **I** settles for $s_{\ell+1} = \{n_{q+1}, n_{q+2}, \dots, n_r\}$ and sets $p_{\ell+1} = m_{r+1}$. This is surely a winning strategy for **I** in $\mathfrak{G}_1(\mathcal{F})$ as the sequence $\langle p_0, s_0, p_1, s_1, \dots\rangle$ is a legal play in $\mathfrak{G}(\mathfrak{Fr}, [\omega]^{<\omega}, \mathcal{F})$ according to the strategy $\$$ and therefore $\{n_i : i \in \omega\} = \bigcup_i s_i \notin \mathcal{F}$.

Now we deal with player **II** . If $\$$ is a winning strategy for **II** in the game $\mathfrak{G}(\mathfrak{Fr}, [\omega]^{<\omega}, \mathcal{F})$, then **II** behaves as follows in the game $\mathfrak{G}_1(\mathcal{F})$: at stage k, **II** imagines that **I** has played $\langle m_0, m_1, \cdots, m_k\rangle$ in the game $\mathfrak{G}(\mathfrak{Fr}, [\omega]^{<\omega}, \mathcal{F})$ and plays one by one the elements of $\$(\langle m_0, m_1, \cdots, m_k\rangle)$ without noticing **I** 's moves until done, and then remembers **I** 's last move, m_{k+1}. At the end, the outcome is one of the game $\mathfrak{G}(\mathfrak{Fr}, [\omega]^{<\omega}, \mathcal{F})$ and thus **II** wins.
Finally, it is proved in [**2**] that **II** never has a winning strategy in $\mathfrak{G}_1(\mathcal{F})$; and therefore by the previous paragraph **II** doesn't have a winning strategy in $\mathfrak{G}(\mathfrak{Fr}, [\omega]^{<\omega}, \mathcal{F})$ either. Alternatively, it is also straightforward to show that actually a winning strategy for **II** in $\mathfrak{G}_1(\mathcal{F})$ yields a winning strategy for **II** in $\mathfrak{G}(\mathfrak{Fr}, [\omega]^{<\omega}, \mathcal{F})$. □

<u>PROOF OF THEOREM 2.12:</u> The fact that this game is dual to $\mathfrak{G}(\mathfrak{Fr}, [\omega]^{<\omega}, \mathcal{F})$ is purely accidental I believe.

By playing two games simultaneously, **II** can produce two outcomes whose union is cofinite, therefore one of them must be in $\mathcal{F}^+$ and thus **I** has no winning strategy. More precisely, let $m_0 = \overline{m}_0$ be **I** 's first move as we again identify a cofinite set $[m, \infty)$ with m. **II** replies with $\{m_0\}$, and then **I** responds with m_1 in the first game. Now in the second game **II** replies with $[m_0, m_1]$ and waits for **I** 's response $\overline{m}_1$; then **II** comes back to the first game and replies with $[m_1, \overline{m}_1]$. Continuing this way, **II** produces the outcome $A = \bigcup_i [m_i, \overline{m}_i]$ in the first game, and $B = \bigcup_i [\overline{m}_i, m_{i+1}]$ in the second. One of these sets is in $\mathcal{F}^+$ and therefore **I** lost one of the games.

Now for player **II** . If $\mathcal{F}$ is meager, then **II** has definitely an easy time winning the game; indeed there must be a sequence $\pi_0 < \pi_1 < \cdots$ such that each member of $\mathcal{F}$ meets all but finitely many of the intervals $[\pi_k, \pi_{k+1})$. Therefore at stage ℓ, **II** plays one of the intervals $[\pi_k, \pi_{k+1})$ with $k > \ell$.
So we must show this is the only way that **II** can have a winning strategy. So fix such a winning strategy \$ for **II** . Define a sequence of integers by $\pi_0 = 1$, and given π_k, choose

$$\pi_{k+1} = max\{\$(m_0, m_1, \cdots, m_i) : i \leq k \text{ and } m_j \leq \pi_k\} + 1.$$

then each member of $\mathcal{F}$ must meet all but finitely many of the intervals $[\pi_k, \pi_{k+1})$. Otherwise, say $Y \in \mathcal{F}$ misses the intervals $[\pi_{k_\ell}, \pi_{k_\ell+1})$ for $\ell \in \omega$, then **I** wins by playing exactly these k_ℓ. □

PROOF OF THEOREM 2.13: Follows immediately by Lemma 3.1 and Theorem 2.3. □

PROOF OF THEOREM 2.14: Follows immediately by Lemma 3.1 and Theorem 2.4. □

PROOF OF THEOREM 2.15: If **II** had a winning strategy in this game, it would be a winning strategy in the game $\mathfrak{G}(\mathfrak{Fr}, [\omega]^{<\omega}, \mathcal{F})$, which is impossible by Theorem 2.11.

If $\mathcal{F}$ is meager, then **I** uses the same strategy as for $\mathfrak{G}(\mathfrak{Fr}, [\omega]^{<\omega}, \mathcal{F})$; if on the other hand $\mathcal{F}$ is not a P-filter, then **I** chooses a witness $\langle X_n : n \in \omega \rangle$ and plays $\bigcap_{i<k} X_i$ at stage k which provides a winning strategy.
Now, as a strategy for player **I** is nothing else but an $\mathcal{F}$-tree of finite sets, it should be rather clear that **I** has no winning strategy if and only if any $\mathcal{F}$-tree has a branch in $\mathcal{F}$, that is if and only if $\mathcal{F}$ is a non-meager P-filter by Lemma 1.3. □

PROOF OF THEOREM 2.16: As far as player **I** is concerned, the proof is entirely similar to that of Theorem 2.7.

So we consider the situation for **II** . First fix $\mathcal{F}$-universal sets $\langle X_n : n \in \omega \rangle$ ω-diagonalizing $\mathcal{F}$. Player **II** fixes a surjection $\sigma : \omega \to \omega$ such that the preimage of every n is infinite and at stage k, after **I** produced a set $Y_k \in \mathcal{F}$, **II** responds with $s_k \in X_{\sigma(k)} \cap [Y_k \setminus k]^{<\omega}$. At the end of the play, **II** has produced $S = \bigcup_k s_k$ which contains infinitely many members of each X_n, and is therefore in $\mathcal{F}^+$.
Now let \$ be a winning strategy for **II** and we define a tree $\mathcal{T} \subseteq {}^{<\omega}([\omega]^{<\omega})$ such that the successors of each node $\overline{s} \in \mathcal{T}$ form an $\mathcal{F}$-universal set and the collection $\langle X_{\overline{s}} : \overline{s} \in \mathcal{T} \rangle$ ω-diagonalizes $\mathcal{F}$.

Let $X_\emptyset = \{\$(X) : X \in \mathcal{F}\}$, an $\mathcal{F}$-universal set, and for each $s \in X_\emptyset$, choose $X^s_\emptyset \in \mathcal{F}$ such that $\$(X^s_\emptyset) = s$. In general, given $X^t_{\overline{s}} \in \mathcal{F}$, for $\overline{s} = \langle s_0, s_1, \cdots, s_i \rangle \in {}^{<\omega}([\omega]^{<\omega})$, define

$$X_{\overline{s}\,^\wedge t} = \{\$(X^{s_0}_\emptyset, X^{s_1}_{s_0}, \cdots, X^t_{\overline{s}}, X)) : X \in \mathcal{F}\}$$

a $\mathcal{F}$-universal set, and for each $u \in X_{\overline{s}\,^\wedge t}$, choose $X^u_{\overline{s}\,^\wedge t} \in \mathcal{F}$ such that

$$\$(X^{s_0}_\emptyset, X^{s_1}_{s_0}, \cdots, X^t_{\overline{s}}, X^u_{\overline{s}\,^\wedge t}) = u.$$

Now every branch of $\mathcal{T}$ constitutes an outcome of a play of the game, and therefore the ($\mathcal{F}$-universal) sets $X_{\overline{s}}$ for $\overline{s} \in \mathcal{T}$ must ω-diagonalize $\mathcal{F}$, as otherwise there is an $X \in \mathcal{F}$ and a branch of the tree which completely misses X, and thus not in $\mathcal{F}^+$. □

PROOF OF THEOREM 2.17: Follows immediately from Lemma 3.1 and Theorem 2.8. □

PROOF OF THEOREM 2.18: Follows immediately from Lemma 3.1 and Theorem 2.9. □

PROOF OF THEOREM 2.19: Player **II** definitely has no hope for a winning strategy; if there is $X \in \mathcal{F}^+ \setminus \mathcal{F}$, then **II** is doomed, and otherwise $\mathcal{F} = \mathcal{F}^+$ and the result follows by Theorem 2.15.

So player **I** has a winning strategy if $\mathcal{F}^+ \neq \mathcal{F}$, and otherwise, the result follows again from Theorem 2.15. □

PROOF OF THEOREM 2.20: The proof is entirely similar to that of Theorem 2.16 and is left to the reader.

PROOF OF THEOREM 2.21: Follows immediately from Lemma 3.1, Theorem 3.2 and Theorem 2.6. □

PROOF OF THEOREM 2.22: Follows immediately from Lemma 3.1, Theorem 3.2 and Theorem 2.7. □

4. Conclusion

The combinatorial properties of a filter being meager, a P-filter or a Q-filter have probably been the most popular; another related property which has been around for some times is that of a filter $\mathcal{F}$ being 'rapid', i.e. for any partition of ω into finite sets $\langle s_k : k \in \omega \rangle$, $\mathcal{F}$ contains a set X such that $X \cap s_k$ has size at most k for each k. If we modify the games above so that at stage k player **II** responds with a finite set of size at most k (or bounded by a fixed "unbounded" function), then the characterizations of the winning strategies for either player will involve modifications of this property of being rapid, such as weakly rapid and so on. The details are left to the interested reader.

However, we have little information on the following interesting variation of the games:

Problem: Characterize winning strategies for the games in which player **II** responds with members of $\mathcal{F}^*$.

REFERENCES

1. P. Aczel, *Quantifiers, Games, and Inductive Definitions*, in Proc. Third Scandinavian Logic Symposium, ed. S. Kanger, North Holland, 1975, 1-14.
2. T. Bartoszynski and M. Scheepers, *Filter and Games*, to appear, 1993.
3. A. Blass, *Selective Ultrafilters and Homogeneity*, Annals of Pure and Applied Logic **38** (1988), 215-255.
4. D. Booth, *Ultrafilters on a Countable Set*, Annals Math. Logic **2** (1970), 1-24.
5. F. Galvin, Unpublished manuscript on ultrafilter games, $\approx$ 1980.
6. S. Grigorieff, *Combinatorics on Ideals and Forcing*, Annals Math. Logic **3** (1971), 363-394.
7. W. Just, A.R.D. Mathias, K. Prikry and P. Simon, *On the existence of large P-ideals*, to appear.
8. K. Kunen, *Some Points in βN*, Math. Proc. Cambridge Phil. Soc. **80** (1976), 385-398.
9. C. Laflamme, *Zapping Small Filters*, Proc. Amer. Math. Soc. **114** (1992), 535-544.
10. ———, *Strong Meager Properties for Filter*, to appear in Fund. Math., 1995.
11. S. Shelah, *Proper Forcing*, Lecture Notes in Mathematics, Springer-Verlag, Berlin 1982.
12. M. Talagrand, *Compacts de fonctions mesurables et filtres non mesurables*, Studia Math. **67** (1980), 13-43.

DEPARTMENT OF MATHEMATICS AND STATISTICS, UNIVERSITY OF CALGARY, CALGARY, ALBERTA, CANADA T2N 1N4

E-mail address: laflamme@acs.ucalgary.ca

Contemporary Mathematics
Volume **192**, 1996

Analytic Non-Borel Sets Modulo Null Sets

R. Daniel Mauldin

Abstract

We give a negative answer to the following question: Can every analytic set, A, in R^n be written as $A = B \cup N$, where B is a Borel set and $N \subset S$, where S is an F_σ set with Lebesgue measure zero?

The purpose of this note is to give a negative answer to the following question of Johnson: Can every analytic set, A, in R^n be written as $A = B \cup N$, where B is a Borel set and $N \subset S$, where S is an F_σ set with Lebesgue measure zero [KGJ]? Without loss of generality and for simplicity, we give a negative answer where R^n is replaced by the unit square.

Our aim is to show that the answer can be obtained from a construction given in [RDM]. This construction was used there to show that the Baire order of the set of functions on the unit interval which are continuous except for a Lebesgue null set is ω_1.

Let I denote the interval [0,1]. If $E \subset I \times I$, and $x \in I$, let $E(x) = \{y : (x,y) \in E\}$. Let λ denote Lebesgue measure. First we show that there is a universal analytic set in the square which represents each Borel set in I a fairly large number of times. The crucial fact is the following proven in [RDM].

Theorem 1. *There is a Borel measurable map h from I onto the Hilbert cube I^ω such that for each x, $h^{-1}(x)$ is not a subset of an F_σ set of measure zero.*

We note that the map h obviously cannot be continuous and the map h constructed in [RDM] is of Borel class 3. I do not know what the lowest possible class is. Of course, we can and do replace I^ω by I.

Theorem 2. *There is an analytic subset A of the unit square, $I \times I$, such that if Eis a Borel subset of I, then $\{x : A(x) = E\}$ is not a subset of an F_σ set of measure zero.*

---

[0]AMS(MOS) subject classifications(1980). Primary 28A05, 54C50, Secondary 26A21, 04A15

Key words and phrases. Analytic set, Borel set, Lebesgue null set.
The Research supported by NSF Grant DMS-9303888.

Proof. Let U be an analytic subset of $I \times I$ which is universal for the Borel subsets of I. Let $A =^{\cdot} \{(x,y) : y \in U(h(x))\}$. The set A has the required properties.

The same type of argument shows that for each $\alpha < \omega_1$, there is a Borel subset B_α of $I \times I$ such that if E is a Borel subset of I of class $\leq \alpha$, then $\{x : B_\alpha(x) = E\}$ is not a subset of an F_σ set of measure zero.

Corollary 3. *There is no Borel set B such that $A \backslash B$ is a subset of an F_σ set with Lebesgue measure zero.*

Proof. Suppose such a Borel set B exists and let $A \backslash B \subset \cup_n K_n$, where each K_n is closed and has measure zero. Let $F_n = \{x : \lambda(K_n(x)) > 0\}$. Then each F_n is an F_σ set with measure zero.

Let $\alpha < \omega_1$ be the exact Borel class of B and let E be a Borel subset of I such that E does not differ from any Borel set of class α by a subset of an F_σ set of measure zero. (We can get such a set because the Baire order of the functions continuous almost everywhere is ω_1. In fact, in [RDM] it is shown that this is equivalent to the Baire order being ω_1.) Now, there is some x such that $A(x) = E$ and x is not in any F_n. Then $E \setminus B(x)$ is a subset of an F_σ set of measure zero. This contradiction establishes the theorem.

Question: For what σ-ideals can one derive similar results?

This question has been considered recently by Kechris and Solecki[K-S] and by Balcerzak[B]

References

[B] M. Balcerzak, Can ideals without the ccc be interesting? Top. Appl. 55(1994), 251-260.

[K-S] A.S Kechris, S.Solecki, Approximation of analytic sets by Borel sets in definable countable chain conditions, Israel J. Math., to appear.

[KGJ] K.G. Johnson, The σ-algebra generated by the Jordan sets in R^n, Real Analysis Exchange, 19(1)(1993-94), 278-282.

[RDM] R.D.Mauldin, The Baire order of the functions continuous almost everywhere, Proc. Amer. Math. Soc., 41(1973), 535-540.

Department of Mathematics
University of North Texas
Denton, Texas 76203

Contemporary Mathematics
Volume **192**, 1996

Laver's Forcing and Outer Measure

Janusz Pawlikowski

ABSTRACT. We show that preservation of outer measure by Laver's forcing can be in a natural way obtained from Sierpiński's inductive analysis of analytic sets.

1. Let $\mathbb{L}$ be Laver's forcing (see [**L**]) and let μ be the Lebesgue measure in ${}^{\omega}2$. We shall prove the following theorem of Woodin [**W**] , rediscovered later by Judah and Shelah [**JS**] ([**W**] is unpublished and the proof in [**JS**] is somewhat difficult to follow).

THEOREM. *For $A \subseteq {}^{\omega}2$, $\mu^*(A) = a \Rightarrow \mathbb{L} \Vdash \mu^*(A) = a$.*

2. We first fix some notation.

DEFINITION. Write Ω for ${}^{<\omega}\omega$. Let $T \in \mathbb{L}$. For $t \in T$ let $T_t = \{s \in T : s \subseteq t \vee t \subseteq s\}$. For $\tau \in \Omega$ let $T(\tau)$ be the image of τ under the canonical isomorphism of Ω and T and let $T\langle\tau\rangle = T_{T(\tau)}$ (so, $T(\emptyset)$ is the stem of T).

For $S, T \in \mathbb{L}$ and $n \in \omega$ write $S \leq_n T$ iff $\forall \tau \in {}^{n}\omega\ S(\tau) = T(\tau)$. Note that if $T_{n+1} \leq_n T_n$ $(n \in \omega)$ then $T = \bigcap_n T_n$ belongs to $\mathbb{L}$ and $T \leq_n T_n$.

For open $D \subseteq \mathbb{L}$ let

$$\widetilde{D} = \{T \in \mathbb{L} : \forall S \leq_0 T\ S \notin D\} \cup D,$$
$$D^* = \{T \in \mathbb{L} : \exists n\ \forall \tau \in \Omega\ |\tau| \geq n \Rightarrow T\langle\tau\rangle \in \widetilde{D}\}.$$

Note that D^* is open and $D \subseteq D^*$.

LEMMA. *If $D \subseteq \mathbb{L}$ is open and nonempty below $S \leq T \in D^*$ then $\exists s \in S\ T_s \in D$.*

1991 *Mathematics Subject Classification.* 03E15 03E35.

Key words and phrases. Laver's forcing, preservation of outer measure, analytic sets, random reals.

These notes were drafted in Fall, 1988, while I was visiting University of Colorado at Boulder. They were used for seminar talks in Wrocław (Fall, 1990) and Jerusalem (Spring, 1991). I am grateful to Rich Laver and Jan Mycielski of Boulder and to Haim Judah of Bar-Ilan for making my visits to Colorado and Israel possible. I am also indebted to Tomek Bartoszyński and Martin Goldstern, whose question motivated me to dig out these notes and transform them into Tex. Partially supported by KBN grant PB 2 1017 91 01.

PROOF. Let $R \in D$, $R \leq S$. Choose $s \in R$ long enough to guarantee $R_s \leq_0 T_s \in \widetilde{D}$. Since D is open, $R_s \in D$, hence by $R_s \leq_0 T_s \in \widetilde{D}$ we have $T_s \in D$. □

3. The following lemma is implicit in [**L**].

LEMMA.
1. *Suppose $D \subseteq \mathbb{L}$ is open. Then $\forall i\ \forall T\ \exists S \leq_i T\ S \in D^*$.*
2. *Suppose $D_n \subseteq \mathbb{L}$ ($n \in \omega$) are open. Then $\forall i\ \forall T\ \exists S \leq_i T\ S \in \bigcap_n D^*_n$.*

PROOF. (1) Let $S^{-1} = T$ and for $k \in \omega$ let $S^k = \bigcup_{\sigma \in {}^{i+k}\omega} R^\sigma$, where R^σ's are chosen so that $R^\sigma \leq_0 S^{k-1}\langle\sigma\rangle$ and $R^\sigma \in D$ whenever possible. Note that $S^k \leq_{i+k} S^{k-1}$. Let $S = \bigcap_k S^k$. Then $S \leq_i T$ and $\forall k \geq 0\ \forall \sigma \in {}^{i+k}\omega\ S\langle\sigma\rangle \in \widetilde{D}$.

(2) By (1) find $\langle S_n : n \in \omega\rangle$ such that $S_n \in D^*_n$ and

$$\cdots \leq_{i+n+1} S_n \leq_{i+n} \cdots \leq_{i+1} S_0 \leq_i T.$$

Let $S = \bigcap_n S_n$. □

4. The next lemma is motivated by Sierpiński's proof that every analytic set is a union and intersection of $\aleph_1$ Borel sets (see [**K**]).

LEMMA. *Let $A_\sigma \subseteq {}^\omega 2$ ($\sigma \in \Omega$) be such that $\forall\sigma\ \mu^*(A_\sigma) \leq a$ and $\forall\sigma\ A_\sigma \subseteq \liminf_n A_{\sigma^\frown n}$. Then*

$$\mu^*\Big(\bigcap_{\Sigma \leq_0 \Omega} \bigcup_{\sigma \in \Sigma} A_\sigma\Big) \leq a.$$

PROOF. Without loss of generality A_σ's are Borel. Define A^α_σ ($\alpha < \omega_1, \sigma \in \Omega$) as follows

$$A^0_\sigma = A_\sigma,$$
$$A^{\alpha+1}_\sigma = \liminf_n A^\alpha_{\sigma^\frown n},$$
$$A^\lambda_\sigma = \bigcup_{\alpha<\lambda} A^\alpha_\sigma, \text{ for limit } \lambda.$$

Note that for a fixed σ, $A^\alpha_\sigma \subseteq \liminf_n A^\alpha_{\sigma^\frown n}$, the sets A^α_σ increase with α and have measure $\leq a$. So, there exist $\alpha_\sigma < \omega_1$ such that $\forall\beta \geq \alpha_\sigma\ \mu(A^{\beta+1}_\sigma \setminus A^\beta_\sigma) = 0$. Let $\alpha = \sup_\sigma \alpha_\sigma$. Then $\mu(A^{\alpha+1}_\emptyset \cup \bigcup_\sigma (A^{\alpha+1}_\sigma \setminus A^\alpha_\sigma)) \leq a$. We are done by the following claim.

CLAIM.

$$\bigcap_{\Sigma \leq_0 \Omega} \bigcup_{\sigma \in \Sigma} A_\sigma \subseteq A^{\alpha+1}_\emptyset \cup \bigcup_\sigma (A^{\alpha+1}_\sigma \setminus A^\alpha_\sigma).$$

PROOF. Let $x \notin A^{\alpha+1}_\emptyset \cup \bigcup_\sigma (A^{\alpha+1}_\sigma \setminus A^\alpha_\sigma)$. Then, by $x \notin A^{\alpha+1}_\emptyset$, $\exists^\infty n\ x \notin A^\alpha_{\langle n\rangle}$. Together with $x \notin \bigcup_\sigma (A^{\alpha+1}_\sigma \setminus A^\alpha_\sigma)$ this gives $\exists^\infty n\ x \notin A^{\alpha+1}_{\langle n\rangle}$. By a similar argument, if $x \notin A^{\alpha+1}_{\langle n\rangle}$ then $\exists^\infty m\ x \notin A^{\alpha+1}_{\langle n,m\rangle}$. Thus $\exists^\infty n\ \exists^\infty m\ x \notin A^{\alpha+1}_{\langle n,m\rangle}$. In this way we construct $\Sigma \leq_0 \Omega$ such that $x \notin \bigcup_{\sigma\in\Sigma} A^{\alpha+1}_\sigma$. It follows by $A_\sigma \subseteq A^{\alpha+1}_\sigma$ that $x \notin \bigcup_{\sigma\in\Sigma} A_\sigma$. □

5. We need the following definition.

DEFINITION. Suppose that $\mathbb{P}$ is a poset and c is a $\mathbb{P}$ name for a clopen subset of ${}^\omega 2$. Let

$$\operatorname{dom}(c) = \{p \in \mathbb{P} : p \text{ decides the value of } c\}.$$

For $p \in \operatorname{dom}(c)$ let $c(p)$ be the clopen that p chooses for c. Let

$$\mu(c) = \sup\{\mu(c(p)) : p \in \operatorname{dom}(c)\}.$$

Note that $\operatorname{dom}(c)$ is an open dense subset of $\mathbb{P}$ and if $p \le q$ are in $\operatorname{dom}(c)$ then $c(p) = c(q)$.

LEMMA. *Let c be an $\mathbb{L}$ name for a clopen set. If $R \le T$, $R \in \operatorname{dom}(c)$ and $T \in \operatorname{dom}(c)^*$, then $\exists s \in R\ c(T_s) = c(R)$.*

PROOF. Since $\operatorname{dom}(c)$ is nonempty below R, by Lemma 2, for some $s \in R$ we have $T_s \in \operatorname{dom}(c)$. Then $c(R) = c(R_s) = c(T_s)$. □

6. Now comes the basic estimation.

LEMMA. *Let c_n $(n \in \omega)$ be $\mathbb{L}$ names for clopen sets. If $T \in \bigcap_n \operatorname{dom}(c_n)^*$ then*

$$\mu^*(\{x \in {}^\omega 2 : T \Vdash x \in \bigcup_n c_n\}) \le \sum_n \mu(c_n).$$

PROOF.

$$\begin{aligned} T \Vdash x \in \bigcup_n c_n &\Leftrightarrow \forall S \le T\ \exists n\ \exists R \le S\ x \in c_n(R) \\ &\Leftrightarrow \forall S \le T\ \exists n\ \exists s \in S\ x \in c_n(T_s) \\ &\Leftrightarrow x \in \bigcap_{S \le T} \bigcup_{s \in S} \bigcup_n c_n(T_s). \end{aligned}$$

(The second $\Rightarrow$ is by Lemma 5.) We conclude by Lemma 4. □

7. We are almost ready for the proof of Theorem 1.

DEFINITION. For a poset $\mathbb{P}$ and a real $\epsilon > 0$ let $\mathcal{I}^\epsilon_{\mathbb{P}}$ be the set of all sequences $c = \langle c_n : n \in \omega\rangle$ such that each c_n is a $\mathbb{P}$ name for a clopen of measure $\le \epsilon/2^{n+1}$.

PROOF OF THEOREM 1. Suppose that $\mathbb{L} \not\Vdash \mu^*(A) = a$. Then there exist $T \in \mathbb{L}$, $\epsilon < a$ and $c \in \mathcal{I}^\epsilon_{\mathbb{L}}$ such that $T \Vdash A \subseteq \bigcup_n c_n$. By Lemma 3, we can assume that $T \in \bigcap_n \operatorname{dom}(c_n)^*$. Since $A \subseteq \{x \in {}^\omega 2 : T \Vdash x \in \bigcup_n c_n\}$, Lemma 6 yields $\mu^*(A) \le \epsilon$, which is a contradiction. □

8. Unfortunately I am not able to prove that if the iterands of a countable support iteration of proper posets preserve outer measure (i.e. behave as $\mathbb{L}$ in Theorem 1) then so does the limit. So, let us introduce a property ★, which implies preservation of outer measure, and which (by a general preservation theorem of Shelah [**S**]) is itself preserved by countable support iterations of proper posets.

As usual let ZFC^* stand for a sufficiently large part of ZFC. If $M \models ZFC^*$, let $\operatorname{Ra}(M)$ be the set of reals which are random over M.

DEFINITION. Say that $\mathbb{P} \models \bigstar$ if, given a countable $N \prec H_\lambda$ such that $\mathbb{P} \in N$, given a rational $\epsilon > 0$, $c \in \mathcal{I}^\epsilon_{\mathbb{P}} \cap N$ and $p \in \mathbb{P} \cap N$, for any

$$x \in \mathrm{Ra}(N) \setminus \bigcap\{\bigcup_n c_n(p_n) : \langle p_n : n \in \omega\rangle \in (\prod_n \mathrm{dom}(c_n)) \cap N \text{ is a descending sequence below } p\},$$

there exists (N, P)-generic $q \leq p$ such that $q \Vdash x \in \mathrm{Ra}(N[G]) \setminus \bigcup_n c_n$.

LEMMA. *Suppose that* $\mathbb{P} \models \bigstar$. *Then, for* $A \subseteq {}^\omega 2$, $\mu^*(A) = a \Rightarrow \mathbb{P} \Vdash \mu^*(A) = a$.

PROOF. Suppose that $\mathbb{P} \models \bigstar$, $A \subseteq {}^\omega 2$, $\mu^*(A) = a$ and $\mathbb{P} \not\Vdash \mu^*(A) = a$. Then there exist $p \in \mathbb{P}$, a rational $\epsilon < a$ and a sequence $c \in \mathcal{I}^\epsilon_{\mathbb{P}}$ such that $p \Vdash A \subseteq \bigcup_n c_n$. Find a countable model $N \prec H_\lambda$ such that $\mathbb{P}$, p and c are in N. Find below p a descending sequence $\langle p_n : n \in \omega\rangle \in (\prod_n \mathrm{dom}(c_n)) \cap N$. Now, for $x \in \mathrm{Ra}(N) \setminus \bigcup_n c_n(p_n)$ we have by $\bigstar$, $p \not\Vdash x \in \bigcup_n c_n$. Hence $p \not\Vdash x \in A$ and thus $x \notin A$. It follows that $\mathrm{Ra}(N) \cap A \subseteq \bigcup_n c_n(p_n)$, hence $\mu^*(A) \leq \sum_n \mu(c_n(p_n)) \leq \epsilon$, which is a contradiction. □

9. We shall show the following theorem.

THEOREM. $\mathbb{L} \models \bigstar$.

For the proof we need two additional lemmas.

10. First we recall a Solovay-folklore lemma. We shall identify ${}^\omega 2$ and $\mathcal{P}(\omega)$.

LEMMA. *Suppose that N is a countable transitive model of ZFC^*, $p_0 \in N$, $\phi(\cdot,\cdot)$ is a Δ_0 formula with parameters $p_1, \ldots, p_n \in N$. Then*

1. *$\{x \subseteq \omega : \exists y \subseteq p_0\ \phi(x, y)\}$ is a Solovay set of reals over N (in the terminology of* [**Je**]*), i.e., there exists a formula ϕ^* with parameters in N such that for any $x \subseteq \omega$ with $N[x] \models ZFC^*$,*

$$\exists y \subseteq p_0\ \phi(x, y) \Leftrightarrow N[x] \models \phi^*(x);$$

2. *there exists a Borel set B coded in N such that for any $x \in \mathrm{Ra}(N)$,*

$$x \in B \Leftrightarrow \exists y \subseteq p_0\ \phi(x, y).$$

PROOF. (1) Let $W \in N$ be a transitive set such that $p_0, p_1, \ldots, p_n, \omega \in W$. There is a formula ψ such that for $x \subseteq \omega$,

$$\exists y \subseteq p_0\ \phi(x, y) \Leftrightarrow \exists y \subseteq W\ \langle W, x, y, \in\restriction W\rangle \models \psi(x, y, p_0, \ldots, p_n).$$

(Here x and y work as predicates; contexts like '$x \in z$' are replaced by '$\exists v \in z\ \forall w(x(w) \Leftrightarrow w \in v$'.) Let $\mathcal{C}$ be the poset adding a 1–1 function from W onto ω (i.e., $\mathcal{C}$ is the set of 1–1 functions from finite subsets of W into ω ordered by the reversed inclusion).

Fix $x \subseteq \omega$ such that $N[x] \models ZFC^*$ and let $c : W \to \omega$ be $\mathcal{C}$ generic function over $N[x]$. Let (†) stand for

$$\exists y \subseteq W\ \langle W, x, y, \in\restriction W\rangle \models \psi(x, y, p_0, \ldots, p_n)$$

and (‡) for

$$\exists y \subseteq \omega\ \langle \omega, c[x], y, c[\in\restriction W]\rangle \models \psi(c[x], y, c(p_0), \ldots, c(p_n)).$$

Then

$$\begin{aligned}(\dagger) &\Leftrightarrow (\ddagger)\\ &\Leftrightarrow N[x][c] \models (\ddagger)\\ &\Leftrightarrow N[x][c] \models (\dagger)\\ &\Leftrightarrow N[x] \models \mathcal{C} \Vdash (\dagger).\end{aligned}$$

(The second equivalence is by absoluteness of Σ^1_1 formulas, the fourth one is by homogeneity of $\mathcal{C}$.)

Thus

$$\exists y \subseteq p_0\ \phi(x,y) \Leftrightarrow N[x] \models \mathcal{C} \Vdash \exists y \subseteq p_0\ \phi(x,y).$$

(2) Follows from (1), see [**Je**]. □

11. The second lemma is as follows.

LEMMA. *Suppose that $N \prec H_\lambda$ is countable, $\epsilon > 0$ is rational, $c \in \mathcal{I}^\epsilon_{\mathbb{L}} \cap N$ and $T \in \mathbb{L} \cap N$. Then there exists a Borel set B, coded in N, such that $\mu(B) \leq \epsilon$ and for any $x \in \mathrm{Ra}(N) \setminus B$ there is $(N,\mathbb{L})$-generic $S \leq T$, such that $S \Vdash x \in \mathrm{Ra}(N[G]) \setminus \bigcup_n c_n$.*

PROOF. We drop sub(p)scripts standing by $\mathcal{I}$. Let $\mathcal{D}$ be the family of all open dense subsets of $\mathbb{L}$. Define $C \subseteq {}^\omega 2$ by $x \notin C$ iff

$$\begin{gathered}\exists S \leq T\ \exists\, f : \mathcal{I} \cap N \to \omega,\ f(c) = 0,\\ \forall D \in \mathcal{D} \cap N\ \exists R \in D^* \cap N\ S \leq R,\\ S \Vdash \forall\, d \in \mathcal{I} \cap N\ \forall\, m \geq f(d)\ x \notin d_m.\end{gathered}$$

Easily, if $x \notin C$ then S witnessing this fact is $(N,\mathbb{L})$-generic and $S \Vdash x \in \mathrm{Ra}(N[G]) \setminus \bigcup_n c_n$. (To see genericity find for a given $D \in \mathcal{D} \cap N$, $R \in D^* \cap N$ such that $S \leq R$. By density, D is nonempty below S, so, by Lemma 2, there is $s \in S$ with $R_s \in D \cap N$.)

We are done by the following two claims.

CLAIM 1. $\mu^*(C) \leq \epsilon$.

PROOF. Let $\mathcal{D} \cap N = \{D_n : n \in \omega\}$. Repeat the construction from Lemma 3 in such a way that each $S_n \in N$. Let $S^\dagger = \bigcap_n S_n$. Then for each $D \in \mathcal{D}$ there is $R \in D^* \cap N$ with $S \leq R$ (in particular $S \in D^*$).

For $f : \mathcal{I} \to \omega$ let

$$C_f = \{x \in {}^\omega 2 : S^\dagger \Vdash \exists\, d \in \mathcal{I} \cap N\ \exists\, m \geq f(d)\ \ x \in d_m\}.$$

Let

$$C^\dagger = \{x \in {}^\omega 2 : \forall\, f : \mathcal{I} \to \omega\ f(c) = 0 \Rightarrow x \in C_f\}.$$

Note that $C \subseteq C^\dagger$ (show $x \notin C^\dagger \Rightarrow x \notin C$!).

By Lemma 6, for any $f : \mathcal{I} \to \omega$ we have

$$\begin{aligned}\mu^*(C_f) &\leq \sum_{d \in \mathcal{I} \cap N} \sum_{m \geq f(d)} \mu(d_m)\\ &= \sum_{d \in \mathcal{I} \cap N} \epsilon \cdot 2^{-f(d)}.\end{aligned}$$

It follows that

$$\mu^*(C^\dagger) \leq \inf_f \mu^*(C_f) \leq \epsilon$$

(the only restriction on f is that $f(c) = 0$). $\square$

CLAIM 2. *There is a Borel set B, coded in N, such that $B \cap \mathrm{Ra}(N) = C \cap \mathrm{Ra}(N)$.*

PROOF. Work with the transitive collapse N^* of N in order to apply Lemma 10. To see that

$$S \Vdash \forall\, d \in \mathcal{I} \cap N\ \forall m \geq f(d)\ x \notin d_m$$

is a Δ_0 proposition about x, S and f (with parameters in N^*) note that

$$\begin{aligned} S \nVdash x \notin d_m &\Leftrightarrow \exists S' \leq S\ S' \Vdash x \in d_m \\ &\Leftrightarrow \exists S' \leq S\ x \in d_m(S') \\ &\Leftrightarrow \exists s \in S\ x \in d_m(S_s) \\ &\Leftrightarrow \exists R \in \mathrm{dom}(d_m)^* \cap N\ \exists s \in S\ S \leq R\ \&\ x \in d_m(R_s). \end{aligned}$$

(For the third $\Rightarrow$ use $S \in \mathrm{dom}(d_m)^*$ and Lemma 5.) $\square$

12. We are ready for the proof of Theorem 9.

PROOF OF THEOREM 9. Fix a countable $N \prec H_\lambda$, rational $\epsilon > 0$, $c \in \mathcal{I}_{\mathbb{L}}^{\epsilon} \cap N$, $T \in \mathbb{L} \cap N$, a descending sequence $\langle T_n : n \in \omega \rangle \in (\prod_n \mathrm{dom}(c_n)) \cap N$ below T and $x \in \mathrm{Ra}(N) \setminus \bigcup_n c_n(T_n)$. Apply Lemma 11 uniformly to $\langle c_{n+i} : i \in \omega \rangle$'s and T_n's and get a sequence $\langle B_n : n \in \omega \rangle$ of Borel sets, which is coded in N, such that for each n, $\mu(B_n) \leq \epsilon/2^n$ and if $x \notin B_n$ then there is $S_n \leq T_n$, $(N, \mathbb{L})$-generic, such that $S_n \Vdash x \in \mathrm{Ra}(N[G]) \setminus \bigcup_i c_{n+i}$. Since $\bigcap_n B_n$ is a null Borel set, which is coded in N, $x \in \mathrm{Ra}(N)$ yields $x \notin \bigcap_n B_n$. Fix n with $x \notin B_n$. Then $S_n \Vdash x \in \mathrm{Ra}(N[G]) \setminus \bigcup_{m \geq n} c_m$. Since for $m \leq n$, $c_m(T_m) = c_m(T_n) = c_m(S_n)$, from $x \notin \bigcup_{m \leq n} c_m(T_m)$ we get $S_n \Vdash x \notin \bigcup_{m \leq n} c_m$. Thus $S_n \Vdash x \in \mathrm{Ra}(N[G]) \setminus \bigcup_m c_m$. $\square$

13. Similar arguments work for some other tree forcings, e.g., Miller's rational perfect forcing, Shelah's $\mathbb{Q}_{f,g}$ forcing etc..

References

[Je] T. Jech, *Set Theory*, Academic Press, New York, 1978.
[JS] H. Judah and S. Shelah, *The Kunen-Miller chart*, J. Symb. Logic **55** (1990), 909–927.
[K] K. Kuratowski, *Topology, vol. 1*, Academic Press, New York, 1966.
[L] R. Laver, *On the consistency of Borel's conjecture*, Acta Math. **137** (1976), 151–169.
[S] S. Shelah, *Proper and improper forcing*, forthcoming book.
[W] H. Woodin, *A letter to J. Baumgartner, August 30, 1981*, unpublished.

DEPARTMENT OF MATHEMATICS, UNIVERSITY OF WROCŁAW, PL. GRUNWALDZKI 2/4, 50-384 WROCŁAW, POLAND

E-mail address: pawlikow@math.uni.wroc.pl

For an infinite cardinal number κ, let $H(\kappa)$ denote:

For every $\omega \times \omega$ matrix $(S_n^m : m, n < \omega)$ of subsets of κ such that for each m $\kappa = \bigcup_{n<\omega} S_n^m$, there exists a sequence $(Y_m : m < \omega)$ of finite subsets of ω such that $\kappa = \bigcup_{m<\omega}(\bigcap_{k\geq m}(\bigcup_{n\in Y_k} S_n^k))$.

Proposition 9. *For an infinite cardinal number κ the following are equivalent:*

(1) *ONE does not have a winning strategy in $\mathsf{G}_3(\kappa)$.*
(2) *$H(\kappa)$.*

Proof. The proof that 1 implies 2 is standard. Let $(S_n^m : m, n < \omega)$ be an $\omega \times \omega$ matrix as in the definition of $H(\kappa)$. Consider the strategy for ONE which calls on ONE to choose $(S_n^m : n < \omega)$ in the $m+1$-st inning. This strategy is not a winning strategy for ONE. Consider a sequence of moves by TWO which defeats this strategy. Then this sequence is of the sort that witnesses $H(\kappa)$ for the given matrix.

We now work on the proof of $2 \Rightarrow 1$. Assume $H(\kappa)$ and let F be a strategy of ONE for the game $\mathsf{G}_3(\kappa)$. Using the strategy F we build a family $(U_\sigma : \sigma \in {}^{<\omega}\omega)$ of subsets of κ as follows: We write $F(\kappa) = (U_{(n)} : n < \omega)$, where this enumeration is bijective if possible. For each $n_1 < \omega$, write $F(\{U_{(j)} : j \leq n_1\}) = (U_{(n_1,m)} : m < \omega)$. In general, with U_σ defined for all sequences of length at most k, we define $F(\{U_{(n)} : n \leq n_1\}, \{U_{(n_1,n)} : n \leq n_2\}, \dots, \{U_{(n_1,\dots,n_{k-1},n)} : n \leq n_k\}) = (U_{(n_1,\dots,n_k,n)} : n < \omega)$.

From the family of U_σ's we now define an $\omega \times \omega$ matrix of subsets of κ as follows: Let m and n be nonnegative integers. We define:

$$S_n^m = \bigcap_{\sigma\in {}^{\leq m}m}(\bigcup_{j\leq n} U_{\sigma\frown(j)}).$$

We see that for each m $\kappa = \bigcup_{n<\omega} S_n^m$, and if ℓ is less than n, then S_ℓ^m is a subset of S_n^m.

Apply $H(\kappa)$ to the matrix $(S_n^m : m, n < \omega)$. We find a sequence $(Y_m : m < \omega)$ of finite subsets of ω such that

$$\kappa = \bigcup_{k<\omega}(\bigcap_{m\geq k}(\bigcup_{n\in Y_m} S_n^m)).$$

But then, letting for each m n_m be the maximum of Y_m, we see that indeed $\kappa = \bigcup_{k<\omega}(\bigcap_{m\geq k} S_{n_m}^m)$.

Select a sequence $m_1 < m_2 < \dots < m_k < \dots$ of positive integers such that $n_1 < m_1$, and for each k we have $(\max\{n_j : j \leq m_k + 1\}) + m_k < m_{k+1}$. It then follows that κ is equal to $\bigcup_{j<\omega} \bigcap_{k\geq j} S_{m_{k+1}}^{m_k+1}$.

But look, for each k we have

$$\bigcup_{j\leq m_{k+1}} U_{(m_1,\dots,m_k,j)} \supseteq S_{m_{k+1}}^{m_k+1}.$$

It follows that the sequence of moves

$$(\{U_{(j)} : j \leq m_1\}, \{U_{(m_1,j)} : j \leq m_2\}, \{U_{(m_1,m_2,j)} : j \leq m_3\} \dots)$$

by TWO against the strategy F of ONE defeats F. □

The statement $\mathsf{H}(\kappa)$ is a combinatorial version of Hurewicz's property where instead of κ we have a topological space, and where the entries of the ω by ω matrix are open subsets of the space. This property was introduced by Hurewicz in his paper [4]. One can show that $\mathfrak{b}$ is the minimal cardinality of a subset of the real line which does not have Hurewicz's's property.

4. The cardinal number $\mathsf{add}(\mathcal{M})$.

Let κ be an infinite cardinal number. Define the game $\mathsf{G}_4(\kappa)$ as follows: Two players, ONE and TWO, play ω innings. In the n-th inning ONE cuts κ into countably many (not necessarily disjoint) pieces U^n_k, $k<\omega$, and TWO selects one of these pieces, $U^n_{k_n}$. TWO wins the game if there exists an increasing $h:\omega\to\omega$ such that for each $\alpha<\kappa$, for all but finitely many n we have:

$$\alpha\in\ \cup_{h(n)\leq j<h(n+1)}U^j_{k_j}.$$

Otherwise, ONE wins.

Theorem 10. *For an infinite cardinal number κ, the following are equivalent:*

(1) *ONE does not have a winning strategy in $\mathsf{G}_4(\kappa)$.*

(2) $\kappa<\min\{\mathfrak{b},\mathsf{cov}(\mathcal{M})\}$.

Proof. 1 $\Rightarrow$ 2: First we show that $\kappa<\mathfrak{b}$. Let $\{f_\alpha:\alpha<\kappa\}$ be a given subset of ${}^\omega\omega$. For $n,m<\omega$, put

$$U^n_m=\{\alpha<\kappa:f_\alpha(n)=m\}.$$

Let ONE's strategy be the function which calls on ONE to play the partition

$$\{U^n_m:m<\omega\}$$

during the $n+1$–th inning. By hypothesis this strategy is not a winning strategy for ONE; thus, find a sequence $g=(m_0,m_1,\dots,m_k,\dots)$ in ${}^\omega\omega$ and an increasing h in ${}^\omega\omega$ such that for each $\alpha<\kappa$, for all but finitely many n,

$$\alpha\in\ \cup_{h(n)\leq j<h(n+1)}U^j_{m_j}.$$

Define f by $m_0=f(0)$ and for each n, $f(n+1)=\max\{m_j:j\leq h(n+1)\}+1$. Then we see that for each $\alpha<\kappa$, $f_\alpha\prec f$.

To see that also $\kappa<\mathsf{cov}(\mathcal{M})$, notice that a winning strategy for ONE in the game $\mathsf{G}_2(\kappa)$ is also a winning strategy for ONE in the game $\mathsf{G}_4(\kappa)$.

(2) $\Rightarrow$ (1): Assume that $\kappa<\min\{\mathfrak{b},\mathsf{cov}(\mathcal{M})\}$. Let F be a strategy for ONE. Define $\{U_\sigma:\sigma\in{}^{<\omega}\omega\}$ as follows:

$$F(\kappa)=(U_{(n)}:n<\omega),\text{ and}$$

$$F(U_{(n_1)},\dots,U_{(n_1,\dots,n_k)})=(U_{(n_1,\dots,n_k,m)}:m<\omega).$$

We are looking for a g in ${}^\omega\omega$ such that for some increasing h in ${}^\omega\omega$, for each $\alpha<\kappa$, for all but finitely many n,

$$\alpha\in\ \cup_{h(n)\leq j<h(n+1)}U_{g\lceil_j}.$$

For each $\alpha<\kappa$ define g_α such that $g_\alpha(0)=\min\{m:\alpha\in U_{(m)}\}$, and for each n, $g_\alpha(n+1)$ is the least $k>g_\alpha(n)$ such that:

$$(\forall i\leq g_\alpha(n))(\forall n_1,\dots,n_i\leq g_\alpha(n))(\exists m\in[g_\alpha(n),k))(\alpha\in U_{(n_1,\dots,n_i,m)}).$$

Since we have $\kappa<\mathfrak{b}$, find a strictly increasing f in ${}^\omega\omega$ such that for each α, for all but finitely many n, $g_\alpha(n)<f(n)$. Now regard $f(n)+1$ as the set $\{0,1,\dots,f(n)\}$ endowed with the discrete topology. Look at the topological space

$$X=\Pi_{n<\omega}(f(n)+1).$$

For each $\alpha<\kappa$ let n_α be such that for all $n\geq n_\alpha$, $g_\alpha(n)<f(n)$. For $n\geq n_\alpha$ put

$$D^n_\alpha=\{g\in X:(\forall k\geq n)(\alpha\notin U_{g\lceil_k})\}.$$

Then each D^n_α is closed and because a finite modification of g_α is an element of X, we also see that each D^n_α is nowhere dense.

Since we have $\kappa < \mathsf{cov}(\mathcal{M})$, the set $X \setminus (\bigcup_{\alpha<\kappa} \bigcup_{n<\omega} D^n_\alpha)$ is nonempty. Let g be an element of this difference. Then we have:

$$(\forall \alpha < \kappa)(\exists^\infty_n)(\alpha \in U_{g\lceil_n}).$$

Now that g is fixed we define for each $\alpha < \kappa$ an h_α as follows:

$$h_\alpha(0) = \min\{k < \omega : (\exists j < k)(\alpha \in U_{g\lceil_j})\}, \text{ and for each } n,$$

$$h_\alpha(n+1) = \min\{k > h_\alpha(n) : (\exists j \in [h_\alpha(n), k))(\alpha \in U_{g\lceil_j})\}.$$

Once again we use the fact that $\kappa < \mathfrak{b}$. Choose an r which is strictly increasing, such that

- $0 < r(0)$ and
- for each $\alpha < \kappa$, $h_\alpha \prec r$

Then define h so that $h(k) = r^{exp_{k+1}(2)}(0)$, the $exp_{k+1}(2)$–th iterate of r, computed at 0 (here, $exp_0(2) = 2$ and $exp_{k+1}(2) = 2^{exp_k(2)}$). We shall show that for all but finitely many k, there are ℓ such that

$$h(k) < h_\alpha(\ell) < h_\alpha(\ell+1) < h(k+1),$$

which will complete the proof. Fix M so large that for all $n \geq M$, $h_\alpha(n) < r(n)$. Then consider any $k \geq 3$ with $M \leq h(k)$. We have: $h(k) < h_\alpha(h(k)) < h_\alpha(h(k)+1) < h(k+1)$. Thus, for any $k \geq 3$ such that $h(k) \geq M$, $\ell = h(k)$ works.

We have found appropriate g and h. □

In view of A.W. Miller's theorem that $\mathsf{add}(\mathcal{M}) = \min\{\mathfrak{b}, \mathsf{cov}(\mathcal{M})\}$, we see that $\mathsf{G}_4(\kappa)$ characterizes $\mathsf{add}(\mathcal{M})$. Miller's equation could hardly be called trivial or self-evident. It would be more satisfying to work directly with $\mathsf{add}(\mathcal{M})$ in our theorem, since this would give an interesting alternative proof of Miller's theorem.

For an infinite cardinal number κ let $\mathsf{A}(\kappa)$ denote:

> For every $\omega \times \omega$ matrix $(S^m_n : m, n < \omega)$ of subsets of κ such that for each m $\kappa = \bigcup_{n<\omega} S^m_n$, there exist an increasing sequence $h : \omega \to \omega$ and a sequence $(n_m : m < \omega)$ such that each element of κ is in all but finitely many of the unions $\bigcup_{h(j) \leq m < h(j+1)} S^m_{n_m}$.

Proposition 11. *For an infinite cardinal number κ the following are equivalent:*

(1) *ONE does not have a winning strategy in $\mathsf{G}_4(\kappa)$.*
(2) $\mathsf{A}(\kappa)$.

Proof. The proof that 1 implies 2 proceeds as usual. We are given an $\omega \times \omega$ matrix as in the definition of $A(\kappa)$; pretend that it is a strategy for ONE, apply 1 to find a counterplay by TWO which defeats this strategy. Then this counterplay constitutes a sequence as required by the principle $\mathsf{A}(\kappa)$.

The proof that 2 implies 1 takes more work, part of which was already done in the proof of 2 $\Rightarrow$ 1 of Proposition 4. As there, we begin with a strategy F of ONE. From it we define a family $(U_\tau : \tau \in {}^{<\omega}\omega)$. From this family we define the sets $U_\sigma(m, j)$ as there. Lemma 5 holds of this family. Since $\mathsf{A}(\kappa)$ implies $\mathsf{R}(\kappa)$ and since $\mathsf{R}(\kappa)$ implies $\mathsf{M}(\kappa)$, we then find by Lemma 6 increasing sequences $(j_n : n < \omega)$ and $(m_n : n < \omega)$ of nonnegative integers such that there is for each $\alpha < \kappa$ infinitely many n such that for each such n there is a sequence $\sigma : m_{n+1} - m_n \to j_{n+1}$ with α in $U_\sigma(m_n, j_n)$.

With for each n m_n and j_n now fixed we define $W(k_1, \ldots, k_n; \sigma_1, \ldots, \sigma_n)$ as before and we let $\mathcal{W}_n$ consist of all subsets of κ of this form. Lemma 7 applies, and we find an $\omega \times \omega$ matrix of subsets of κ as in $\mathsf{A}(\kappa)$, by letting the m-th row consist of the elements of $\mathcal{W}_m$, enumerated in some order.

Now the argument starts to deviate from that in Proposition 4: Instead of applying $\mathsf{R}(\kappa)$ to our matrix, we apply $\mathsf{A}(\kappa)$. We find an increasing sequence $h_1 : \omega \to \omega$, and for each m a $W_m \in \mathcal{W}_m$ such that

$$\kappa = \cup_{k<\omega}(\cap_{j\geq k}(\cup_{h_1(j)\leq m<h_1(j+1)} W_m)).$$

Each W_m is of the form $W(k_1^m, \ldots, k_m^m; \sigma_1^m, \ldots, \sigma_m^m)$. Let such a representation for each be fixed from now on. For each n we choose an $\ell_n \in \{k_1^n, \ldots, k_n^n\} \setminus \{\ell_1, \ldots, \ell_{n-1}\}$ and we let τ_n be the σ_i^n corresponding with $\ell_n (= k_i^n)$. Since the ℓ_n–s are pairwise distinct, also the m_{ℓ_n}–s and the j_{ℓ_n}–s are pairwise distinct.

As before one sees from the definitions that for each k,

$$W_k \subseteq U_{\tau_k}(m_{\ell_k}, j_{\ell_k}).$$

Also, for each $\alpha \in \kappa$, for all but finitely many j α is in $\cup_{h_1(j)\leq k<h_1(j+1)} U_{\tau_k}(m_{\ell_k}, j_{\ell_k})$.

Define $g : \omega \to \omega$ such that for each k,

$$g(m_{\ell_k} + i) = \tau_k(i)$$

whenever $i < m_{\ell_k+1} - m_{\ell_k}$. Finally, for each k set $h(k) = m_{\ell_{h_1(j)}}$. Then we have that

$$\kappa = \cup_{k<\omega}(\cap_{j\geq k}(\cup_{h(j)\leq m<h(j+1)} U_{g\lceil_k})).$$

This completes the proof of this Proposition. □

5. The cardinal number unif($\mathcal{M}$).

Let κ be an infinite cardinal number. Define the game $\mathsf{G}_5(\kappa)$ as follows: Two players, ONE and TWO, play ω innings. In the n-th inning ONE cuts κ into countably many *disjoint* pieces U_k^n, $k < \omega$, and TWO selects one of these pieces, say $U_{k_n}^n$. TWO wins the game if each $\alpha < \kappa$ is in no more than finitely many of the selected $U_{k_n}^n$. Otherwise, ONE wins.

Theorem 12. *For an infinite cardinal number κ, the following are equivalent:*

(1) *ONE does not have a winning strategy in $\mathsf{G}_5(\kappa)$.*
(2) $\kappa < \mathsf{unif}(\mathcal{M})$.

Proof. Our proof of this theorem uses a combinatorial characterization of the cardinal number $\mathsf{unif}(\mathcal{M})$, namely: $\mathsf{unif}(\mathcal{M})$ is the least cardinal number κ for which the assertion

> *For every family $\mathcal{F}$ of at most κ elements of ${}^{\omega}\omega$ there exists a sequence $n_0, n_1, \ldots, n_m, \ldots$ such that for each $f \in \mathcal{F}$, the set $\{m < \omega : f(m) = n_m\}$ is finite.*

To see that 1 implies 2, assume that ONE does not have a winning strategy in the game $\mathsf{G}_5(\kappa)$. Let $\mathcal{F}$ be a subset of ${}^{\omega}\omega$ of cardinality κ; enumerate it bijectively as $\{f_\alpha : \alpha < \kappa\}$. For each m and each n define:

$$U_n^m = \{\alpha < \kappa : f_\alpha(m) = n\}.$$

Now consider the strategy of ONE which calls on ONE to play the partition $\{U_n^m : n < \omega\}$ in the $n+1$-th inning. By hypothesis, this is not a winning strategy for

ONE. Consider a sequence of moves $(U^0_{n_0}, \dots, U^m_{n_m}, \dots)$ by TWO which defeats this strategy. Then each $\alpha < \kappa$ is outside all but finitely many of the $U^m_{n_m}$. This means that for each α, for all but finitely many m, $f_\alpha(m) \neq n_m$. Thus, $n_0, n_1, \dots, n_m, \dots)$ has the required properties.

To see that 2 implies 1, consider a strategy F of ONE. Define a family $(U_\tau : \tau \in {}^{<\omega}\omega)$ of subsets of κ as follows:

(1) $U_\emptyset = \kappa$,
(2) $F(U_\emptyset) = \{U_{(n)} : n < \omega\}$, and
(3) for $(n_1, \dots, n_k)$, $F(U_{(n_1)}, \dots, U_{(n_1,\dots,n_k)}) = \{U_{(n_1,\dots,n_k,m)} : m < \omega\}$.

Fix $\alpha < \kappa$ and define f_α such that for each τ in ${}^{<\omega}\omega$, $f_\alpha(\tau)$ is the unique m such that $\alpha \in U_{\tau\frown(m)}$. Since we are assuming that $\kappa < \mathsf{unif}(\mathcal{M})$, we find a g in ${}^{({}^{<\omega}\omega)}\omega$ such that for each α, for all but finitely many τ we have $f_\alpha(\tau) \neq g(\tau)$.

Now fix n_0 and select $n_1, n_2, \dots, n_k, \dots$ such that for each i we have

$$g((n_0, \dots, n_i)) = n_{i+1}.$$

Then TWO wins the F–play generated by the sequence

$$(U_{(n_0)}, \dots, U_{(n_0,\dots,n_k)}, \dots)$$

of moves of TWO. □

It would be aesthetically more pleasing to have a direct proof of the equivalence of assertions 1 and 2; such a proof would give an interesting alternative proof of Miller's characterization of $\mathsf{unif}(\mathcal{M})$.

For an infinite cardinal number κ the symbol $\mathsf{U}(\kappa)$ denotes the following assertion:

> For every $\omega \times \omega$ matrix $(S^m_n : m, n < \omega)$ of subsets of κ such that for each m the sequence $(S^m_n : n < \omega)$ is a partition of κ into pairwise disjoint subsets, there is a sequence $n_0, n_1, \dots, n_m, \dots$ such for each $\alpha \in \kappa$ there are only finitely many m such that $\alpha \in S^m_{n_m}$.

Proposition 13. *For an infinite cardinal number κ the following statements are equivalent:*

(1) *ONE does not have a winning strategy in $\mathsf{G}_5(\kappa)$.*
(2) $\mathsf{U}(\kappa)$.

Proof. The proof that 1 implies 2 is standard. We show that 2 implies 1. Thus, let F be a strategy for ONE in the game $\mathsf{G}_5(\kappa)$. Put $(U(n) : n < \omega) = F(\kappa)$. Assume that U_τ has been defined for all sequences τ of length at most k. Let $(n_1, \dots, n_k)$ be given, and define $(U_{(n_1,\dots,n_k,n)} : n < \omega)$ to be $F(U_{(n_1)}, \dots, U_{(n_1,\dots,n_k)})$.

From the family $(U_\tau : \tau \in {}^{<\omega}\omega)$ we now define a matrix $(S^m_n : m, n < \omega)$ of subsets of κ to which the principle $\mathsf{U}(\kappa)$ will be applied. Let $(\tau_m : m < \omega)$ bijectively enumerate the set ${}^{<\omega}\omega$. Fix $m, n < \omega$, and define: $S^m_n = U_{\tau_m\frown(n)}$. Then $(S^m_n : m, n < \omega)$ is a matrix of the sort in the definition of $\mathsf{B}(\kappa)$. Applying B_κ we find for each m an n_m such that each $\alpha \in \kappa$ is in at most finitely many of the sets $S^m_{n_m}$. Now define a sequence $j_1 < j_2 < \dots$ such that $j_1 = n_0$, and for each k, $j_{k+1} = n_m$ where $\tau_m = (j_1, \dots, j_k)$.

Then the sequence $U_{(j_1)}$, $U_{(j_1,j_2)}, \dots,$ $U_{(j_1,\dots,j_k)}, \dots$ is a sequence of moves by TWO against the strategy F of ONE which also defeats F. □

Added in proof: (February 10, 1995) I recently learned from old unpublished manuscripts that the game–theoretic characterization of $\mathfrak{d}$ was already known to Galvin back in the 1970's.

REFERENCES

[1] T. Bartoszynski and H. Judah, *Set Theory: on the structure of the real line*, preprint of a textbook in progress.

[2] F. Galvin, *Indeterminacy of Point-Open Games*, **Bulletin de L'academie Polonaise des Sciences** 26 (1978), 445 – 448.

[3] W. Hurewicz, *Über die Verallgemeinerung des Borelschen Theorems*, **Mathematische Zeitschrift** 24 (1925), 401 – 421.

[4] W. Hurewicz, *Über Folgen stetiger Funktionen*, **Fundamenta Mathematicae** 9 (1927), 193 – 204.

[5] K. Menger, *Einige Überdeckungssätze der Punktmengenlehre*, **Sitzungsberichte der Wiener Akademie Abt. 2a, Mathematik, Astronomie, Physik, Meteorologie und Mechanik** 133 (1924), 421 – 444.

[6] A.W. Miller, *Some properties of Measure and Category*, **Transactions of the American Mathematical Society** 266 (1981), 93 – 114.

[7] J. Pawlikowski, *Undetermined sets of Point–Open games*, **Fundamenta Mathematicae** 144 (1994), 279 – 285.

[8] F. Rothberger, *Eine Verschärfung der Eigenschaft* C, **Fundamenta Mathematicae** 30 (1938), 50 – 55

[9] M. Scheepers, *Lebesgue measure zero subsets of the real line and an infinite game*, **The Journal of Symbolic Logic**, to appear.

DEPARTMENT OF MATHEMATICS, BOISE STATE UNIVERSITY, BOISE, IDAHO 83725

Contemporary Mathematics
Volume **192**, 1996

ON SOME PROBLEMS IN GENERAL TOPOLOGY

SAHARON SHELAH

§0. INTRODUCTION

This work was done in 1977 and was widely quoted but not submited.

In section 3 it is proved that Arhangelskii's problem has a consistent positive answer: if $V \models CH$, then for some $\aleph_1$-complete $\aleph_2$-c.c. forcing notion P of cardinality $\aleph_2$ we have $\Vdash_P$ "CH and there is a Lindelöf regular topological space of size $\aleph_2$ with clopen basis with every point of pseudo-character $\aleph_0$ (i.e. each singleton is the intersection of countably many open sets)".

Meanwhile this was continued in Hajnal and Juhasz [HJ], Stanley and Shelah [ShSt:167], I.Gorelic [Go] and Ch.Morgan.

In section 4 we prove the consistency of: $CH + 2^{\aleph_1} > \aleph_2 +$ there is no space as above with $\aleph_2$ points" (starting with a weakly compact cardinal).

Section 2 deals with $\beta(\mathbb{N})$, it is proved that the following is consistent with ZFC: $MA + 2^{\aleph_0} = \aleph_2 + (*)$ where

$(*)$ if $A_i^0, A_i^1 \subseteq \omega$ (for $i < \omega_1$) and $A_i^0 \cap A_j^1$ is finite for $i, j < \omega_1$; and $\mathfrak{D}_i^\ell$ is a non-principal ultrafilter over ω such that $A_i^\ell \in \mathfrak{D}_i^\ell$ (for $i < \omega_1$ and $\ell \in \{0, 1\}$), *then* there is a $B \subseteq \omega$ such that $B \in \mathfrak{D}_i^\ell \Leftrightarrow \ell = 0$.

The scheme is as in Baumgartner [B].

Another problem on $\beta(\mathbb{N})$ which I remember was asked and published by E. van Douwen and G. Woods, is answered in §1: is there a discrete $D \subseteq \beta(\mathbb{N})$, of cardinality $\aleph_1$ and $A \subseteq D$ such that $cl(A) \cap cl(D \setminus A) \neq \emptyset$.

I thank U. Abraham for urging the publication (in this form) and for corrections, the referee for corrections and M. Džamonja for corrections. Compared with the old version we added details, explanations, the introduction, added 2.4 and stated 1.2, 4.2; one section was omitted.

§1. A PROBLEM ON $\beta(\mathbb{N})$

1.1 Question. Does there exist a discrete $D \subseteq \beta(\mathbb{N})$ of cardinality $\aleph_1$, and an $A \subseteq D$ such that $cl(A) \cap cl(D \setminus A) \neq \emptyset$?

1991 *Mathematics Subject Classification.* 54A25, 54D15.

Partially supported by the NSF. Research in this paper was done in fall 77 (while the author was at University of Wisconsin, Madison), typed and distributed in winter 78 (while the author was at the University of California, Berkeley), but the author forgot to submit it for publication.

1.2 Answer. Yes, moreover we can let D have any cardinality λ such that $\aleph_1 \leq \lambda = cf(\lambda) \leq 2^{\aleph_0}$ and have $\bigcap\{cl(D') : D' \subseteq D \text{ and } |D'| = |D|\}$ not empty.

1.3 Definition. Let B_i (for $i \leq \omega_1$) be the BA (Boolean algebra) freely generated by $\{x_\alpha : \alpha < i\}, B_i^c$ is the completion of B_i, $B = B_{\omega_1}^c$.

1.4 Claim. *In the space of ultrafilters of $B_{\omega_1}^c$ we can find such a D.*

PROOF: We define by induction on $i \leq \omega_1$, an ultrafilter $\mathfrak{D}_i$ of $B_{\omega_1}^c$ such that

(i) $x_\alpha \in \mathfrak{D}_i \Leftrightarrow \alpha = i$,

(ii) if $a \in B_i^c$, $a \in \mathfrak{D}_i$ then $a \in \mathfrak{D}_j$ for every $j > i$.

This is easy. Let $\mathfrak{D}_{\omega_1} = \bigcup_{i<\omega_1}(\mathfrak{D}_i \cap B_i^c)$. Now, $\{\mathfrak{D}_i : i < \omega_1\}$ is discrete by (i) and $\mathfrak{D}_{\omega_1} \in cl\{\mathfrak{D}_i : i \in S\}$ for any $S \subseteq \omega_1, |S| = \aleph_1$, because $B_{\omega_1}^c = \bigcup_{i<\omega_1} B_i^c$, as B_{ω_1} satisfies the countable chain condition.

1.5 Solution of the problem. Let $X_i \subseteq \omega$ $(i < \omega_1)$ be independent, i.e. any non trivial Boolean combination of the X_i is not empty. Let $f : \mathcal{P}(\omega) \to B_{\omega_1}^c$ be any homomorphism such that $f(X_i) = x_i$ (exists as $B_{\omega_1}^c$ is complete and the X_i are independent). It is not hard to prove f is onto, and $\{f^{-1}(\mathfrak{D}_i) : i < \omega_1\}$ is as required ($\mathfrak{D}_i$ from the claim), and

$$f^{-1}(\mathfrak{D}_{\omega_1}) \in cl(\{f^{-1}(\mathfrak{D}_{2i}) : i < \omega_1\}) \cap cl(\{f^{-1}(\mathfrak{D}_{2i+1}) : i < \omega_1\}).$$

$\square_{1.5}$

§2 A question on $\beta(\mathbb{N}) \setminus \mathbb{N}$

2.1 Claim. *Assuming the consistency of ZFC we prove the consistency of the following assertion with*

$ZFC + 2^{\aleph_0} = \aleph_2 + MA$:

$\otimes$ *if $A_i^0, A_i^1 \subseteq \omega$(for $i < \omega_1$) and $A_i^0 \cap A_j^1$ is finite for $i, j < \omega_1$; $\mathfrak{D}_i^\ell$ is a non-principal ultrafilter over ω such that $A_i^\ell \in \mathfrak{D}_i^\ell$(for $i < \omega_1$ and $\ell \in \{0,1\}$), then there is a $B \subseteq \omega$ such that $B \in \mathfrak{D}_i^\ell \Leftrightarrow \ell = 0$.*

PROOF: We assume $V \models 2^{\aleph_0} = \aleph_1 \wedge 2^{\aleph_1} = \aleph_2$. We repeat the proof of Solovay-Tennenbaum of $Con(ZFC+2^{\aleph_0} = \aleph_2+MA)$, similarly to Baumgartner [B]. That is, we define by induction on $\alpha < \omega_2$ a set of forcing conditions P_α, increasing (under $\subseteq$ and even $\lessdot$) and continuous with $\alpha, |P_\alpha| \leq \aleph_1$, and each P_α satisfies the countable chain condition. We start with $V \models 2^{\aleph_0} = \aleph_1 + 2^{\aleph_1} = \aleph_2 + \diamondsuit_{\{\delta<\omega_2 : cf(\delta)=\aleph_1\}}$.

Now, in addition, at some $\alpha < \omega_2$ we consider a system $\langle A_i^\ell : i < \omega_1, \ell = 0,1\rangle$, $\langle \mathfrak{D}_i^\ell : i < \omega_1, \ell = 0,1\rangle$, which belongs to $V^{P_\alpha}, A_i^\ell \subseteq \omega$, $A_i^\ell \in \mathfrak{D}_i^\ell$, $\mathfrak{D}_i^\ell$ a family of subsets of ω, such that the filter it generates in V^{P_α} (which we denote by the same letter) is $\aleph_1$-saturated (i.e. there are no $C_\alpha \neq \emptyset \text{ } mod \text{ } \mathfrak{D}_i^\ell, C_\alpha \subseteq \omega$ for $\alpha < \omega_1$ such that $C_\alpha \cap C_\beta = \emptyset \text{ } mod \text{ } \mathfrak{D}_i^\ell$ for $\alpha < \beta < \omega_1$). By the usual bookkeeping, every such system appears, and this is possible as $V \models 2^{\aleph_0} = \aleph_1 + 2^{\aleph_1} = \aleph_2$.

Clearly $V^{P_\alpha} \models 2^{\aleph_0} = \aleph_1$. We define in V^{P_α} a set of forcing conditions Q satisfying the c.c.c., whose generic set gives a B such that $B \in \mathfrak{D}_i^0$ and $\omega \setminus B \in \mathfrak{D}_i^1$ (for $i < \omega_1$), and define $P_{\alpha+1} = P_\alpha * Q$. This is sufficient, as if $p \in P_{\omega_2}$ forces that $\langle \underset{\sim}{A}_i^\ell : i < \omega_1 \rangle, \langle \underset{\sim}{\mathfrak{D}}_i^\ell : i < \omega_1, \ell = 0,1 \rangle$ contradict $\otimes$, then for a club of E of ω_2, for every $\alpha \in E$ of cofinality $\aleph_1$, $\langle \underset{\sim}{A}_i^\ell : \ell < \omega_1 \rangle$ is a P_α-name, and $\langle \mathcal{P}(\omega)^{V^{P_\alpha}} \cap \underset{\sim}{\mathfrak{D}}_i^\ell : \ell, i \rangle$ is a P_α-name of an ultrafilter. Then, clearly for stationarily many $\alpha < \omega_2$ of cofinality $\aleph_1$, in V^{P_α} the above holds and as $\diamondsuit_{\{\delta < \omega_2 : cf(\delta) = \aleph_1\}}$ holds we can assume that we have considered the system in question at some α (of course in the bookkeeping we take care of MA too).

So it suffices to prove:

2.2 Claim. *If $V \models 2^{\aleph_0} = \aleph_1$, $A_i^\ell \subseteq \omega$ (for $\ell = 0,1$ and $i < \omega_1$), $A_i^0 \cap A_j^1$ are finite for $i, j < \omega_1$, $\mathfrak{D}_i^\ell$ an $\aleph_1$-saturated filter over ω, $A_i^\ell \in \mathfrak{D}_i^\ell$, then we can find a partial order Q of size $\aleph_1$, satisfying c.c.c., and in V^Q there is an $X \subseteq \omega$ such that $X \in \mathfrak{D}_i^0$ and $\omega \setminus X \in \mathfrak{D}_i^1$ (for $i < \omega_1$; note: $\mathfrak{D}_i^\ell$ stands for the filter it generates).*

2.2A Remark. The sequence $\langle (B_i^0, B_i^1) : i < \omega_1 \rangle$ constructed below should be just generic enough (we do not use this presentation because some people do not like it[1]). More specifically letting $f_{2i+\ell} \in {}^\omega 2$ be defined by

$$f_i^\ell(n) = 0 \Leftrightarrow [\text{the } n\text{th element of } A_i^\ell \text{ belongs to } B_i^\ell]$$

we demand: for every $n < \omega$ and open dense $J \subseteq {}^n({}^\omega 2)$, for some $\alpha_J < \omega_1$, for every $\alpha_0 < \ldots < \alpha_{n-1}$ from (α_J, ω_1) we have $\langle f_{\alpha_0}, \ldots, f_{\alpha_{n-1}} \rangle \in J$.

PROOF: We shall choose sets $B_i^\ell \subseteq A_i^\ell$, $B_i^\ell \notin \mathfrak{D}_i^\ell$, and let $Q = \{(f,g) : f, g$ are finite functions from ω_1 to ω, and $(A_i^0 \setminus B_i^0 \setminus f(i)) \cap (A_j^1 \setminus B_j^1 \setminus g(j)) = \emptyset$ when $i \in Dom(f)$ and $j \in Dom(g)\}$.

(let $Q_i = \{(f,g) \in Q : Dom(f) \cup Dom(g) \subseteq i\}$).

Q is ordered naturally: $(f_1, g_1) \leq (f_2, g_2)$ <u>iff</u> $f_1 \subseteq f_2$ & $g_1 \subseteq g_2$

For a generic $G \subseteq Q$ the set $X = \bigcup \{B_i^0 \setminus f(i) : \exists g((f,g) \in G)\}$ is as required, and $|Q| = \aleph_1$. We should show only that we can choose B_i^ℓ such that Q satisfies the c.c.c. (the density condition is easy: for every $i, j < \omega_1$ by the almost disjoint condition, $A_i^0 \cap A_j^1$ is finite, hence for every $(f,g) \in Q$ and $i < \omega_1$ there is $n^* < \omega$ such that $A_i^0 \subseteq n^*$ is disjoint to A_j^1 for $j \in Dom(g) \cup \{i\}$ and $A_i^1 \setminus n^*$ is disjoint to A_j^0 for $j \in Dom(f) \cup \{i\}$. Let $f' = f \cup \{\langle i, n^* \rangle\}$, $g' = g \cup \{\langle i, n^* \rangle\}$. So, there is a $(f', g') \in Q$ such that $(f,g) \leq (f', g')$ and $i \in Dom(f') \cap Dom(g')$).

Suppose (f_i, g_i) (for $i < \omega_1$) exemplify a contradiction to c.c.c. Then, by the well known techniques, we can assume that there is a $(f,g) \leq (f_i, g_i)$, $Dom(f_i) = Dom(g_i) = w_i$, $Dom(f) = Dom(g) = w$, $w \subseteq w_i$ and the sets $w_i \setminus w$ are pairwise disjoint, $w_i \setminus w = \{\eta_i(0), \ldots, \eta_i(t)\}$, $i < j \Rightarrow \eta_i(t) < \eta_j(0)$, $\eta_i(0) < \eta_i(1) < \ldots < \eta_i(t)$, and for $m \leq t$ we have $f_i(\eta_i(m)) = k_m$ and $g_i(\eta_i(m)) = j_m$. For some δ we can get an "elementary submodel" of the whole system (so $\eta_i(t) < \delta$ for $i < \delta$). Let $t_\zeta, \delta_\zeta, f_i^\zeta, g_i^\zeta$ (for $i < \delta$), $\eta_i^\zeta(m), k_m^\zeta, j_m^\zeta$ (for $\zeta < \omega_1$ and $m \leq t$) enumerate all

[1] at least they did not back in the seventies

possible such systems. Now we shall choose B^ℓ_α by induction on α and then on ℓ with some restrictions: (say $\ell = 0$ for notational simplicity).

Let us try to explain the idea of the proof. If $\max\{\zeta, \delta_\zeta\} < \alpha < \omega_1$, then we think of $(\langle \alpha, k^\zeta_0\rangle, \emptyset) \in Q$ as a candidate to be a part of some (f_i, g_i). Now, either it is compatible with infinitely many (f^ζ_i, g^ζ_i) (for $i < \delta_\zeta$), or the condition on being an elementary submodel eliminates this possibility, and similarly if $Dom(f') = Dom(g') = \{\eta(0), \dots, \eta(m-1)\} \subseteq \alpha$ and (f, g) is compatible with infinitely many (f^ζ_i, g^ζ_i), then either so is $(f' \cup \{\langle \alpha, k^\zeta_m\rangle\}, g')$, or the condition on elementary submodels is violated.

There are countably many such conditions and we can find $2^{\aleph_0}$ pairwise almost disjoint infinite $Y_\xi \subseteq A^\ell_i$ (for $\xi < 2^{\aleph_0}$), such that $A^\ell_\alpha \setminus Y_\xi$ are as required and all but countably many of the Y_ξ's are $= \emptyset \ mod\ \mathfrak{D}^\ell_\alpha$ (as it is $\aleph_1$-saturated), so we have many candidates for B^ℓ_α.

Now we present the construction itself. Assume B^0_i, B^1_0 have been choosen for $i < \alpha$ and we shall define B^0_α, B^1_α. As $\langle B^0_i, B^1_i : i < \alpha\rangle$ is defined, so is Q_α.

Let

$$\begin{aligned} K_\alpha = \{(\zeta, m, n, i, f', g', \eta) \colon\ & \zeta < \alpha \wedge m \le t_\zeta \wedge n < \omega \wedge i < \delta_\zeta \wedge (f', g') \in Q_\alpha \\ & \wedge |Dom(f')| = m \wedge \eta \in {}^m\alpha \wedge \eta \text{ strictly increasing} \\ & \wedge Dom(f') = Dom(g') = \{\eta(0), \dots \eta(m-1)\} \\ & \wedge f'(\eta(s)) = k^\zeta_s (\text{ for } s < m) \\ & \wedge g'(\eta(s)) = j^\zeta_s (\text{ for } s < m)\}. \end{aligned}$$

Clearly, K_α is countable. Let $L_\alpha = \{(Y^0, Y^1) : Y^\ell \subseteq A^\ell_\alpha \text{ for } \ell = 0, 1\}$, and for $(Y^0, Y^1) \in L_\alpha$, let $Q_\alpha[Y^0, Y^1]$ be defined as $Q_{\alpha+1}$, had we chosen (B^0_α, B^1_α) to be $(A^0_\alpha \setminus Y^0, A^1_\alpha \setminus Y^1)$. We say that $(Y^0, Y^1) \in L_\alpha$ satisfies $(\zeta, m, n, i, f', g', \eta) \in K_\alpha$ if:

<u>either</u>

(α) for some $\beta < \delta_\zeta$, $(f^\zeta_\beta, g^\zeta_\beta)$ and $(f' \cup \{< \alpha, k^\zeta_m >\}, g' \cup \{< \alpha, j^\zeta_m >\}, h)$ are compatible conditions in $Q_\alpha[Y^0, Y^1]$ and $\beta e_{\zeta,n} i$ where $j e_{\zeta,n} i$ means:

$$\begin{aligned} (\forall \ell \le t^\zeta)[A^0_{\eta^\zeta_j(\ell)} \cap n = A^0_{\eta^\zeta_i(\ell)} \cap n \ \wedge\ A^1_{\eta^\zeta_j(\ell)} \cap n = A^1_{\eta^\zeta_i(\ell)} \cap n \\ \wedge B^0_{\eta^\zeta_j(\ell)} \cap n = B^0_{\eta^\zeta_i(\ell)} \cap n \ \wedge\ B^1_{\eta^\zeta_j(\ell)} \cap n = B^1_{\eta^\zeta_i(\ell)}]. \end{aligned}$$

<u>or</u>

(β) for some natural number $u < \omega$, for every $(Z^0, Z^1) \in L_\alpha$ satisfying $Z^0 \cap u = Y^0 \cap u \,\&\, Z^1 \cap u = Y^1 \cap u$, clause ($\alpha$) fails even for $Q_\alpha[Z^0, Z^1]$.

Now, L_α is a complete separable metric space (by the metric d defined as $d((Y^0, Y^1), (Z^0, Z^1)) = \min((Y^0 \Delta Z^0) \cup (Y^1 \Delta Z^1))$ where Δ is the symmetric difference i.e. $Y \Delta Z = (Y \setminus Z) \cup (Z \setminus Y)$.)

Clearly:

($*$) for each $(\zeta, m, n, i, f', g', \eta) \in K_\alpha$ the set

$$L^\alpha_{(\zeta, m, n, i, f', g', \eta)} = \{(Y^0, Y^1) \in L_\alpha : (Y^0, Y^1)$$

satisfies $(\zeta, m, n, i, f', g', \eta)\}$ is an open dense set.

As K_α is countable, we can find a $\langle (Y^0_\xi, Y^1_\xi) : \xi < 2^{\aleph_0}\rangle$ such that:

(a) $(Y^0_\xi, Y^1_\xi) \in \cap\{L^\alpha_{(\zeta,m,n,i,f',g',\eta)} : (\zeta, m, n, i, f', g', \eta) \in K_\alpha\}$

(b) for $\zeta < \xi$ the set $Y^0_\zeta \cap Y^0_\xi$ is finite.

(this is like building a perfect set of Cohen generic reals.)

So (by the $\aleph_1$-saturation), for some $\xi < 2^{\aleph_0}$ we have $Y^0_\xi = \emptyset \, mod \, \mathfrak{D}^0_\alpha$ and $Y^1_\xi = \emptyset \, mod \, \mathfrak{D}^1_\alpha$.

We let $(B^0_\alpha, B^1_\alpha) = (A^0 \setminus Y^0, A^1_\alpha \setminus Y^1)$.

So we have finished the inductive definition of the (B^0_α, B^1_α)'s.

Now we show that the construction quarantees that Q satisfies the c.c.c. Suppose $\{(f_i, g_i) : i < \omega_1\}$ is an uncountable antichain. As we explained above, we may assume there are t_ζ, $\delta_\zeta, \ldots$, and $\{(f_i, g_i) : i < \delta_\zeta\}$ is an "elementary subsystem". So in particular, if for some $u < \omega$ and $F^\ell : t_\zeta + 1 \to \mathcal{P}(u) \times \mathcal{P}(u)$ (where $\ell \in \{0,1\}$) we have an $i < \omega_1$ such that $(A^\ell_i \cap u, B^\ell_{\eta^\zeta_i(m)} \cap u) = F^\ell(m)$ for each $\ell < 2$ and $m \leq t_\zeta$, then such i exists already below δ_ζ. So for each $u < \omega$ and $\gamma < \omega_1$ for some $i_{\gamma,u} < \delta_\zeta$ we have $\gamma e_{\zeta,u} i_{\gamma,u}$. But now consider $\gamma > \delta_\zeta$ such that $\eta^\zeta_\gamma(0) > \delta_\zeta$ and $\eta^\zeta_\gamma(0) > \zeta$, and let $m \leq t_\zeta$ be the least ordinal $n^\zeta_\gamma(m)$ for which condition (α) fails for some finite $u < \omega$ and $i = i_{\gamma,u}$ which is $< \delta_\zeta$ (note that if $\alpha = \eta^\zeta_\gamma(t_\zeta)$ and $u = 0$ then condition (α) cannot possibly hold). Let it fail for $u = u_m$. Then condition (β) holds. But if u witnesses (β) then for $\max\{u_m, u\}$ we have a failure for $m-1$, contradiction.

The construction guarantees that Q satisfies the c.c.c, hence we have finished. $\square_{2.2}$

$\square_{2.1}$

2.3 Remark. 1) We can act as in [Wi:C 2], and then use really $\aleph_1$-saturation and all filters are in V.

2) Can we ignore CH and make $2^{\aleph_0}$ larger?

Assume $\lambda = \lambda^{\aleph_1} = cf(\lambda)$ and $\diamondsuit_S$ where $S = \{\delta < \lambda : cf(\delta) = \aleph_1\}$ and $(\forall \alpha < \lambda)[|\alpha|^{\aleph_0} < \lambda]$. We can find a forcing notion P, which is c.c.c. of cardinality λ, and $\Vdash_P$ "$2^{\aleph_0} = \lambda$ and MA and $(*)$ of 2.1 holds."

Why? we use finite support iteration, if $\alpha \notin S, Q_\alpha$ is adding a Cohen real; if $\delta \in S$ and $\diamondsuit_S$ guesses $P, \langle A^\ell_i : i < \omega_1, \ell < 2\rangle$, and $\langle \mathfrak{D}^\ell_\alpha \cap \mathcal{P}(\omega)^{V^{P_\alpha}} : \alpha < \omega_1, \ell < 2\rangle$ as in the proof, we imitate the proof, but for $\langle (B^0_\alpha, B^1_\alpha) : \alpha < \omega_1\rangle$ we use a sequence $\langle T_\alpha : \alpha < \omega_1\rangle$, T_α a perfect set of members of L_α such that for every large enough $\alpha < \omega_1$, all branches of T_α are Cohen generic over V^{P_δ}.

To answer the question of the referee:

2.4 Claim. *The statement $\otimes$ of 2.1 follows from PFA.*

PROOF Consider the forcing $P = Levy(\aleph_1, \aleph_2) * \underset{\sim}{Q}$ where $\underset{\sim}{Q}$ is constructed as in the proof of 2.1 in the universe $V^{Levy(\aleph_1,\aleph_2)}$ (note that forcing with $Levy(\aleph_1, \aleph_2)$ adds no reals, so $\mathfrak{D}_i$ is still an ultrafilter and CH holds) and let $\underset{\sim}{B}$ be the name of the desired set. So $\Vdash_P$ "for $i < \omega_1$, $\ell < 2$, for some $A \in \mathfrak{D}^\ell_i$ we have $[\ell = 0 \Rightarrow \underset{\sim}{B}^* \supseteq A]$ and $[\ell = 1 \Rightarrow \underset{\sim}{B} \cap A =^* \emptyset]$". Apply PFA.

§3. Concerning Arhangelskii's Problem

3.1 Theorem. *The following is consistent with $ZFC + GCH$:*
(∗) *There is a regular space of cardinality $\aleph_2$ which is Lindelöf and has pseudo-character $\aleph_0$.*

Remark. We had first said "a Hausdorff space..." but Kunen noted the proof actually yields a regular space.

PROOF: Assume V satisfies GCH. Let P be the set of tuples of the form $p = \langle A, f, E, T\rangle = \langle A^p, f^p, E^p, T^p\rangle$, where:

(1) A is a countable subset of ω_2,

(2) f is a two place function from A to $\omega + 1$, where we write $f_x(y)$ instead of $f(x, y)$, and $f_x(y) = \omega \Leftrightarrow y = x$ holds.

(3) E is a three place relation on A, but we write $xE_\gamma y$, and demand that for each fixed γ, E_γ is an equivalence relation, and $\gamma \in A \wedge \gamma < \beta \wedge xE_\beta y \Rightarrow xE_\gamma y$, while $x < \gamma \wedge xE_\gamma y \Rightarrow x = y$. We stipulate E_{ω_2} as the equality on A.

(4) T is a countable set. Each member B of T will be called a formal cover, $|B| = \aleph_0$, and B is of the form $\{\tau_n^B : n < \omega\}$, where each τ_n^B is the formal intersection of finitely many $U_x^n (x \in A, n \in \mathbb{Z})$ [the intended meaning is: $U_x^n = \{y : f_x(y) \geq n\}$ for $n \geq 0$, and $U_x^{-n} = \omega_2 \setminus U_x^n$ for $n > 0$; we say $p \vDash$ "$y \in \tau$" in the natural case (i.e. $y \in A$, $\tau = \bigcap_{i<k} U_{x(i)}^{\ell(i)}$, where $k < \omega, \ell(i) \in \mathbb{Z}, x(i) \in A$ and for each $i < k$ we have: $f_{x(i)}(y) \geq |\ell(i)|$ iff $\ell(i) \geq 0$)]. We let, for a formal term τ, $dom(\tau) = \{x \in A : x$ is mentioned in $\tau\}$, $Dom(B) = \bigcup\{dom(\tau) : \tau \in B\}$.

The real restrictions are

(5) if $zE_\gamma y$, and $x < \gamma$ then $f_x(y) = f_x(z)$.

(6) (a) if $B \in T, x \in A$, then $p \vDash$ "$x \in \tau_n^B$" for some $\tau_n^B \in B$.

(b) moreover, for each finite $A^1 \subseteq A$ and $h : A^1 \to \omega$, there is a $\tau \in B$ such that

$$[x \in A^1 \wedge U_x^n \text{ appears in } \tau] \Rightarrow [[0 \leq n \leq h(x)] \vee [n < -h(x)]]$$

(c) moreover, for each $B \in T, x \in A$ and $\gamma \in \{\omega_2\} \cup Dom(B)$ satisfying $\gamma \leq x$, and a finite $A^1 \subseteq A$ and function $h : A^1 \to \omega$, there is a $\tau = \bigcap_{i<k} U_{x(i)}^{\ell(i)} \in B$ such that: $[x(i) < \gamma] \Rightarrow [x \in U_{x(i)}^{\ell(i)}]$ and $[x(i) \geq \gamma \wedge x(i) \in A^1] \Rightarrow [[0 \leq \ell(i) \leq h(y)] \vee [\ell(i) < -h(y)]]$. Note that for $\gamma = \min Dom(B)$, clause (c) reduces to clause (b) and for $\gamma = \omega_2$, clause (c) reduces to clause (a).

[Explanation: The set A approximates the set of points, the function f describes the U_x^n's which will generate the topology as clopen sets, T is a set of "countable covers", i.e. we think of a possible covering which is a counterexample to Lindelöfness and "promise" that a countable subfamily of such cover, will cover the entire space.

In demand (6), clause (a) just says that each $B \in T$ really covers, clause (b) is necessary when we prove e.g. density of $\{p : x \in A^p\}$ for $x \in \omega_2$, (see the proof of Fact B). This is done by an increasing ω-sequence of descriptions of the important new values of f, so this clause tells us a finite information, so does not prevent us from preserving "B is a cover".

Still, why do we need clause (c) of demand (6)? We want that our forcing notion satisfies the $\aleph_2$-c.c., so we use the Δ-system lemma, and during the construction (i.e. the proof of Fact E i.e. the construction of a common upper bound of p_1, p_2) we have finitely many commitements on new values of f, we want to make $x \in \bigcup_{n<\omega} \tau_n^B$ for $B \in T^{p_j}$, $x \in A^{p_i} \setminus A^{p_j}$, so $f_y(x)$ is determined for all $y \in A^{p_i} \cap A^{p_j}$ and for finitely many $y \in A^{p_j} \setminus A^{p_i}$, and clause (6)(c) guarantees we can deal with this. We have above avoided "justifying" the use of the equivalence relation E^p, it is needed when in the proof of Fact D (Lindelöfness holds), to the union of a generic enough sequence $\langle p_n : n < \omega \rangle$ for the union q we add to the T^q a covering thus defeating a possible counterexample, we need E to verify condition (6)(c).]

If there are several p's in consideration, we shall write $A^p, f^p, \dots$ or $p^\ell = \langle A^\ell, f^\ell, E^\ell, T^\ell \rangle$. Now we define the order on $P : p \le q$ iff $A^p \subseteq A^q, f^p = f^q \restriction A^p, E^p = E^q \restriction A^p, T^p \subseteq T^q$. In V^P we define the following topology on ω_2. For $x \in \omega_2, n < \omega$, let $U_x^n = \{y : f_x^p(y) \ge n$ for some p in the generic set $\}$, and if $n > 0$, we let $U_x^{-n} = \omega_2 \setminus U_x^n$. Now, $\{U_x^n : x \in \omega_2, n \in \mathbb{Z}\}$ will be closed and open, and the topology X is the minimal one which satisfies this. So the set of finite intersections of U_x^n's forms a basis. By clause (2) in the definition of P, and the Fact B below, we know $\bigcap_{n<\omega} U_x^n = \{x\}$, so as each U_x^n is clopen, the space is Hausdorff and even regular, and has pseudo-character $\aleph_0$.

Fact A. P is $\aleph_1$-complete; in fact, any ascending ω-sequence has a naturally defined union.

In fact, we already use

Fact B. For every $p \in P$ and $z \in \omega_2$ there is a $q \in P$ such that $q \ge p$ and $z \in A^q$.

Moreover, if $z \notin A^p$, for any finite subset A^* of A^p and function h^* from A^* to $\mathbb{Z}$ we can demand $f_x^q(z) = h^*(x)$ for $x \in A^*$.

PROOF OF B: The non-trivial part is to satisfy clause (6). We first define $f_x(z)$ for $x \in A^p$ to satisfy (6)(c) when z here stands for x there. So let $\{(B_k, A_k, h_k, \gamma_k) : k < \omega\}$ be a list of all tuples (B, A, h, γ) such that $\gamma \in \{\omega_2\} \cup Dom(B)$, $\gamma \le z$, $B \in T^p$, $A \subseteq A^p$ is finite and $h : A \to \omega$. Now we define by induction on k a finite set $D_k \subseteq A^p$ and $f_x(z)$ for $x \in D_k$.

For $k = 0$, $D_0 = A^*$, $\bigwedge_{x \in D_0} f_x(z) = h^*(x)$. If we have defined D_k, let us define h'_k, A'_k as follows: $A'_k = A_k \cup D_k$ and $h'_k(x)$ is $h_k(x)$ if $x \in A_k$ and $f_x(z)$ if $x \in D_k \setminus A_k$; choose $\tau \in B_k$ as exists by (6)(c) (with A'_k in place of A^1 and h'_k in place of h), let $D_{k+1} = D_k \cup Dom(\tau)$, and define $f_x(z)$ for $k \in Dom(\tau) \setminus D_k$ as ℓ if U_z^ℓ appears in τ with $\ell \ge 0$ and as 0 otherwise.

We can at last complete the definition of $h_x(z)$ for $x \in A \setminus \bigcup_{k<\omega} D_k$. Lastly define $f_z(y)$ for $y \in A^p \cup \{z\}$ as ω if $y = z$ and 1 if $y \ne z$. If we let $q = \langle A \cup \{z\}$, f^q

is f expanded as described above, E^p (i.e. x is E_γ-equivalent only to itself), $T^p\rangle$ then q is O.K. $\square_B$

Similarly, we can prove

Fact C.

(1) For every $p \in P$, $z \in \omega_2 \setminus A^p$ and $\gamma \leq y$ in $A^p, \gamma \leq z$, there is a $q \in P$ such that $p \leq q$ and $q \vDash$ "$zE_\gamma y$". Moreover, for a given finite $A^1 \subseteq A^p \setminus \gamma$, and a function $h^1 : A^1 \to \omega$, we can demand $q \vDash$ "$f_x(z) = h^1(x)$ for $x \in A$".

(2) The following is a dense subset of P, closed under unions:

$$\mathcal{I} = \Big\{ q \in P : \underline{\text{for every}}\ \gamma \in A^q \cup \{\omega_2\} \text{ and finite } A \subseteq A^q \text{ and function } h : A \to \omega \text{ and } x \in A^q \text{ satisfying } \gamma \leq x\ \underline{\text{there is}}\ x' \in A^q \text{ such that } \gamma \leq x', x' E_\gamma x \text{ and } (\forall y \in A \setminus \gamma)[f_x(y) = h(y)] \Big\}.$$

PROOF OF C: 1) like the proof of Fact B.

2) For any $p \in P$, choose $p_n \in P$, $p_0 = p$, $p_n \leq p_{n+1}$, each time use part (1) with a suitable bookkeeping and take the union by Fact A. $\square_B$

Fact D. The space is Lindelöf.

PROOF OF D: Let $\underset{\sim}{\sigma}$ be a name of a cover and $\underset{\sim}{\sigma}_x$ a member of it to which x belongs, and *w.l.o.g.* $\underset{\sim}{\sigma}_x$ is a member of the basis (i.e. is a τ). Now, for each $p \in P$ and $x \in A$, there is a $q \geq p$, such that $q \Vdash$ "$\underset{\sim}{\sigma}_x = \tau$" for some specific τ. By Fact A for every $p \in P$ there is $q \in P$ such that

$$p \leq q \in \mathcal{I} \ \wedge \bigwedge_{x \in A^p} \bigvee \{q \Vdash \text{“}\underset{\sim}{\sigma}_x = \tau\text{”} : \tau \text{ as in demand (4)}\}.$$

So for every $p \in P$ we can find $\langle p_n : n < \omega \rangle$ such that $p \leq p_0$, $p_n \leq p_{n+1}$, $p_n \in \mathcal{I}$ and $x \in A^{p_n} \Rightarrow p_{n+1} \Vdash \underset{\sim}{\sigma}_x = \tau^x$". Let $q = \bigcup_{n<\omega} p_n$, now $q \in P$ is an upper bound of $\{p_n : n < \omega\}$ and $q \in \mathcal{I}$ by Fact C(2). Now

$$B^* = \{\tau : q \Vdash \text{“}\underset{\sim}{\sigma}_x = \tau\text{” for some } x \in A^q \text{ and } \tau \text{ as in (4) (for } q)\},$$

satisfies the requirements on B in clause (4). Now B^* also satisfies the requirements of part (6) in the definition: clause (a) holds as

$$x \in A^q \Rightarrow \bigvee_n x \in A^{p_n} \Rightarrow \bigvee_n (p_{n+1} \text{ forces a value to } \underset{\sim}{\sigma}_x).$$

Clause (c) holds by the above as $q \in \mathcal{I}$. So $q^* \overset{\text{df}}{=} \langle A^q, f^q, E^q, T^q \cup \{B^*\}\rangle \in P$ [why? check; the main point is (6)(c) which holds as $q \in \mathcal{I}$]. Also $q \leq q^*$, so q^* forces that B^* is (essentially) a countable subcover, as required. $\square_D$

Fact E. P satisfies the $\aleph_2$-chain condition. (Hence in V^P, G has power $> 2^{\aleph_0}$).

PROOF: Let $p_i \in P (i < \omega_2)$. It is well known that we can assume that for some p and α_i (for $i < \omega_2$): α_i is increasing, $A^{p_i} \cap \alpha_i = A^p, A^{p_i} \subseteq \alpha_{i+1}, p \leq p_i$. Now, like in the proof of Fact B, we can prove p_0, p_1 can be extended to a condition $\langle A^{p_0} \cup A^{p_1}, f^{p_0} \cup f^{p_1}, E^{p_0} \cup E^{p_1}, T^{p_0} \cup T^{p_1} \rangle$. $\square_E$

$\square_{3.1}$

Remarks.

(1) The proof works with $\aleph_0$ replaced by any λ such that $\lambda^{<\lambda} = \lambda, 2^\lambda = \lambda^+$; countable is replaced by "of cardinality $\leq \lambda$" and τ are still finite formal intersections, $Rang(f) \leq \omega = 1$. We get λ-Lindelöf but still pseudo-character $\aleph_0$.

(2) It is well known that there is no Lindelöf space of pseudo-character $\aleph_0$ of power $\geq$ " first measurable".

§4. More on Arhangelskii's problem

We prove:

4.1 Theorem. *If $(ZFC + GCH + \exists$ a weakly compact) is consistent, then so is the following: $ZFC + CH + not(*)$ where*

$(*)$ *There is a regular space of cardinality $\aleph_2$ which is Lindelöf and has pseudo-character $\aleph_0$.*

Question. (CH) Is there a Lindelöf space with pseudo-character $\aleph_0$ and with $2^{\aleph_1}$ points? More than $\aleph_2$ points?

PROOF: Let $V \models$ "$GCH \wedge \kappa$ weakly compact."

Let P_i $(i < \kappa^+)$ be the forcing for adding a Cohen subset to ω_1 (so P_i is $\aleph_1$-complete).

Q_2^κ is the Levy collapse of κ to $\aleph_2$ (i.e. every $\theta \in (\aleph_1, \kappa)$ is collapsed to $\aleph_1$, and each condition is countable).

$Q_1 = \Pi_{i<\kappa^+} P_i = \{p \in \Pi_{i<\kappa^+} P_i : p$ has countable support (i.e. $p(i) = \emptyset$ for all except countably many i)$\}$.

$Q = Q_2^\kappa \times Q_1$.

In V^Q we know κ is $\aleph_2$, CH holds, $2^{\aleph_1} = \kappa^+$ (and only the cardinals $\theta \in (\aleph_1, \kappa)$ were collapsed.)

We prove that in V^Q there is no Lindelöf space X with countable pseudo-character, such that $|X| = \kappa$.

If there is such an X, we can assume its set of points is ω_2, and $x \in U_x^n, U_x^n$ open, $\bigcap_n U_x^n = \{x\}$ for all $x \in \omega_2$ (by the countable pseudo-character).

So the topology is the minimal one to which all U_x^n belong; this is O.K. as if we decrease family of open sets, the Lindelöfness is preserved.

We *w.l.o.g.* identify X with $\langle (x, n, U_x^n) : x < \omega_2, n < \omega \rangle$. So the topology is the minimal one to which all U_x^n belong; this is O.K. as if we decrease the family of open sets, the Lindelöfness is preserved. Note

$\oplus$ if $X \in V_1 \subseteq V_2$ and $U \subseteq X$ with $U \in V_1$, then $V_1 \models$ "U is open in X" iff $V_2 \models$ "U is open in X".

(because U is open iff $\forall x \in U \exists x_1, \dots, x_k \exists n_1, \dots, n_k (x \in \bigcap_{\ell=1}^{k} U_{k_\ell}^{n_\ell} \subseteq U)$)

Claim A. *In V^Q (but CH is used), for some closed unbounded $C \subseteq \kappa$, we have $\alpha \in C \wedge cf(\alpha) > \aleph_0 \Rightarrow (X \restriction \alpha$ is not Lindelöf, moreover, there is a $g : \alpha \to \omega$, so that $\{U_x^{g(x)} : x \in \alpha\}$ has no countable subcover).*

PROOF OF A: C will be the family of $\alpha < \omega_2$ such that:

if $Dom(h)$ is a countable bounded subset of α, $Range(h) \subseteq \omega, \omega_2 \neq \bigcup\{U_x^{h(x)} : x \in Dom(h)\}$ *then* there is a $\beta < \alpha$, $\beta \notin \bigcup\{U_x^{h(x)} : x \in Dom(h)\}$.

Clearly, C is closed and by CH it is unbounded.

If $cf(\alpha) > \aleph_0$, we can omit the "bounded", as every countable subset of α is bounded.

For $\alpha \in C$ such that $cf(\alpha) > \aleph_0$, define $g : \alpha \to \omega$ as follows : $g(x)$ is the first $n < \omega$ such that $\alpha \notin U_x^n$ (exists as $\bigcap_{n<\omega} U_x^n = \{x\}$).

Clearly $\{U_x^{g(x)} : x < \alpha\}$ cover α. Suppose $\{U_x^{g(x)} : x < \alpha\}$ has a countable subcover $\{U_x^{g(x)} : x \in Y\}, |Y| \leq \aleph_0$. Let $h = g \restriction Y$ and we get a contradiction to the definition of C (because α witnesses that the union is not ω_2). $\square_A$

Claim B. *Suppose V satisfies CH and X and $U_x^n (x \in X)$ are as above. Suppose X is Lindelöf, P is an $\aleph_1$-complete forcing, but in V^P, the space X is not Lindelöf.*

Then also in V^{P_0} (remember that P_0 is adding one Cohen subset to ω_1), the space is not Lindelöf.

PROOF OF B: Suppose $\underset{\sim}{\tau}$ is a P-name for a cover contradicting Lindelöfness, *wlog* the cover consists of old sets. Let $p \in P$.

We define $p_\eta \in P$ for $\eta \in {}^{\omega_1 >}\omega$ by induction on the length $\ell(\eta)$ of η, an old open set U_η (where $\ell(\eta)$ is a successor) such that:

(1) $p_{\eta \restriction \alpha} \leq p_\eta$ ($\leq$ means "weaker than")
(2) $p_\eta \Vdash_P$ "$U_\eta \in \underset{\sim}{\tau}$" (for $\ell(\eta)$ a successor)
(3) $X = \bigcup_{n<\omega} U_{\eta\frown\langle n \rangle}$.

For $\ell(\eta) = 0, \eta = \langle\,\rangle, p_\eta = p$, for limit only (1) applies and we use $\aleph_1$-completeness. If $\eta \in ({}^\alpha\omega)$ and $p_{\eta \restriction (\beta+1)}$ is defined for $\beta < \alpha$, let

$$F_\eta = \{U : U \text{ an open set of } X \text{ (in the universe } V) \text{ and for some } q \geq p_\eta, q \Vdash \text{"} U \in \underset{\sim}{\tau} \text{"}\}.$$

Clearly, F_η is a cover.

F_η is a cover, but X is Lindelöf, so for some countable $F' \subseteq F_\eta$ we have $X = \bigcup F'$. Let $F' = \{U_n^\eta : n < \omega\}$ (maybe with repetitions). Let $U_{\eta\frown\langle n \rangle} \overset{\text{def}}{=} U_n^\eta$, and let $p_{\eta\frown\langle n \rangle} \geq p_\eta$ be chosen so that $p_{\eta\frown\langle n \rangle} \Vdash_P$ "$U_{\eta\frown\langle n \rangle} \in \underset{\sim}{\tau}$."

Now we show that in V^{P_0} the space X is not Lindelöf. For a generic $g \in ({}^{\omega_1}2)$ let σ be the family $\{U_{g\restriction(\alpha+1)} : \alpha < \omega_1\}$. It is easily seen that σ is a cover of X. Suppose $X \subseteq \bigcup\{U_{g\restriction(\alpha+1)} : \alpha < \beta\}$ for some $\beta < \omega_1$. Without loss of generality, β is a limit ordinal. Then $p_{g\restriction\beta} \Vdash_P$ "$\sigma \subseteq \underset{\sim}{\tau}$", in contradiction with the choice of $\underset{\sim}{\tau}$. $\Box_B$

Continuation of the proof of 4.1: W.l.o.g. $y \in U^n_x$ is determined in $Q_3 = Q_2^\kappa \times \Pi_{i<\kappa} P_i$. [Why? As Q satisfies the κ^+-c.c. there are κ maximal antichains $\mathcal{I}_{y,n,x}$ of elements forcing a truth value to "$y \in U^n_x$". So $|\bigcup_{y,n,x} \mathcal{I}_{y,n,x}| \leq \kappa$, so for some α we have $\bigcup \mathcal{I}_{y,n,x} \subseteq Q_2^\alpha \times \prod_{i<\kappa} P_i$.] Also in V^{Q_3}, the space X (i.e. the space defined by letting a subset be open iff it is forced to be open, see $\oplus$ above) is Lindelöf, noting that a cover in V^{Q_3} is also a cover in V^Q and those two universes have the same ω-sequences of members of V^{Q_3} as no ω-sequences are added by $\Pi_{i\in[\kappa,\kappa^+)} P_i$). So, forcing by $P_0 = P_\kappa$ over V^{Q_3} does not contradict Lindelöfness, by Claim B.

As κ is weakly compact, for some stationary set S we have

$$S \subseteq \{\alpha < \kappa : \alpha \text{ strongly inaccessible}\},$$

and in addition, for each $\alpha \in S$, we can split the forcing Q to $Q'_\alpha \times Q''_\alpha$, both Q'_α and Q''_α are $\aleph_1$-complete forcings, and $Q'_\alpha = Q_2^\alpha \times \Pi_{i<\alpha} P_i$ so that:

$$Q'_\alpha \text{ determines "} y \in U^n_x \text{" for } \ y, x < \alpha, n < \omega,$$

and in $V^{Q'_\alpha}$, α becomes $\aleph_2$. Also, the part of the space X that we get after forcing with Q'_α, X_α that is $X \restriction \alpha$, is Lindelöf of pseudo-character $\aleph_0$, as exemplified by the U^n_x, and adding a P_0-generic does not contradict Lindelöfness. (Here we use the weak compactness of κ i.e. Π^1_1-indescribability of κ.)

Now, by claim A, for some such α, in $V^{Q'_\alpha \times Q''_\alpha}$ the space X_α is no longer Lindelöf. Therefore, forcing by Q''_α abolishes Lindelöfness. Also Q'_α, Q''_α are $\aleph_1$-complete, so Q''_α is $\aleph_1$-complete in $V^{G'_\alpha}$, hence (Claim B) P_0 forcing abolishes Lindelöfness, a contradiction. $\Box_{4.1}$

4.2 Remark. Note that during the proof we did not use the regularity of X.

References

[B] J.Baumgartner, **All $\aleph_1$-dense sets of reals can be isomorphic**, *Fund. Math.*, 1973 vol. 79 (101–106).

[Go] I. Gorelic, **The Baire category and forcing large Lindelöf spaces with points G_δ**, *Proceedings of Amer. Math. Soc.*, 1993 vol. 118.

[HJ] A. Hajnal and I.Juhasz, **Lindelöf spaces à la Shelah**, *Coll. Math. Soc. J. Bolyai, Topology*, 1978 vol. 23.

[ShSt:167] S. Shelah and L. Stanley, **S-forcing. IIa. Adding diamonds and more applications: coding sets, Arhangelskii's problem and $\mathcal{L}[Q_1^{<\omega}, Q_2^1]$**, *Israel Journal of Mathematics*, 1986 vol. 56 (1–65).

[Wi:C2] E. Wimmer, **The Shelah P-point independence theorem**, *Israel Journal of Mathematics*, 1982 vol. 43 (28–48).

Institute of Mathematics
The Hebrew University
Jerusalem, Israel

Rutgers University
Department of Mathematics
New Brunswick, NJ USA

Contemporary Mathematics
Volume **192**, 1996

REMARKS ON $\aleph_1$-CWH NOT CWH FIRST COUNTABLE SPACES

SAHARON SHELAH

Abstract. CWH, CWN stand for collectionwise Hausdorff and collectionwise normal respectively. We analyze the statement "there is a $\lambda - CWH$ not CWH first countable (Hausdorff topological) space". We prove the existence of such a space under various conditions, show its equivalence to: there is a λ-CWN not CWN first countable space and give an equivalent set theoretic statement; the nicest version we can obtain is in 4.8.

§0 INTRODUCTION

This paper is concerned with several "almost, but not quite"-type questions for first countable Hausdorff spaces, e.g.:

(1) Is there an $\aleph_1$-metrizable but not metrizable such space?
(2) Is there such space which is $\aleph_1$-CWH but not CWH?
(3) Similarly for CWN.

We feel that these questions are of considerable interest, especially the first one which is somewhat of a classical problem in set-theoretic topology.

Generalizations to $\lambda > \aleph_1$ are considered, and the analogy with CWN is explored. Here, CWH stands for collectionwise Hausdorff, and CWN for collectionwise normal. For the purposes of this paper, a space is always a first countable Hausdorff space.

Note that λ-metrizable $\Rightarrow$ λ-CWH.

Our motivation is to prove that, to a certain degree, the above three questions are equivalent. Note that in the class of Moore spaces, the equivalence of metrizability and CWN is well known (see [Fl84]).

We give below a summary of some of the results.

1991 *Mathematics Subject Classification.* 54A25, 54A35, 54D20.

I would like to thank Alice Leonhardt for the beautiful typing.
My thanks also go to Franklin Tall for showing me the problem in 1992, to the participants of the seminar in Jerusalem in July '92 and Fall '93 for their remarks, and to Mirna Džamonja for many corrections and writing up the first part of the introduction.

In the first section it is shown that instances of non trivial cardinal arithmetic (for example $\neg SCH$) imply examples for (1). For this we note that for singular λ of countable cofinality, a λ-metrizable not metrizable space can be constructed from the assumption that there are λ^+ functions $\eta_\alpha \in {}^\omega\lambda$ such that for $\beta < \lambda^+$ there are pairwise disjoint end segments for $\langle \eta_\alpha : \alpha < \beta \rangle$.

The strength of this assumption is discussed, for example if it were to fail for all relevant λ, then for every singular μ we would have $\mathrm{pp}(\mu) = \mu^+$. This in turn has many implications, like that $\diamondsuit^*(\{\delta < \mu^+ : \mathrm{cf}(\delta) \neq \mathrm{cf}(\mu)\})$ holds for μ singular strong limit.

Also in the first section is a construction the method of which is going to play a major role in the rest of the paper. We wish to construct an $\aleph_1$-CWH not CWH space from the existence of a strong limit λ of uncountable cofinality with $2^\lambda = \lambda^+$. We succeed to obtain "not CWH" at this stage, but we need something more to prove the "$\aleph_1$-CWH".

Spaces which are $\aleph_1$-CWH but not CWH are further discussed in §2. It is proved that MA + ¬CH implies the existence of an $\aleph_1$-metrizable but not metrizable space. Consideration is given to a combinatorial property $INCWH(\lambda)$, defined in 2.4. If one can show that there is an $\lambda > \aleph_1$ such that $INCWH(\lambda)$ holds, then the original problem (1) is solved. It is further shown that $INCWH(\lambda)$ implies the existence of an $(< \lambda)$-CWH but not λ-CWH space, and more. The proof uses the construction mentioned in the discussion of the first section. Further variants of $INCWH(\lambda)$ are introduced.

Discussion of the variants of freeness continues in the third section. It is shown that the version introduced earlier, $INCWH(\lambda)$ can be further weakened, to a property called $INCWH^4(\lambda)$, to obtain a principle equivalent to the existence of a space X with λ points which is $(< \lambda)$-CWH, but not λ-CWH.

Having thus hopefully convinced the reader that freeness has a lot to do with the original problem, we give a detailed discussion of the general concept of freeness and its connection to topological properties we discussed earlier. This is the subject of the fourth section. The method of constructing topological spaces introduced in the first section is further explored now. Some equivalences are given and in particular, the CWN-spaces enter the arena. Freely stated, Theorem 4.8 shows that basically, the existence of a space which is (λ)-CWH, $^*CWN$ respectively, but not λ-CWH, $^*CWN$ respectively are equivalent, and in addition equivalent to the existence of a $(< \lambda)$-free not free family of functions with domains countable sets of ordinals and range $\subseteq \omega$. (The theorem gives in fact more but does not distinguish λ_1, λ_2 if $\lambda_1^{\aleph_0} = \lambda_2^{\aleph_0}$.)

In the fifth section of the paper, we continue the investigation of variants of freeness.

The author had a flawed proof of the existence of spaces as above in ZFC, for some $\lambda > \aleph_1$, in June of 1992; still we decided that there is some interest in the correct part and some additions.

Further work on the variants of freeness, as well as their connection with metrizability, is in preparation.

* * *

We shall deal mainly with first countable topological spaces.
All spaces will be Hausdorff.

0.1 Definition. 1) A space X is metrizable if the topology on X is induced by a metric.
2) A space X is $(<\lambda)$-metrizable if for each $Y \subseteq X$ such that $|Y| < \lambda$, the induced topology on Y is metrizable. Let μ-metrizable mean $(<\mu^+)$-metrizable.
3) A space X is CWH (collectionwise Hausdorff) <u>if for every</u> subspace Y on which the induced topology is discrete (i.e. every subset is open) <u>there is</u> a sequence $\langle u_y : y \in Y \rangle$ of pairwise disjoint open subsets of X, such that for every $y \in Y$ we have $y \in u_y$.
4) A space X is $(<\lambda)$-CWH if for every $Y \subseteq X$ of cardinality $<\lambda$, Y (with the induced topology) is CWH.
$\mu - CWH$ means $(<\mu^+) - CWH$.
5) A space X is CWN (collectionwise normal) if: whenever $\langle Y_i : i < \alpha \rangle$ is a sequence of pairwise disjoint subsets of X, and each Y_i is clopen in $X \restriction (\bigcup_{j<\alpha} Y_j)$, then we can find pairwise disjoint open $\langle \mathcal{U}_i : i < \alpha \rangle$ in X such that $Y_i \subseteq \mathcal{U}_i$.
6) A space is $(<\lambda)-^*$ CWN if every subspace with $<\lambda$ points is CWN (we use the $*$ because another notion is: there is a bound $\alpha < \lambda$ such that all relevant subspaces are of size $<\alpha$).
$\mu -^* CWN$ means $(<\mu^+) -^* CWN$.

0.2 Question. (ZFC) 1) Are there $\aleph_1$-metrizable not metrizable (first countable Hausdorff topological) spaces?
2) Are there $\aleph_1 - CWH$ not CWH first countable spaces?
We shall also consider analogous questions with $\aleph_1$ replaced by any $\lambda > \aleph_0$.
<u>Note</u>: λ-metrizable $\Rightarrow \lambda - CWH$. Also, metrizable $\Rightarrow CWN \Rightarrow CWH$.

0.3 Observation. Assume X is a space with character $\chi \leq \lambda$ (i.e. every point has a neighborhood basis of cardinality $\leq \chi$).
<u>Then</u>:

(a) X is $\lambda - CWH$ <u>iff</u> for every subset Y of cardinality $\leq \lambda$ on which the induced topology is discrete there is a sequence $\langle u_y : y \in Y \rangle$ of pairwise disjoint open subsets of X such that $y \in u_y$.
(b) In (a), for any fixed $\chi \leq \mu \leq \lambda$, we can restrict ourselves (on both sides) to discrete subsets of cardinality μ.

Proof. (a) The implication $\Leftarrow$ is immediate. For the implication $\Rightarrow$ assume that $Y \subseteq X, |Y| \leq \lambda$ and $X \restriction Y$ is the discrete topology. Let $\langle \mathcal{U}_i^y : i < i^y \leq \chi \rangle$ be a neighborhood basis in X for $y \in Y$; choose for $y^1, y^2 \in Y$ and $i_1 < i^{y^1}, i_2 < i^{y^2}$ a point $z[y^1, y^2, i_1, i_2]$ which is in $\mathcal{U}_{i_1}^{y^1} \cap \mathcal{U}_{i_2}^{y^2}$, if this intersection is non-empty. By the assumption $X \restriction Y_1$ is CWH, where
$Y_1 = Y \cup \{z[y^1, y^2, i_1, i_2]\colon y^1 \in Y, y^2 \in Y, i_1 < i^{y^1}, i_1 < i^{y^2}\}$.
(b) Follows from the proof of (a). $\square_{0.3}$

§1 Analysis of "$\aleph_1 - CHW$ but not CHW"

1.1 Lemma. *1) Assume*

$(*)_\lambda$ *$cf(\lambda) = \aleph_0 < \lambda$, $\eta_\alpha \in {}^\omega\lambda$ for $\alpha < \lambda^+$, and for each $\beta < \lambda^+$, we can find pairwise disjoint end segments for $\langle \eta_\alpha : \alpha < \beta \rangle$ (e.g. $\exists h_\beta : \beta \to \omega$ such that*

$$\alpha_1 < \alpha_2 < \beta \wedge k > h_\beta(\alpha_1) \wedge k > h_\beta(\alpha_2) \Rightarrow \eta_{\alpha_1} \restriction k \neq \eta_{\alpha_2} \restriction k).$$

Then 1) the space ${}^{\omega>}\lambda \cup \{\eta_\alpha : \alpha < \lambda^+\}$ with the topology given below is

(α) *first countable and Hausdorff*

(β) *$\lambda - CWH$, even λ-metrizable*

(γ) *not $\lambda^+ - CHW$.*

The topology is the obvious one: each $\eta \in {}^{\omega>}\lambda$ is isolated, and for each $\alpha < \lambda^+$, the neighborhood basis of η_α is $\{\{\eta_\alpha \restriction \ell : k < \ell \leq \omega\} : k < \omega\}$.
2) Hence, the space is not metrizable but is λ-metrizable.

Proof. Straightforward. $\square_{1.1}$

1.2 Conclusion. 1) If the answer to 0.2(1) or 0.2(2) is "no", then $(*)_\lambda$ of 1.1 is not true for any λ.
2) If $(*)_\lambda$ of 1.1 fails for all λ, then

$$(*) \quad cf(\lambda) = \aleph_0 < \lambda \Rightarrow pp(\lambda) = \lambda^+$$

(by [Sh:355],1.5A).
3) If 2)'s conclusion holds, then for every λ singular we have $pp(\lambda) = \lambda^+$. (By [Sh:371],1.10 or [Sh:371],1.10A(6) or [Sh:355],2.4(1)), hence for every pair $\theta < \mu$ we have $cf([\mu]^{\leq\theta}, \subseteq) = \text{cov}(\mu, \theta^+, \theta^+, 2) \leq \mu^+$ (by [Sh:400],1.8), in fact

$$cf([\mu]^{\leq\theta}, \subseteq) = \begin{cases} \mu & \text{if} \quad cf(\mu) > \theta \\ \mu^+ & \text{if} \quad cf(\mu) \leq \theta. \end{cases}$$

4) If 3)'s conclusion holds then:

$(*)$ if λ is singular strong limit then

(a) $2^\lambda = \lambda^+$
hence

(b) $\diamondsuit^*_{S_\lambda}$ where $S_\lambda = \{\delta < \lambda^+ : cf(\delta) \neq cf(\lambda)\}$, and $\diamondsuit^*_S$ means that there is a $\langle \mathcal{P}_\delta : \delta \in S \rangle$ satisfying $\mathcal{P}_\alpha \subseteq [\alpha]^\lambda$, $|\mathcal{P}_\alpha| = \lambda$ such that

$$(\forall X \subseteq \lambda^+)(\exists \text{ club } C)\left[\bigwedge_{\delta \in S \cap C} (X \cap \delta) \in \mathcal{P}_\delta\right]$$

((b) holds by [Sh:108]; see there on earlier work of Gregory).
Note that clearly $\diamondsuit^*_S$ & $S_1 \subseteq S \Rightarrow \diamondsuit^*_{S_1}$.
(5) Not only $pp(\lambda) > \lambda^+$ and $\lambda > \aleph_0 = cf(\lambda)$ implies $(*)_\lambda$ (from 1.11); but assume

we have $\langle \lambda_n : n < cf(\lambda) = \aleph_0 \rangle$, $\sum_n \lambda_n = \lambda$, $\lambda_n = cf(\lambda_n)$, $\mathrm{tcf}(\Pi\lambda_n/J^{bd}_\omega) = \lambda^+$, as exemplified by $\bar{f} = \langle f_\alpha : \alpha < \lambda^+ \rangle$ such that

> $\oplus$ if $\aleph_0 < cf(\delta) \leq \kappa < \lambda$, then there is a closed unbounded $A \subseteq \delta$ and $n_\alpha < cf(\lambda)$ for $\alpha \in A$ such that $n_\alpha, n_\beta < n < cf(\lambda) \Rightarrow f_\alpha(n) < f_\beta(n)$.

Then using $\oplus$ we get $(*)_\lambda$ of 1.1, so we get a $\kappa - CWH$, κ-metrizable first countable space which is not metrizable nor CWH (so not λ^+-metrizable and not $\lambda^+ - CWH$). In fact $\lambda > cf(\lambda) = \aleph_0$, $cf([\lambda]^{\aleph_0}, \subseteq) > \lambda^+$ is sufficient too (see [Sh:355],§6).

1.3 Construction. Assume $\lambda = \beth_{\omega_1}$ (or just λ is a strong limit, $cf(\lambda) \neq \aleph_0$), $2^\lambda = \lambda^+$ and S is a stationary subset of λ^+ such that

$$S \subseteq \{\delta < \lambda^+ : cf(\delta) = \aleph_0 \text{ and } \lambda^2 \text{ divides } \delta\}.$$

We shall build a space with the set of points $\{x_\alpha, y_\alpha : \alpha < \lambda^+\}$. Each x_α will be isolated in X and each y_α will have a countable neighborhood basis in X. We shall have $\{u_{\alpha,n} : n < \omega\}$ as a neighborhood base of y_α with $u_{\alpha,n}$ decreasing in n and $u_{\alpha,n} = \{y_\alpha\} \cup \{x_\beta : f_\alpha(\beta) > n\}$ where f_α is a function from λ^+ to ω which we shall define below.
Note that each y_α is isolated in the space restricted to $\{y_\alpha : \alpha < \lambda^+\}$.
The only thing left is to define f_α for $\alpha < \lambda^+$.
We set $f_\alpha(\beta) = 0$ except in some specified cases. For the space to be Hausdorff it is enough to have:
for $\alpha < \beta$ there is an $m = m(\alpha, \beta) < \omega$ such that
$\neg(\exists\gamma)[f_\alpha(\gamma) \geq m \;\&\; f_\beta(\gamma) \geq m]$. We shall make a stronger condition:

$(*)_0$ $\alpha < \beta \Rightarrow (\exists^{\leq 1}\gamma)[f_\alpha(\gamma) \geq 1 \;\&\; f_\beta(\gamma) \geq 1]$.

Remember that, as remarked in 1.2(4), it is reasonable to assume

> $(*)_1$ $\diamondsuit_S$ holds since $2^\lambda = \lambda^+$ and $cf(\lambda) > \aleph_0$ (or by the proof of 1.2 for arbitrarily large $\mu < \lambda$, $(*)_\mu$ of 1.1 holds). So there is a $\langle g_\alpha : \alpha \in S \rangle$ with $g_\alpha : \alpha \to \omega$, such that
>
> $$(\forall g \in {}^{\lambda^+}\omega)(\exists^{\text{stat}}\alpha \in S)(g_\alpha = g \restriction \alpha).$$

Now, if the space is CWH then there is a $g : \lambda^+ \to \omega$ such that $\langle u_{\alpha, g(\alpha)} : \alpha < \lambda^+ \rangle$ are pairwise disjoint.

We define by induction on α a limit $< \lambda^+$ the value of $f_i(j)$ for $i, j < \alpha$. Denote $f_i^\alpha = f_i \restriction \alpha$ and denote the sequence $\langle f_i^\alpha : i < \alpha \rangle$ by $\bar{f}^\alpha$, so if $i < \alpha$, then f_i^α is a sequence in ${}^\alpha\omega$ and f_i^α is an initial segment of f_i^β when $\alpha < \beta$. Usually we just give value zero to $f_i(j)$.
If $\alpha = \omega$, or α is limit there are no problems.
If $cf(\alpha) > \aleph_0$, and $\bar{f}^\alpha$ is defined, we define $\bar{f}^{\alpha+\omega}$ by letting all the new values be equal to zero. If $\alpha \in S$, and g_α looks as a candidate for g, i.e. $\langle u^\alpha_{i, g_\alpha(i)} : i < \alpha \rangle$ are pairwise disjoint, where $u^\alpha_{i,k} =: \{x_\beta : \beta < \alpha \;\&\; f_i(\beta) > k\}$, and if for some m

it happens that $\text{otp}(\{\beta < \alpha : g_\alpha(\beta) = m\}) = \alpha$, then choose for the minimal such $m = m_\alpha$

(a) $\beta_n^\alpha < \beta_{n+1}^\alpha \cdots < \alpha = \bigcup_n \beta_n^\alpha$

(b) $g_\alpha(\beta_n^\alpha) = m$

and define $\bar{f}^{\alpha+\omega}$ (extending $\bar{f}^\alpha$) by

$$f_\alpha^{\alpha+\omega}(\alpha + n) = n$$

and

$$f_{\beta_n^\alpha}^{\alpha+\omega}(\alpha + n) = m + 1$$

(other values of $\bar{f}^{\alpha+\omega}$ are zero). If g_α fails the conditions above or if $\alpha \notin S$ but $cf(\alpha) = \aleph_0$, choose $m_\alpha = -1$, β_n^α satisfying conditions (a) above and extend $\bar{f}^\alpha$ as just described.
So, if g_α satisfies all the conditions above, we cannot extend g_α to a g which is as required (for CWH) and defined on λ^+ : if $g(\alpha) = k$ we get

$$x_{\alpha+k+18} \in (u_{\alpha,k+1} \cap u_{\beta_{k+18}^\alpha, g_\alpha(\beta_{k+18}^\alpha)}).$$

So the space is not CWH (hence not metrizable). For simplicity, we can request that $\beta_n^\alpha \notin \bigcup_{\gamma \in S} [\gamma, \gamma + \omega)$. Hence, $(*)_0$ holds, so the space is Hausdorff.
Suppose the space is not $\aleph_1 - CWH$. So for some $\mathcal{U} \in [\lambda^+]^{\aleph_1}$

$$X \restriction \{x_\alpha, y_\alpha : \alpha \in \mathcal{U}\}$$

is not CWH.
So without loss of generality if

$$\alpha \in S \cap \mathcal{U}$$

then

$$\alpha + n \in \mathcal{U} \text{ and}$$
$$\beta_n^\alpha \in \mathcal{U}.$$

So

$\otimes$ for every $g : \mathcal{U} \to \omega$ (candidate to give the separation), <u>we get</u>: for some $\alpha \in S \cap \mathcal{U}$, $(\exists^\infty n)\, g(\beta_n^\alpha) \leq m_\alpha$.

$\square_{1.3}$

1.4 Comments. (0) Unfortunately, we have not proved "X is $\aleph_1 - CWH$".
(1) The space constructed in 1.3 has neighborhood bases consisting of countable sets, like the ones considered in the earlier consistency results from [JShS:320]. However, above a superrcompact *pp* is trivial, no such phenomenon arrises here.
(2) But $\Vdash_{\mathrm{Levy}(\aleph_1,\lambda^+)}$ "X is not $\aleph_1 - CWH$" may fail unless we put more restrictions on the β^α_n. See (3).
(3) <u>If</u> we build X as above, let $\mathbb{P} = \mathrm{Levy}(\aleph_1, \lambda^+)$ and there is a $\mathbb{P}$-name $\underset{\sim}{g}$ such that

$$\Vdash_{\mathbb{P}} \text{"}\underset{\sim}{g} : \lambda^+ \to \omega \text{ witnesses that } X \text{ is } CWH\text{"},$$

<u>then</u> X is $\aleph_1 - CWH$.
[Why? given a $Y \in [\lambda^+]^{\aleph_1}$, we can find $\langle p_i : i < \omega_1\rangle$ increasing in $\mathbb{P}$ such that $\bigwedge_{\alpha\in Y} \bigvee_i p_i \Vdash$ "$\underset{\sim}{g}(\alpha) = $ something"].
(4) It is well known that if $\lambda = cf(\lambda) > \aleph_0$ and $S \subsetneq \{\delta < \lambda : cf(\delta) = \aleph_0\}$ is stationary not reflecting, then there is a $(< \lambda) - CWH$ not CWH space. A space like this can be constructed in a fashion similar to that of 1.3. Namely, we can choose for each $\delta \in S$ an increasing sequence $\langle \alpha^\delta_n : n < \omega\rangle$ which converges to δ. We require $\alpha^\delta_n \notin S$, say α^δ_n a successor ordinal. We define

$$f(\beta,\delta) = \begin{cases} n & \text{if} \quad \delta \in S \text{ and } \beta = \alpha^\delta_n \\ 0 & \text{otherwise.} \end{cases}$$

We set $X = \{y_\delta : \delta \in S\} \cup \{x_\beta : \beta \in \lambda\}$ set x_β isolated and let $u_{\delta,n} = \{y_\delta\} \cup \{x_\beta : \beta < \delta \;\&\; f(\beta,\delta) \geq n\}$ be a neighborhood base for $y_\delta (n < \omega)$.

To see that X is not CWH, do as above, and to see it is $(< \lambda) - CWH$, use induction on $\alpha < \lambda$. since S is not reflecting, at $\alpha \in S$ we can choose a cofinal sequence which avoids S, and apply the induction hypothesis.

1.5 Definition. We say that the space X is $\lambda - WCWH$ if for any discrete set of λ points, some subset of cardinality λ can be separated by disjoint open sets.
X is $WCWH$ if X is $|X|$-$WCWH$.

1.5A Remark. By a theorem of Foreman and Laver for first countable spaces we have the consistency of: $\aleph_1 - WCWH \Rightarrow \aleph_2 - WCWH$.

Namely in [FoLa88], starting with a huge embedding $j : V \to M$ with critical point κ and $j(\kappa) = \lambda$, the following is obtained:

There is a forcing notion $\mathbb{P} * \underset{\sim}{\mathbb{R}}$ such that $\mathbb{P}$ is κ-c.c.,$|\mathbb{P}| = \kappa$, $V[G_{\mathbb{P}}] \models$ "$\kappa = \omega_1$", $\mathbb{R} \in V[G_{\mathbb{P}}]$ is λ-c.c., of cardinality λ and $(< \kappa)$-closed and $V[G_{\mathbb{P}*\underset{\sim}{\mathbb{R}}}] \models$ "$\lambda = \omega_2$". In addition, there is a regular embedding $h : (\mathbb{P} * \underset{\sim}{\mathbb{R}}) \to j\mathbb{P}$ with $h(p) = p$ for all $p \in \mathbb{P}$ and the master condition property holds for h, $j\mathbb{P}$, $\mathbb{P} * \underset{\sim}{\mathbb{R}}$. Finally, if G is $(\mathbb{P} * \underset{\sim}{\mathbb{R}})$-generic, then in $V[G], j\mathbb{P}/h''(G)$ is κ-centered.

The consistency of $\aleph_1 - WCWH \Rightarrow \aleph_2 - WCWH$ for first countable spaces clearly follows from the above result of [FoLa88]. For the convenience of the reader we include the following easy Claim 1.5B which shows this implication. In fact, M. Foreman informs us that from other results in [FoLa88], the implication is even easier.

1.5B Claim. *Suppose X is a first countable topological space and $|X| = \kappa^+$, while $Y_0 \subseteq X$ is a discrete subspace of X, with $|Y_0| = \kappa^+$. If $\mathbb{P}$ is a κ-centered forcing notion such that*

$$\Vdash_{\mathbb{P}} \text{“There is a } Y \subseteq Y_0 \text{ with } |Y| = |X| \text{ and } Y \text{ is separated in } X\text{”},$$

then
in V, there is a $Y \subseteq Y_0, |Y| = |Y_0|$ and Y is separated in X.

Proof. Without loss of generality, the set of points of Y_0 is κ^+, and we denote $\lambda = \kappa^+$. We may fix a set $\{\underset{\sim}{x}_\gamma : \gamma < \lambda\}$ of $\mathbb{P}$-names such that

$$\Vdash_{\mathbb{P}} \text{“}\{\underset{\sim}{x}_\gamma : \gamma < \lambda\} \text{ is separated } \subseteq \lambda \text{ and has cardinality } \lambda\text{”}.$$

We can also assume that there are no repetitions among the $\underset{\sim}{x}_\gamma$, and that $\underset{\sim}{x}_\gamma \geq \gamma$. Suppose that in V, the neighborhood bases for points in Y_0 are given by

$$\langle \langle u^n_y : n < \omega \rangle : y \in Y_0 \rangle .$$

So, without loss of generality $\{u^{\underset{\sim}{n}(\gamma)}_{\underset{\sim}{x}_\gamma} : \gamma < \lambda\}$ are pairwise disjoint, in $V^{\mathbb{P}}$.

Now, let $\mathbb{P} = \bigcup\limits_{i<\kappa} \mathbb{P}_i$ where each $\mathbb{P}_i$ is directed.

For each $\alpha < \lambda$, there is a condition p_α forcing a value to $\underset{\sim}{x}_\alpha, \underset{\sim}{n}(\alpha)$ say $\beta_\alpha, m(\alpha)$. So, there is an $i(*) < \kappa$ such that $A = \{\alpha : p_\alpha \in \mathbb{P}_{i(*)}\}$ is unbounded in λ. Therefore, $\{\beta_\alpha : \alpha \in A\}$ is separated by

$$\{u^{m(\alpha)}_{\beta_\alpha} : \alpha \in A\}.$$

(So, having that any two members of $\mathbb{P}_i$ are compatible, or that out of any λ elements of $\mathbb{P}$ there are λ pairwise compatible, i.e. $\mathbb{P}$ is λ-Knaster, suffices). $\square_{1.5.B}$

On the other hand, e.g.

1.6 Claim. *There is a first countable Hausdorff space X which is $(2^{\aleph_0})^+ - WCWH$ but is not $WCWH$.*

Proof. Let $\lambda = \sum\limits_{n<\omega} \lambda_n$, $\lambda_n^{\aleph_0} < \lambda_{n+1}$. Let $\langle \eta_\alpha : \alpha < \lambda^+ \rangle$, $\eta_\alpha \in {}^\omega\lambda$, $\alpha < \beta$ and $\eta_\alpha <_{J^{bd}_\omega} \eta_\beta$.

We define the topological space X on ${}^{\omega>}\lambda \cup \{\eta_\alpha : \alpha < \lambda^+\}$ as in 1.1.

Proof that X is not $\lambda^+ - WCWH$: if $\mathcal{U} \in [\lambda^+]^{\lambda^+}$, $\langle \eta_\alpha : \alpha \in \mathcal{U} \rangle$ cannot be separated as $|\{\eta_\alpha \restriction \ell : \ell < \omega, \alpha \in \mathcal{U}\}| \leq \lambda$.

If $\mathcal{U} \in [\lambda^+]^{(2^{\aleph_0})^+}$, without loss of generality $\mathrm{otp}(\mathcal{U}) = (2^{\aleph_0})^+$; set $\mathcal{U} = \{\alpha_\zeta : \zeta < (2^{\aleph_0})^+\}$. Now for some $Y \in [(2^{\aleph_0})^+]^{(2^{\aleph_0})^+}$ and n, $\langle \eta_{\alpha_\zeta} \restriction [n,\omega) : \zeta \in Y\rangle$ is strictly increasing (not just modulo J^{bd}_ω but in every coordinate (see [Sh:111] or [Sh:355], §1, also [Sh:355],§6; see also [Sh:400],§5, [Sh:430], §6)). $\square_{1.6}$

1.7 Remark. We can prove other Claims similar to 1.6 (see the references above).

§2 On not CWH, $\aleph_1 - CWH$ Spaces

2.1 Definition. For an ordinal γ let us define

$(*)^1_\gamma$ there is an $S \subseteq \{\delta < \gamma : cf(\delta) = \aleph_0\}$ and, for $\delta \in S$, a sequence $\langle \beta^\delta_n : n < \omega\rangle$ strictly increasing with limit δ, and a $m_\delta < \omega$, such that $(\forall g \in {}^\gamma\omega)(\exists \delta \in S)(\exists^\infty n)[g(\beta^\delta_n) \leq m_\delta]$.

2.2 Claim. *(1) If the answer to 0.2 is no (or much less), then for some $\gamma < \omega_2$, $(*)^1_\gamma$ holds.*
(2) If $MA + \neg CH$, then $\gamma < 2^{\aleph_0} \Rightarrow \neg(*)^1_\gamma$.
(3) Without loss of generality, in $(*)^1_\gamma$, each β^δ_n is a successor ordinal.

Proof. 1) By the proof of 1.3 and 1.2. take $\gamma = otp(u)$, where u is like at the end of 1.3.
(2) Check. Use the natural forcing $\{p : p$ is a finite function from γ to $\omega\}$ with $p \leq q$ iff $p \subseteq q$ & $(\forall\delta)(\delta \in S \cap \mathrm{Dom}(p) \rightarrow (\forall n)[\beta^\delta_n \in \mathrm{Dom}(q)\backslash\mathrm{Dom}(p) \rightarrow q(\beta^\delta_n) > n])$.
(3) Check. $\square_{2.2}$

2.2A Conclusion. If $MA + \neg CH$ <u>then</u> the answer to 0.2 is yes. In fact, there is an $\aleph_1$-metrizable (hence $\aleph_1$-CWH) not CWH (hence not metrizable) first countable space.

Proof. By 2.2(1) and 2.2(2).

2.3 Claim. *If $(*)^1_\gamma$ for some $\gamma < \omega_2$, then $(*)^1_{\omega_1}$.*

Proof. Choose $\gamma^* < \omega_2$ minimal such that $(*)^1_{\gamma^*}$. Clearly $\gamma^* \geq \omega_1$.

If $\gamma^* = \omega_1$ we are done. So assume $\gamma^* > \omega_1$, and we shall get a contradiction. We fix an $S \subseteq \gamma^*$ and m_δ, $\langle \beta^\delta_n : n < \omega\rangle$ for $\delta \in S$, which exemplify $(*)^1_{\gamma^*}$. Note that for every $\gamma < \gamma^*$ there is a $g_\gamma \in {}^\gamma\omega$ such that:

$\otimes$ if $\delta \in S \cap \gamma$ <u>then</u> $\{n : g_\gamma(\beta^\delta_n) \leq m_\delta\}$ is finite.

Case 1. $\gamma^* = \gamma + 1$ and $\gamma \notin S$.
Extend g_γ by $\{\langle \gamma, 0\rangle\}$.

Case 2. $\gamma^* = \gamma + 1$ and $\gamma \in S$.
Define $g \in {}^{\gamma^*}\omega$:

$\underline{\text{if}}\ \beta \in \gamma, \beta \notin \{\beta^\gamma_n : n < \omega\}$	$\underline{\text{then}}\ g(\beta) = g_\gamma(\beta)$
$\underline{\text{if}}\ \beta = \gamma$	$\underline{\text{then}}\ g(\beta) = 0$
$\underline{\text{if}}\ \beta = \beta^\gamma_n$	$\underline{\text{then}}\ g(\beta) = \text{Max}\{g_\gamma(\beta), n + 8, m_\gamma + 8\}$.

So g gives a contradiction.

Case 3. $cf(\gamma^*) = \aleph_0$.

Let $\gamma^* = \bigcup_{n<\omega} \gamma_n$, $\gamma_0 = 0$, $\gamma_n < \gamma_{n+1}$, and each γ_{n+1} is a successor of a successor ordinal.

Let $g = \cup\{g_{\gamma_{n+1}} \restriction [\gamma_n, \gamma_{n+1}) : n < \omega\}$ - it gives a contradiction.

Case 4. $cf(\gamma^*) = \omega_1$.

Let $\langle \gamma_i : i < \omega_1 \rangle$ be increasing continuous with limit γ^*, $\gamma_0 = 0$, γ_{i+1} a successor of a successor ordinal.
Let $S^* =: \{\gamma_i : \gamma_i \in S$ (so i is a limit ordinal)$\}$.

Subcase A. γ^*, $\langle <\beta^\gamma_n : n < \omega >: \gamma \in S^* \rangle$, $\langle m_\gamma : \gamma \in S^* \rangle$ do not exemplify $(*)^1_{\gamma^*}$.
So some $g^* \in {}^{\gamma^*}\omega$ shows this. Define g by:

$$\underline{\text{if}}\ \beta \in [\gamma_i, \gamma_{i+1})\ \underline{\text{then}}\ g(\beta) = \text{Max}\{g_{\gamma_{i+1}}(\beta), g^*(\beta)\}$$

So g gives a contradiction.

Subcase B. $\langle <\beta^\gamma_n : n < \omega >: \gamma \in S^* \rangle$, $\langle m_\gamma : \gamma \in S^* \rangle$ exemplify $(*)^1_{\gamma^*}$.
If S^* is not stationary, then we get a contradiction as in case 3, noting that in this case we can without loss of generality assume $\bigwedge_i \gamma_i \notin S^*$. Therefore we may note that S^* is stationary, even though this will not be used in the rest of the proof. Let $\gamma^* = \bigcup_{i<\omega_1} a_i$ with a_i countable increasing continuous, such that $a_0 = \emptyset$,

$a_i \cap \{\gamma_j : j < \omega_1\} = \{\gamma_j : j < i\}$, $a_i \subseteq \gamma_i$ and $\gamma_j \in a_i \wedge j \in S^* \Rightarrow \bigwedge_n \beta^{\gamma_j}_n \in a_i$.

For $i \in S^*$ let $u_i =: \{n < \omega : \beta^{\gamma_i}_n \in a_i\}$.
Note

> $\otimes$ if $i \in S^*$ and $j < i$, then $\{n \in u_i : \beta^{\gamma_i}_n \in a_j\}$ is finite, as it is included in $\{n < \omega : \beta^{\gamma_i}_n < \gamma_j\}$. [Why? Remember $a_j \subseteq \gamma_j$].

Let $S^{**} = \{i \in S^* : u_i$ is infinite and i is a limit ordinal$\}$. So we already know

> $\oplus$ for every $g \in {}^{\gamma^*}\omega$, for some $i \in S^*$, for infinitely many $n < \omega$ we have $g(\beta^{\gamma_i}_n) \leq m_{\gamma_i}$.

We claim

$\oplus^+$ for every $g \in {}^{\gamma^*}\omega$ for some $i \in S^{**}$, for infinitely many $n \in u_i$ we have $g(\beta_n^{\gamma_i}) \leq m_{\gamma_i}$.

Otherwise, for some $g^* \in {}^{\gamma^*}\omega$ this fails and we define g:

<u>let</u> $\beta \in a_{i+1} \backslash a_i$ (there is one and only one such i),
<u>then</u> $g(\beta) = \text{Max}\{g^*(\beta), m_{\gamma_i} + 8, m_{\gamma_i+1} + 8\}$

As g gives a contradiction to $\oplus$, clearly $\oplus^+$ holds.

Now let h be a one to one function from ω_1 onto γ^* such that for i limit, h maps $\{j : j < i\}$ onto a_i.
Let for $i \in S^{**}$, $\{j_n^i : n < \omega\}$ enumerate $\{j < i : h(j) \in \{\beta_n^{\gamma_i} : n \in u_i\}\}$, and $m_i^* = m_{\gamma_i}$ for $i \in S^{**}$.
Now $\langle < j_n^i : n < \omega >: i \in S^{**}\rangle$, $\langle m_i^* : i \in S^{**}\rangle$ exemplify that γ^* could have been chosen to be $= \omega_1$, as required. $\square_{2.3}$

We define the combinatorial property we actually use

2.4 Definition. 1) $INCWH(\lambda) = INCWH^1(\lambda)$ means:

λ is regular $> \aleph_0$ and for some stationary $S \subseteq \{\delta < \lambda : cf(\delta) = \aleph_0\}$ we have $\langle m_\delta, < \beta_n^\delta : n < \omega >: \delta \in S\rangle$ such that:
$m_\delta < \omega$, $\beta_n^\delta < \beta_{n+1}^\delta < \delta = \bigcup_{n<\omega} \beta_n^\delta$, each β_n^δ is a successor <u>and</u>:

(a) for every $g \in {}^\lambda\omega$, for some $\delta \in S$, for infinitely many n we have $g(\beta_n^\delta) \leq m_\delta$.

$(b)_\lambda$ for every $\mathcal{U} \subseteq \lambda$, $|\mathcal{U}| < \lambda$, for some $g \in {}^{\mathcal{U}}\omega$, for every $\delta \in S \cap \mathcal{U}$, for every $n < \omega$ large enough, $g(\beta_n^\delta) > m_\delta$.

2) We can replace m_δ by $\langle m_n^\delta : n < \omega\rangle$, requesting $g(\beta_n^\delta) \leq m_\delta^n$ in (a) and $g(\beta_n^\delta) > m_\delta^n$ in $(b)_\lambda$. In this way we obtain an apparently weaker property, which we call $INCWH^2(\lambda)$.
For other versions of the principle, as well as the connections between the various versions, see §3.

2.4A Discussion. 1) If $INCWH(\lambda)$, then there is a space (as in 1.3) which is Hausdorff first countable with λ points, not metrizable, not even CWH, but every subspace of smaller cardinality is metrizable. So, the notation, "$INCWH$" is derived from "incompactness for CWH", where incompactness is understood in the model-theorethic sense.
2) So if we prove $(\exists \lambda > \aleph_1) INCWH(\lambda)$ we have solved the original problem 0.2.
3) $(b)_\kappa$ means that we require $|\mathcal{U}| < \kappa$. Note that $(b)_{\aleph_1}$ holds trivially and that $\mu \leq \kappa$ & $(b)_\kappa \Rightarrow (b)_\mu$.

More formally

2.5 Claim. *If $INCWH(\lambda)$ <u>then</u> $SINCWH(\lambda)$ (even exemplified by a $(<\lambda)$-metrizable space), where:*

2.6 Definition. $SINCWH(\lambda)$ means that there is a first countable T_2-space X with λ points which is $(<\lambda)-CWH$ (i.e. for every discrete subset of cardinality $<\lambda$ we can choose pairwise disjoint open neighborhoods separating the subspace) but not $\lambda - CWH$.

Proof of 2.5. Assuming $INCWH(\lambda)$ we build a space X witnessing $SINCWH(\lambda)$. The points of X are y_α $(\alpha<\lambda)$ and $x_{\alpha,\beta}$ $(\beta<\alpha<\lambda)$ with $x_{\alpha,\beta}$ isolated and y_α which have neighborhood bases $\langle u_{\alpha,n}: n<\omega\rangle$:

$$\underline{\text{if}}\quad \alpha\in S \qquad u_{\alpha,n}=\{y_\alpha\}\cup\{x_{\alpha,\beta}:\text{for some } k>n \text{ we have } \beta=\beta^\alpha_k\}$$

$$\underline{\text{if}}\ \alpha\notin S \qquad u_{\alpha,n}=\{y_\alpha\}\cup\{x_{\delta,\alpha}:\alpha<\delta\in S,\ \text{for some } k \text{ we have } \alpha=\beta^\delta_k,\ \text{and } k\le m_\delta\}.$$

Here, S is a fixed stationary $\subseteq\{\delta<\lambda: cf(\delta)=\aleph_0\}$ which exemplifies $INCWH(\lambda)$, together with $\langle m_\delta,\langle\beta^\delta_n:n<\omega\rangle:\delta\in S\rangle$.

<u>Checking of "X not CWH"</u>

Let $Y=\{y_\alpha:\alpha<\lambda\}$.

Note that $X\restriction Y$ is a discrete subspace of X. Note that $\{u_{\alpha,n}:n<\omega\}$ is the neighborhood basis of y_α. Suppose that there is $\langle u_{\alpha,g(\alpha)}:\alpha<\lambda\rangle$, a sequence of pairwise disjoint sets, for some $g\in{}^\lambda\omega$. As $u_{\alpha,g(\alpha)}\cap u_{\beta,g(\beta)}=\emptyset$ for $\alpha\neq\beta(<\lambda)$ clearly for $\alpha\in S$ and $\beta=\beta^\alpha_k$ we get $k>g(\alpha)\Rightarrow g(\beta)>m_\alpha$ (since otherwise $x_{\alpha,\beta}\in u_{\alpha,g(\alpha)}\cap u_{\beta,g(\beta)}$, why? $x_{\alpha,\beta}\in u_{\alpha,g(\alpha)}$ as $k>g(\alpha)$ and $x_{\alpha,\beta}\in u_{\beta,g(\beta)}$ as $g(\beta)\le m_\alpha$).

So g contradicts (a) of $INCWH(\lambda)$.

<u>Checking of "X is $(<\lambda)-CWH$"</u>

Let $Z\subseteq X,|Z|<\lambda$ and $X\restriction Z$ is discrete. Let $Z_0=\{x_{\alpha,\beta}:\beta<\alpha<\lambda\}\cap Z$, $Z_1=\{y_\alpha:\alpha\in\lambda\backslash S\}\cap Z$, $Z_2=\{y_\alpha:\alpha\in S\}\cap Z$, so $\langle Z_1,Z_2,Z_3\rangle$ is a partition of Z. Let $\mathcal{U}=\{\alpha\in S:y_\alpha\in Z_2\}$, so $|\mathcal{U}|<\lambda$ and $\mathcal{U}\subseteq\lambda$ hence by the assumption, there is a $g_0\in{}^{\mathcal{U}}\omega$ as in $(b)_\lambda$.

We define u_z, a neighborhood of z for $z\in Z$ (remembering Z is discrete):

if $z=x_{\alpha,\beta}\in Z_0, u_z=\{x_{\alpha,\beta}\}$

if $z=y_\alpha\in Z_1, u_z=u_{\alpha,n(\alpha)}$ where
$n(\alpha)=\text{Min}\{n:n\ge g(\alpha)+8 \text{ and } u_{\alpha,n}\cap Z_0=\emptyset\}$

if $z=y_\delta\in Z_2, u_z=u_{\delta,n(\delta)}$ where
$n(\delta)=\text{Min}\{n:n\ge m_\delta+8 \text{ and } u_{\delta,n}\cap Z_0=\emptyset\}$.

Now check that $\{u_z:z\in Z\}$ separates the points of Z. $\square_{2.5}$

Note that 2.5 also follows from 3.6 + 3.8 below.

2.7 Claim. *Assume λ and $\langle m_\delta, \langle \beta^\delta_n : n < \omega \rangle : \delta \in S \rangle$ are as in 2.4, but we require λ just to be an ordinal, and weaken $(b)_\lambda$ to*

$(b)_\kappa$ *for every $\mathcal{U} \subseteq \lambda$, $|\mathcal{U}| < \kappa$, for some $g \in {}^{\mathcal{U}}\omega$*
for every $\delta \in S \cap \mathcal{U}$, for every n large enough $g(\beta^\delta_n) > m_\delta$.

Then if λ satisfies this weakened version of $INCWH$, then for some regular μ, $\kappa \leq \mu \leq \lambda$ we have $INCWH(\mu)$.

Proof. If we allow μ in the definition of $INCWH(\mu)$ to be an ordinal: straightforward (and suffices for our main interest). Namely, we choose a $\mathcal{U}$ such that

(α) $\mathcal{U} \subseteq \lambda$,
(β) there is no $g \in {}^\lambda\omega$ such that for every $\delta \in S \cap \mathcal{U}$ for every n large enough $g(\beta^\delta_n) > m_\delta$,
(γ) under (α) + (β) the order type of $\mathcal{U}$ is minimal.

Clearly $\mathrm{otp}(\mathcal{U}) \leq \lambda$ and $\mathrm{otp}(\mathcal{U}) \geq \kappa$. By the same proof as 2.3, $\mathrm{otp}(\mathcal{U})$ is a regular cardinal, we call it μ and with the a_i's as in the proof of 2.3, we get $INCWH(\mu)$. $\square_{2.7}$

2.8 Conclusion. If $\lambda = cf(\lambda) > \aleph_0$ and $\diamondsuit_{\{\delta<\lambda : cf(\delta)=\aleph_0\}}$, then for some regular uncountable $\lambda' \leq \lambda$ (but not necessarily $\lambda' > \aleph_1$!), we have $INCWH(\lambda')$.

Proof. By the proof of 1.3 there is a sequence $\langle m_\delta, \langle \beta^\delta_n : n < \omega \rangle : \delta \in S \rangle$ exemplifying *(a)* of 2.4, with $S = \{\delta < \lambda : cf(\delta) = \aleph_0\}$. We know that $(b)_{\aleph_1}$ holds. Now use 2.7.

2.9 Observation. If $S_1 \subseteq S_2 \subseteq \{\delta < \lambda : cf(\delta) = \aleph_0\}$, $\langle m_\delta, \langle \beta^\delta_n : n < \omega \rangle : \delta \in S_1 \rangle$ witness $INCWH(\lambda)$, then we can find a $\langle m'_\delta, \langle \gamma^\delta_n : n < \omega \rangle : \delta \in S_2 \rangle$ witnessing $INCWH(\lambda)$.

Proof. By [Sh:351], proof of 4.4(2) we can find $\langle \eta_\delta : \delta \in S_2 \backslash S_1 \rangle$ such that:
η_δ is an increasing ω-sequence of successor ordinals with limit δ such that

$$\eta_{\delta_1}(n_1) = \eta_{\delta_2}(n_2) \Rightarrow n_1 = n_2 \ \ \& \ \ \eta_{\delta_1} \restriction n_1 = \eta_{\delta_2} \restriction n_2.$$

Now we define $m^2_\delta, \beta^{2,\delta}_n$ for $\delta \in S_2, n < \omega$: if $\delta \in S_1$ then $m^2_\delta = m_\delta$, $\beta^{2,\delta}_n = 2\beta^\delta_n$ and if $\delta \in S_2 \backslash S_1$ then $m_\delta = 3$, $\beta^{2,\delta}_n = 2\eta_\delta(n) + 1$.
Now check. $\square_{2.9}$

2.10 Remark. We can replace in our discussion $\aleph_0$ as the character of the spaces under consideration by θ. Towards this we define a family of spaces.

2.11 Definition. $X \in \mathcal{T}^\ell_\theta$ if X is a Hausdorff space with each point x having a neighborhood basis $\{u_{x,\alpha} : \alpha < \alpha^*\}$ such that:

(a) $\ell = 0$ and $\alpha^* \leq \theta$ or
(b) $\ell = 1, \alpha^* \leq \theta$ and $\langle u_{x,\alpha} : \alpha < \alpha^* \rangle$ is decreasing, or
(c) $\ell = 2, \alpha^* = \theta$, and $\langle u_{x,\alpha} : \alpha < \alpha^* \rangle$ is decreasing.

If $\ell = 2$, we may omit it.

2.12 Definition. We define also the principles
$INCWH(\lambda, \theta) = INCWH^1(\lambda, \theta)$ and $INCWH^2(\lambda, \theta)$ as in 2.4 (with ω replaced by θ).
E.g. $INCWH^2(\lambda, \theta)$ means that there are S and $\langle \varepsilon^\delta_i : i < \theta \rangle, \langle \beta^\delta_i : i < \theta \rangle$ for $\delta \in S$ such that:

$(A)(i)$ $S \subseteq \{\delta < \lambda : \mathrm{cf}(\delta) = \theta\}$
(ii) β^δ_i is a successor ordinal (this is not a serious obstracle, we in fact want $\beta^\delta_i \notin S$)
(iii) $i < j < \theta, \delta \in S \Rightarrow \beta^\delta_i < \beta^\delta_j < \delta$
(iv) $\delta = \bigcup_{i<\theta} \beta^\delta_i$
(v) $\varepsilon^\delta_i < \theta$

$(B)(i)$ for every $g \in {}^\lambda\theta$ for some $\delta \in S$ we have $\theta = \sup\{i < \theta : g(\beta^\delta_i) \leq \varepsilon^\delta_i\}$
(ii) for $X \in [\lambda]^{<\lambda}$ there is $g \in {}^\lambda\theta$ such that for every $\delta \in S \cap X$ we have $\theta > \sup\{i : g(\beta^\delta_i) \leq \varepsilon^\delta_i\}$.

2.13 Claim.

(α) *if $\lambda > cf(\lambda) = \theta$, $pp(\lambda) > \lambda^+$ (or the parallel of 1.2(5)), then*

$\otimes$ *there is an $X \in \mathcal{T}^2_\theta$, $|X| = \lambda^+$, X is $\lambda - CWH$, X has a discrete subspace of size λ^+, but for some $X' \subseteq X$, $|X'| = \lambda$, $cl(X') = X$ (so $|cl(X')| > \lambda$) (this is a strong form of X is not $\lambda^+ - CWH$).*

(β) *if $\lambda > cf(\lambda) = \theta$, λ is a strong limit and $2^\lambda = \lambda^+$, then: $INCWH(\lambda', \theta)$ for some $\lambda' = cf(\lambda') \in [\theta^+, \lambda^+]$.*

Proof. Similar to the above, replacing $\aleph_0$ with θ. $\square_{2.13}$

§3 Variants of Freeness

3.1 Definition. 1) $INCwh(\lambda) = INCwh^1(\lambda)$ is defined as in 2.4 except that $\langle \beta^\delta_n : n < \omega \rangle$ is not required to be increasing with limit δ, just $[n \neq m \Rightarrow \beta^\delta_n \neq \beta^\delta_m]$.
2) $INCwh^2(\lambda)$ is defined as in (1) but we use $\langle m^\delta_n : n \in \omega \rangle$ rather than a single m_δ. (Compare 2.4(2).)

3.2 Claim. *0)* $INCWH^\ell(\lambda) \Rightarrow INCwh^\ell(\lambda)$, $INCWH^1(\lambda) \Rightarrow INCWH^2(\lambda)$, $INCwh^1(\lambda) \Rightarrow INCwh^2(\lambda)$.
1) $INCwh^2(\mathfrak{b})$ *(where*
$\mathfrak{b} = \text{Min}\{|F| : f \subseteq {}^\omega\omega$ *and for no* $g \in {}^\omega\omega$ *do we have for every* $f \in F$ *that* $f <^* g\}$.
As usual, $f <^* g$ *means that* $\{n : f(n) \geq g(n)\}$ *is finite).*
2) Assume $\lambda \leq 2^{\aleph_0}$ *and for* $\alpha < \lambda$, f_α *is a partial function from* ω *to* ω, $\text{Dom}(f_\alpha)$ *is infinite and* $U \subseteq \lambda$ & $|U| < \lambda \Rightarrow (\exists f \in {}^\omega\omega) \bigwedge_{\alpha \in U} f_\alpha \leq^* f$ *but for no* $f \in {}^\omega\omega$, $\bigwedge_{\alpha<\lambda} f_\alpha <^* f$, *<u>then</u>* $INCwh^2(\lambda)$.
3) It does not matter in 3.1 if we demand "β^δ_n *is a successor ordinal".*

Proof. 0) Check.
1) By 2).
2), 3) Check. $\square_{3.2}$

3.2A Questions.
1) Are there examples like in 2.6 for λ singular e.e., does $SINCWH(\lambda)$ hold for λ singular?
2) Suppose that in 3.2(2) we allow for each α a filter F_α on $\text{Dom}(f_\alpha)$ generated by $\aleph_0$ sets and we require $\bigwedge_{\alpha \in U} \{\beta \in \text{Dom}(f_\alpha) : f_\alpha(\beta) < f(\beta)\} \in F_\alpha$; is this equivalent to $INCwh^2(\lambda)$?

3.3 Claim. *Assume* $INCwh^2(\kappa)$, $\lambda > \kappa$ *and* λ *is regular,* $S \subseteq \{\delta < \lambda : cf(\delta) = \aleph_0\}$ *is stationary and* $\diamondsuit_S$ *holds.*
<u>Then</u> (1) there is a $\langle\langle m^\delta_n, \beta^\delta_n : n < \omega\rangle : \delta \in S\rangle$ *as in 2.4(2), but only (a) and* $(b)_\kappa$ *hold.*
2) For some regular $\lambda' \in [\kappa, \lambda]$, *we have* $INCWH^2(\lambda')$.
3) We can replace $INCwh^2(\lambda')$, $INCWH^2(\lambda')$ *by* $INCwh^1(\lambda)$, $INCWH^1(\lambda')$ *respectively.*

Proof. Now (2) follows from (1) as in 2.7 and we leave (3) to the reader. The proof of 3.3(1) is like the construction of 1.3 with one twist. Let $h : \lambda \to \kappa$ be such that for every $\zeta < \kappa$, the set $h^{-1}(\{\zeta\})$ has cardinality λ. Let $\langle\langle m^\zeta_n, {}^*\beta^\zeta_n : n < \omega\rangle : \zeta \in S^*\rangle$ witness $INCwh^2(\kappa)$.

Let $\langle g_\delta : \delta \in S\rangle$ witness $\diamondsuit_S$ i.e. $g_\delta \in {}^\delta\omega$ and for every $g \in {}^\lambda\omega$ for stationarily many $\delta \in S$ we have $g_\delta = g \restriction \delta$.
For each $\delta \in S$ we define a function $g^*_\delta \in {}^\kappa\omega$:

$$g^*_\delta(\zeta) = \text{Min}\Big\{m : \text{for arbitrarily large } \alpha < \delta \text{ we have} : m = g_\delta(\alpha) \text{ and } h(\alpha) = \zeta\Big\}, \text{ if defined.}$$

If for some $\zeta < \kappa$, $g^*_\delta(\zeta)$ is not defined (i.e. there is no such m) - we do nothing. If $g^*_\delta \in {}^\kappa\omega$ is defined we know that for some $\zeta(\delta) \in S^*$,

$$(\exists^\infty n)(g^*_\delta({}^*\beta^{\zeta(\delta)}_n) \le m^{\zeta(\delta)}_n).$$

(Such a $\zeta(\delta)$ exists by the choice of $\langle\langle m^\zeta_n, {}^*\beta^\zeta_n : n < \omega\rangle : \zeta \in S^*\rangle$). We fix such a δ.

Now, choose γ^δ_n such that:

(a) $\gamma^\delta_n < \delta, h(\gamma^\delta_n) = {}^*\beta^{\zeta(\delta)}_n, g_\delta(\gamma^\delta_n) = g^*_\delta({}^*\beta^{\zeta(\delta)}_n)$

(b) $\delta = \bigcup_{n<\omega} \gamma^\delta_n$ and $\gamma^\delta_n < \gamma^\delta_{n+1}$.

$\langle\gamma^\delta_n : n < \omega\rangle$ exist by the definition of q^*_δ.

We claim $\left\langle\langle m^{\zeta(\delta)}_n, \gamma^\delta_n : n < \omega\rangle : \delta \in S\right\rangle$ witness the conclusion. Looking at Definition 2.4, we see that the preliminary properties hold. Ordinals γ^δ_n are not necessarily successors, but this does not matter by 3.2(3).
We have to prove clause (a) of 2.4, we well as $(b)_\kappa$ of 2.7.

Proof of (a). Let $g \in {}^\lambda\omega$. For each $\zeta < \kappa$, the set $\{\alpha < \lambda : h(\alpha) = \zeta\}$ has cardinality λ, so

$$g^*(\zeta) = \mathrm{Min}\{m : (\exists^\lambda \alpha)[h(\alpha) = \zeta \wedge g(\alpha) = m]\}$$

is well defined. Let

$$A =: \{(\zeta, m) : (\exists^\lambda \alpha < \lambda)[g(\alpha) = m \wedge h(\alpha) = \zeta] \text{ and } \zeta < \kappa, m < \omega\}.$$

Then

$$E =: \Big\{\delta < \lambda : \text{for every } (\zeta, m) \in A, \text{ for unboundedly many } \alpha < \delta \text{ we have } g(\alpha) = m, h(\alpha) = \zeta, \text{ and for every } (\zeta, m) \in (\kappa \times \omega)\backslash A, \text{ we have } \delta > \sup\{\alpha < \lambda : g(\alpha) = g^*(\zeta) \wedge h(\alpha) = \zeta\}\Big\}$$

is a club of λ.
For stationarily many $\delta \in S$, $g_\delta \subseteq g$ so there is such a $\delta \in E \cap S$.
Now check: $g^*_\delta = g^*$ (g^*_δ was defined earlier). The rest is also easy to check.

Proof of $(b)_\kappa$ i.e. $(<\kappa)$-freeness. Let $U \subseteq \lambda$, $|U| < \kappa$, hence $V = \{h(\alpha) : \alpha \in U\}$ is a subset of κ of cardinality $< \kappa$, so by the choice of $\langle m^\delta_n, {}^*\beta^\delta_n : n < \omega, \delta \in S^*\rangle$ there is a $f^* : V \to \omega$ exemplifying $(b)_\kappa$ for $\langle m^\delta_n, {}^*\beta^\delta_n : n < \omega, \delta \in S^*\rangle$.
Choose $f : U \to \omega$ by $f(\alpha) = f^*(h(\alpha))$, now f exemplifies $(b)_\kappa$ for $\left\langle\langle m^{\zeta(\delta)}_n, \gamma^\delta_n : n < \omega\rangle : \delta \in S\right\rangle$. $\square_{3.3}$

3.4 Conclusion. $(\exists\lambda \geq \mu) INCWH^\ell(\lambda)$ is equivalent to $(\exists\lambda \geq \mu) INCwh^\ell(\lambda)$ (for $\ell = 1, 2$).

3.5 Definition. 1) $INCWH^3(\lambda)$ means: there are $S \subseteq \lambda$ and $f : \lambda \times \lambda \to \omega$ such that: **IF** we define the spaces as before, i.e.
- the points of $X = X_{f,S}$ are $y_\alpha, x_{\alpha,\beta}, (\alpha < \beta < \lambda)$
- each $x_{\alpha,\beta}$ is isolated
- the sets

$$\begin{aligned} u_{\alpha,n} = \{y_\alpha\} &\cup \{x_{\alpha,\beta} : f(\alpha,\beta) \geq n, \alpha < \beta, \alpha \notin S, \beta \in S\} \\ &\cup \{x_{\beta,\alpha} : f(\beta,\alpha) \geq n, \beta < \alpha, \beta \notin S, \alpha \in S\} \end{aligned}$$

for $n < \omega$ is a neighborhood base at y_α, **THEN**:

(a) $\alpha < \beta < \lambda$, $u_{\alpha,n} \cap u_{\beta,m} \neq \emptyset \Rightarrow x_{\alpha,\beta} \in u_{\alpha,n} \cap u_{\beta,m} \Rightarrow \beta \in S \wedge \alpha \notin S$
(b) for every $\alpha < \beta < \lambda$ for some n we have: $u_{\beta,n} \cap u_{\alpha,n} = \emptyset$,

and

(c) the space X is not CWH but is $(< \lambda) - CWH$.
Note that (a) follows directly from the definition of $u_{\alpha,n}$'s.

2) $INCWH^4(\lambda)$ means: there is a symmetric two-place function f from $\lambda \times \lambda$ to $F =: \{v : v \subseteq \omega \times \omega$ is finite, and $(n,m) \in v, n' \leq n, m' \leq m \Rightarrow (n',m') \in v\}$ which is not free (i.e. for any $g : \lambda \to \omega$ for some $\alpha < \beta$ we have $(g(\alpha), g(\beta)) \in f(\alpha,\beta))$, but is λ-free (i.e. for every $A \subseteq \lambda$, $|A| < \lambda$, there is a $g : A \to \omega$ with no such $\alpha < \beta$ which are from A).

The point is that (and also see 3.7)

3.6 Claim. *1)* $INCWH^1(\lambda) \Rightarrow INCWH^2(\lambda) \Rightarrow INCWH^3(\lambda) \Rightarrow INCWH^4(\lambda)$.

Proof. 1) $INCWH^1(\lambda) \Rightarrow INCWH^2(\lambda)$ is obvious from the definition.
$INCWH^2(\lambda) \Rightarrow INCWH^3(\lambda)$ follows from the proof of 3.7 below.
Suppose that X is defined as in the definition of $INCWH^3(\lambda)$, using some $f^* : \lambda \times \lambda \to \omega$ and S which exemplify $INCWH^3(\lambda)$.
We define $f : \lambda \times \lambda \to F$ by:
if $\beta < \alpha < \lambda$, then:

$$f(\alpha,\beta) = f(\beta,\alpha), \text{ and}$$

$$f(\beta,\alpha) = \{(n,m) : u_{\alpha,n} \cap u_{\beta,m} \neq \emptyset\}.$$

To check that f is as required we simply use the fact that X is $(< \lambda) - CWH$ but not CWH. $\square_{3.6}$

3.7 Claim. $SINCWH(\lambda) \Leftrightarrow INCWH^4(\lambda)$.

Proof $\Rightarrow$. Let the space X exemplify $SINCWH(\lambda)$. Let $\{y_\alpha : \alpha < \lambda\} \subseteq X$ exemplify "X not $\lambda - CWH$" i.e. it is discrete not separated and $\alpha \neq \beta \Rightarrow y_\alpha \neq y_\beta$.

Let $u_{\alpha,n} \supseteq u_{\alpha,n+1}, \{u_{\alpha,n} : n < \omega\}$ be a neighborhood basis of y_α. Now for each α, n, β, m choose if possible $x_{\alpha,n,\beta,m} \in u_{\alpha,n} \cap u_{\beta,m}$. Let $f(\alpha,\beta) = \{(n,m) : x_{\alpha,n,\beta,m}$ is defined$\}$. This f exemplifies $INCWH^4(\lambda)$ (remember in the definition of freeness, in 3.5(2), we consider only $\alpha < \beta$).

$\Leftarrow$ We define the space X with the points $y_\alpha, x_{\alpha,\beta}$ $(\alpha < \beta < \lambda)$ in which each $x_{\alpha,\beta}$ is isolated and the neighborhood basis for y_α is given by (for $n \in \omega$)

$$\begin{aligned} u_{\alpha,n} = \{y_\alpha\} \cup \{x_{\alpha,\beta} : \alpha < \beta \text{ and } \exists m \geq n((n,m) \in f(\alpha,\beta))\} \\ \cup \{x_{\beta,\alpha} : \beta < \alpha \text{ and } \exists m \geq n((m,n) \in f(\alpha,\beta)\}. \end{aligned}$$

Here, f is the function which exemplifies $INCWH^4(\lambda)$. We show that X is $(< \lambda)-CWH$ and not CWH. Suppose that X is CWH and that $u_{\alpha,g(\alpha)} (\alpha \in \lambda)$ exemplify this. Let $\alpha < \beta$ be such that $(g(\alpha), g(\beta)) \in f(\alpha,\beta)$. Then $x_{\alpha,\beta} \in u_{\alpha,g(\alpha)} \cap u_{\beta,g(\beta)}$, contradiction. On the other hand, if $A \subseteq \lambda$ and $|A| < \lambda$, let $g : A \to \omega$ be such that for no $\alpha < \beta$ from A, do we have $(g(\alpha), g(\beta)) \in f(\alpha,\beta)$. Then for $\alpha < \beta \in A$, we have $u_{\alpha,g(\alpha)} \cap u_{\beta,g(\beta)} = \emptyset$. $\square_{3.7}$

We finish this section by the following

3.8 Claim. *In 2.5 we can weaken* $INCWH^1(\lambda)$ *to* $INCWH^2(\lambda)$.

Proof. Suppose that $\langle\langle m_n^\delta, \beta_n^\delta : n < \omega\rangle : \delta \in S\rangle$ exemplify $INCWH^2(\lambda)$ and define the space X as in 2.5, except that the neighborhood basis for y_α when $\alpha \notin S$ is given by (for $n < \omega$)

$$\begin{aligned} u_{\alpha,n} = \{y_\alpha\} \cup \{x_\delta, \alpha : &\alpha < \delta \in S \text{ and for some } k \\ &\text{we have } \alpha = \beta_k^\delta \text{ and } n \leq m_n^\delta\}. \end{aligned}$$

$\square_{3.8}$

Comment. The $INCWH^\ell(\lambda)$ are not so artificial: $SINCWH(\lambda)$ is equivalent to $INCWH^4(\lambda)$.

§4 General Set Theoretic Spectrum of Freeness

4.0 Definition. For $\lambda > cf(\lambda) = \theta$ let $(*)_\lambda$ mean: there is a $\{\eta_\alpha : \alpha < \lambda^+\} \subseteq {}^\theta\lambda$ which is λ-free in the sense of 1.1(1)$(*)_\lambda$: for any α for some $g \in {}^\theta\alpha$, $\langle\{\eta_\beta(i) : i \in [g(\beta),\theta)\} : \beta < \alpha\rangle$ are pairwise disjoint and for simplicity we require $\eta_{\alpha_1}(\zeta_1) = \eta_{\alpha_2}(\zeta_2) \Rightarrow \zeta_1 = \zeta_2$

4.1 Definition. We define various versions of the spectrum of freeness.
1) For θ a regular cardinal and $\sigma \geq 1$ (if $\sigma = 1$ we omit it) let:

$$SP_{\theta,\sigma} = \Bigg\{\lambda : \text{there is a family } H \text{ such that :}$$

(a) every $h \in H$ is a partial function from ordinals to θ

(b) $h \in H \Rightarrow |\mathrm{Dom}(h)| = \theta$

(c) every $H' \subseteq H$ of cardinality $< \lambda$ is σ-free which means that it can be represented as a union $\bigcup_{i<i(*)} H'_i$ where $i(*) < 1 + \sigma$, and each H'_i is free. For H'_i to be free means that there is a g, a function from ordinals to θ such that $(\forall h)(\exists \xi < \theta)[h \in H'_i \rightarrow (\forall \alpha \in \mathrm{Dom}(h)[h(\alpha) \leq g(\alpha) \vee h(\alpha) \leq \xi]$

(d) H is not σ-free, $|H| = \lambda\Bigg\}$

2)

$$SPd_{\theta,\sigma} = \{\lambda : \text{there is an } H \text{ satisfying (a)-(d) above and (e) each } h \in H \text{ is one to one}\}.$$

3)

$$SPw_{\theta,\sigma} = \Bigg\{\lambda : \text{there is a family } H \text{ such that :}$$

(a) if $(h, \bar{u}) \in H$ then h is a function from ordinals to θ

(b) if $(h, \bar{u}) \in H$, then $\bar{u} = \langle u_\varepsilon : \varepsilon < \theta\rangle$ is a decreasing sequence of subsets of $\mathrm{Dom}(h)$

(c) every pair (H', Z'), with $Z' \subseteq$ ordinals, $|Z'| < \lambda$ and $H' \subseteq H$ of cardinality $< \lambda$ is σ-free, which means $H' \times Z'$ can be represented as $\bigcup_{i<i(*)} H'_i \times Z'_i$ where $i(*) < 1 + \sigma$ and each (H'_i, Z'_i) is free. This means that there are functions g, f with $g : H'_i \rightarrow \theta$ and f from ordinals to θ such that for every $(h, \bar{u}) \in H'_i$, for some $\zeta < \theta$ for every $\alpha \in u_\zeta \cap Z'_i \cap \mathrm{Dom}(h)$ we have $h(\alpha) \leq \max\{f(\alpha), g(h, \bar{u})\}$.

(d) (H, λ) is not σ-free, $|H| = \lambda\Bigg\}$

The reader can restrict himself to the case $\sigma = 1$ (also in Definition 4.2(3)).

4.1A Observation. 0) In Definition 4.1(1), if each $h \in H$ converges to θ (i.e. $\forall \zeta < \theta |\{\alpha : h(\alpha) < \zeta\}| < \theta$), in clause (c) of 4.1(1) we can just demand $(\forall h)[h \in H' \to \theta > |\{\alpha : h(\alpha) > g(\alpha)\}|]$.
1) In Definition 4.1(1) without loss of generality $\bigcup_{h \in H} \mathrm{Dom}(h) \subseteq \lambda$ and in 4.1(3) without loss of generality $\bigcup_{(h,\bar{u}) \in H} \mathrm{Dom}(h) \subseteq \lambda$. Also, without loss of generality $\mathrm{Dom}(g) = \lambda$.
2) Note $\theta^+ \cap SP_\theta = \emptyset$
[why? if $H = \{h_\zeta : \zeta < \zeta^* \leq \theta\}$, let $\bigcup_\zeta \mathrm{Dom}(h_\zeta) = \{\alpha_i : i < \theta\}$, and let

$$g(\alpha_i) = \sup\{h_\zeta(\alpha_i) : \zeta < i,\ \alpha_i \in \mathrm{Dom}(h_\zeta)\}.$$

This g exemplifies that H is free.] This also follows from 4.1(B1) and 4.2(2).
3) $SP_\theta \cap [\theta^+, 2^\theta] \neq \emptyset$ [this follows from 4.1A(4) below (and 4.3(1))].
4) We let

$$\begin{aligned}\mathfrak{b}[\theta] = \ & \mathrm{Min}\{|F| : F \subseteq {}^\theta\theta, \text{ and for no } g \in {}^\theta\theta \text{ do we have} \\ & (\forall f \in F)(\exists \zeta < \theta)(f \restriction [\zeta, \theta) < g \restriction [\zeta, \theta))\}\end{aligned}$$

if $\sigma \leq \theta^+$ then clearly $\mathfrak{b}[\theta] \in SP_{\theta,\sigma}$.
5) In Definition 4.1(3) without loss of generality for $(h, \bar{u}) \in H$, we have $\bigcap_{\zeta < \theta} u_\zeta = \emptyset$.
Also without loss of generality, for $(h, \bar{u}) \in H$ we have $u_\zeta = \{\alpha \in \mathrm{Dom}(h) : h(\alpha) \geq \zeta\}$ (we say: $\bar{u}$ is standard for h).
6) Suppose that H is as in 4.1(3) and $[(h, \bar{u}) \in H \Rightarrow |u_0| < \lambda = \mathrm{cf}(\lambda)]$ or $\sup\{|u_0| : (h, \bar{u}) \in H\} < \lambda$; and assume that for every $(h, \bar{u}) \in H$ we know that $\bar{u}$ is standard. Then clause (c) means:
For every $H' \subseteq H$ with $|H'| < \lambda$, there are sets H'_i for $i < i(*) < 1 + \sigma$ such that $H' = \bigcup_{i < i(*)} H'_i$ and for each $i < i(*)$, there is a function f_i from ordinals to θ with the following property.
For every $(h, \bar{u}) \in H'_i$

$$\exists \xi < \theta \exists \zeta < \theta \forall \alpha \in u_\xi [h(\alpha) \leq \max\{\zeta, g_i(\alpha)\}].$$

7) Note also that we can without loss of generality assume that $Z' \subseteq \bigcup_{(h,\bar{u}) \in H'} \mathrm{Dom}(h)$, for 4.1(3)c).
8) We usually restrict our attention to the case $\sigma \leq \theta^+$. Actually, the main interest is in $\sigma = 1$.
On the other hand, for $\sigma \leq \theta$, every H which satisfies (a) and (b) of Definition 4.1(1) is σ-free iff it is free.
[Why? If H is free than it is definitely σ-free. If $H = \bigcup_{i < i(*)} H_i$ for some $i(*) < 1 + \sigma$, and H_i is free as exemplified by g_i, <u>then</u>

$$g \stackrel{\text{def}}{=} \sup\{g_i : i < i(*)\}$$

is a function from ordinals to θ which exemplifies H is free.]

4.1B Notation. For $a \subseteq \theta \times \theta$: we say that a is pie if $(\zeta_1, \xi_1) \neq (\zeta_2, \xi_2) \in a \Rightarrow \neg(\zeta_1, \xi_1) \leq (\zeta_2, \xi_2)$ coordinatewise.
$\mathrm{Pie}(\theta \times \theta) = \{a : a \subseteq \theta \times \theta$ and a is pie (hence finite)$\}$
$C\ell(a) = \{(\zeta, \xi) \in (\theta \times \theta) : (\exists x \in a)(x \leq (\zeta, \xi)$ coordinatewise$\}$, for $a \subseteq \theta \times \theta$.

4.2 Definition. 1) For θ a regular cardinal and $\sigma \geq 1$ (if $\sigma = 1$ we omit it) let:

$SQ_{\theta,\sigma} = \Big\{\lambda$: there is a family H such that :

(a) every $h \in H$ is a partial function defined on the ordinals

(b) $h \in H \Rightarrow |\mathrm{Dom}(h)| = \theta$, $\mathrm{Rang}(h) \subseteq \mathrm{Pie}(\theta \times \theta)$

(c) every $H' \subseteq H$ of cardinality $< \lambda$ is σ-free which means that it can be represented as a union $\bigcup_{i<i(*)} H'_i$ where $i(*) < 1 + \sigma$, and each H'_i is free. For H'_i to be free means that there is a g, a function from ordinals to θ such that $(\forall h)(\exists \zeta < \theta)[h \in H'_i \to (\forall \alpha \in \mathrm{Dom}(h))[(g(\alpha), \zeta) \in C\ell(h(\alpha))]$

(d) H is not σ-free, $|H| = \lambda\Big\}$.

2)

$SQd_{\theta,\sigma} = \Big\{\lambda$: there is an H satisfying (a)-(d) above and

(e) each $h \in H$ is simple, which means: there is an enumeration $\mathrm{Dom}(h) = \{\alpha_\zeta : \zeta < \theta\}$ with no repetitions, such that for each $\zeta < \theta$ for some $\beta_\zeta, \gamma_\zeta < \theta$ we have $C\ell(h(\alpha_\zeta)) = \{(\zeta_1, \zeta_2) : (\zeta_1, \zeta_2) \not\leq (\beta_\zeta, \gamma_\zeta)\}$ and $\langle \gamma_\zeta : \zeta < \theta \rangle$ is strictly increasing and $\bigcup_{\xi<\zeta} \beta_\xi < \gamma_\zeta\Big\}$.

Note that $(\zeta_1, \zeta_2) \not\leq (\beta_\zeta, \gamma_\zeta)$ means that either $\zeta_1 > \beta_\zeta$ or $\zeta_2 > \gamma_\zeta$.

3)

$$SQw_{\theta,\sigma} = \Big\{\lambda : \text{there is a family } H \text{ such that :}$$

(a) if $(h, \bar{u}) \in H$ then h is a function from ordinals to $\text{Pie}(\theta \times \theta)$

(b) if $(h, \bar{u}) \in H$, then $\bar{u} = \langle u_\varepsilon : \varepsilon < \theta \rangle$ is a decreasing sequence of subsets of $\text{Dom}(h)$

(c) every pair (H', Z'), with $Z' \subseteq$ ordinals, $|Z'| < \lambda$ and $H' \subseteq H$ of cardinality $< \lambda$ is σ-free, which means $H \times Z$ can be represented as $\bigcup\limits_{i<i(*)} H'_i \times Z'_i$ where $i(*) < 1 + \sigma$ and each (H'_i, Z'_i) is free. This means that there are functions g, f with $g : H'_i \to \theta$ and f from ordinals to θ such that for every $(h, \bar{u}) \in H'_i$, for some $\zeta < \theta$ for every $z \in u_\zeta \cap Z'_i \cap \text{Dom}(h)$ we have $(g(h), f(z)) \in C\ell(h(z))$

(d) (H, λ) is not σ-free, $|H| = \lambda$

(e) $(h, \bar{u}) \in H \Rightarrow \bigcap\limits_{\varepsilon<\theta} u_\varepsilon = \emptyset \Big\}$.

4.2(A) Remark. 1) In 4.2(3)c), we can assume that $Z' \subseteq \bigcup\limits_{h \in H'} \text{Dom}(h)$.
2) As in 4.1, we consider normally only the case $\sigma \le \theta^+$.
3) $SPx_{\theta,\sigma}$ can be understood as a particular case of $SQx_{\theta,\sigma}$, where $\text{Rang}(h)$ is restricted to $\{(\zeta, \zeta) : \zeta < \theta\}$. Here, $x \in \{w, d\}$ or x is omitted.

4.2B Fact. 1) $\lambda \in SP_{\theta,\sigma}$ implies that $\lambda \in SQ_{\theta,\sigma}$
$\lambda \in SPd_{\theta,\sigma}$ implies that $\lambda \in SQd_{\theta,\sigma}$, and .
$\lambda \in SPw_{\theta,\sigma}$ implies that $\lambda \in SQw_{\theta,\sigma}$.
2) $\lambda \in SQd_{\theta,\sigma}$ implies that $\lambda \in SP_{\theta,\sigma}$.

Proof. 1) If H exemplifies that $\lambda \in SP_{\theta,\sigma}$, let $H^\otimes = \{h^\otimes : h \in H\}$, where for $h \in H$, $h^\otimes$ is a function with domain $\text{Dom}(h)$ and

$$h^\otimes(\alpha) = \{(h(\alpha), h(\alpha))\}.$$

Similarly for $SPd_{\theta,\sigma}$.
If H exemplifies that $\lambda \in SQw_{\theta,\sigma}$, let $H^\otimes = \{(h^\otimes, \bar{u}) : (h, \bar{u}) \in H\}$.
2) See §5, Remark 5.8.

4.2C Notation. For a function h from a subset of ordinals to $\mathrm{Pie}(\theta \times \theta)$, we say that h converges to θ, if

$$(\forall \beta < \theta)(\exists \alpha)(\forall \gamma \in \mathrm{Dom}(h) \setminus \alpha)$$
$$[(\varepsilon_1, \varepsilon_2) \in h(\gamma) \Rightarrow \varepsilon_1 > \beta \text{ and } \varepsilon_2 > \beta].$$

4.2D Observation. 0) In Definition 4.2(1), if each $h \in H$ converges to θ, in clause (c) of 4.2(1) we can just demand
$(\forall h)[h \in H' \to \theta > |\{\alpha : \exists(\varepsilon_1, \varepsilon_2) \in h(\alpha)[\varepsilon_1 > g(\alpha) \vee \varepsilon_2 > g(\alpha)]\}|]$.
1) In Definition 4.2(1) without loss of generality $\bigcup_{h \in H} \mathrm{Dom}(h) \subseteq \lambda$ and in 4.2(3) without loss of generality $\bigcup_{(h,\bar{u}) \in H} \mathrm{Dom}(h) \subseteq \lambda$. Also, without loss of generality, $\mathrm{Dom}(g) = \lambda$.
2) Note $\theta^+ \cap SQ_\theta = \emptyset$
[why? if $H = \{h_\zeta : \zeta < \zeta^* \leq \theta\}$, $\bigcup_\zeta \mathrm{Dom}(h_\zeta) = \{\alpha_i : i < \theta\}$, let $g(\alpha_i) = \sup\{\max\{h_\zeta(\alpha_i)_1, h_\zeta(\alpha_i)_2 : \zeta < i,\ \alpha_i \in \mathrm{Dom}(h_\zeta) \text{ and } h_\zeta(\alpha_i) = (h_\zeta(\alpha_i)_1, h_\zeta(\alpha_i)_2)\}\}$.]
3) $SQ_\theta \cap [\theta^+, 2^\theta] \neq \emptyset$ [this follows from 4.1A(3) and 4.2B(1)]. Actually, $\mathfrak{b}[\theta] \in SQ_\theta$.
4) In 4.2(3)c), if $|u_0| < \lambda = \mathrm{cf}(\lambda)$ for $(h, \bar{u}) \in H$, we obtain the following property. For every $H' \subseteq H$ of cardinality $< \lambda$, there are sets H'_i for $i < i(*) < 1 + \sigma$, such that there are functions $\{g_i : i < i(*)\}, g_i : H'_i \to \theta$ satisfying: if $(h, \bar{u}) \in H'_i$, then $(\exists \zeta < \theta)(\exists \xi < \theta)(\forall \alpha \in u_\zeta)[(g_i(\alpha), \xi) \in C\ell(h(\alpha))]$.

4.3 Claim. *1) If there is an H as in (a), (b) of 4.1(1) which is $(< \mu) - \sigma$-free not $\lambda - \sigma$-free <u>then</u> there is a $\lambda' \in [\mu, \lambda] \cap SP_{\theta,\sigma}$. Similarly for Definitions 4.1(2), 4.1(3) (see also Claim 4.3A).*
2) If $pp_{\Gamma(\theta)}(\lambda) > \lambda^+$, $\lambda > cf(\lambda) = \theta$ (or just $(*)_\lambda$ of 4.0) and $\lambda \geq \sigma$ <u>then</u> $SP_{\theta,\sigma} \cap [\lambda^+, \lambda^\theta] \neq \emptyset$.[1]

Proof. 1) Straightforward.
2) Let $\{\eta_\alpha : \alpha < \lambda^+\} \subseteq {}^\theta\lambda$ be λ-free, without loss of generality $\langle \{\eta_\alpha(\zeta) : \alpha < \lambda^+\} : \zeta < \theta \rangle$ are pairwise disjoint and let

$$H = \Big\{ h : \text{for some } \alpha < \lambda^+ \text{ and } a \subseteq \lambda^+, \mathrm{otp}(a) = \theta, \mathrm{Dom}(h) = a,$$
$$h \text{ is strictly increasing and for } \beta \in a$$
$$h(\beta) = \sup\{\varepsilon : \eta_\alpha(\varepsilon) = \eta_\beta(\varepsilon)\} \Big\}.$$

Now H is not free: if $g : \lambda^+ \to \theta$, then for some $\varepsilon^* < \theta$, $A = \{\alpha < \lambda^+ : g(\alpha) = \varepsilon^*\}$ is of cardinality λ^+. Choose by induction on $\zeta < \lambda^+$ an ordinal $\alpha^*_\zeta < \lambda^+$ increasing with ζ such that

[1] $\Gamma(\theta)$ refers to the class of θ-complete ideals on θ which $\supseteq J^{bd}_\theta$. pp_Γ is pp taken only over the ideals in Γ

$$\bigcup\{\text{Rang}(\eta_\alpha) : \alpha \in A \cap \alpha^*_{\zeta+1} \backslash \alpha^*_\zeta\} = \bigcup\{\text{Rang}(\eta_\alpha) : \alpha \in A \backslash \alpha^*_\zeta\}.$$

Next choose $\alpha \in A\backslash\alpha^*_\theta$ and $\beta_\zeta \in A \cap [\alpha^*_\zeta, \alpha^*_{\zeta+1})$ for $\zeta < \theta$ such that $\eta_{\beta_\zeta}(\zeta) = \eta_\alpha(\zeta)$ (note that the existence of such β_ζ follows from the definition of α^*_ζ and the assumption $\zeta \neq \xi \Rightarrow \eta_\alpha(\zeta) \neq \eta_\beta(\xi)$ for $\zeta, \xi < \theta$ and $\alpha, \beta < \lambda^+$) and let $a = \{\beta_\zeta : \zeta < \theta\}$, $h \in {}^a\theta$, $h(\beta_\zeta) = \sup\{\varepsilon : \eta_\alpha(\varepsilon) = \eta_{\beta_\zeta}(\varepsilon)\} \geq \zeta$, so $h \in H$. As $\beta_\zeta \in A$, $g(\beta_\zeta) = \varepsilon^*$ =constant, so if $\xi < \theta$, $\{\beta \in \text{Dom}(h) : h(\beta) \geq g(\beta), \xi\}$ includes $\{\beta_\zeta : \xi, \varepsilon^* < \zeta < \theta\}$, which is a contradiction.

On the other hand, H is λ^+-free. For suppose $H' \subseteq H$, $|H'| \leq \lambda$. For $h \in H'$ choose α_h, a_h witnessing $h \in H$. Then $b = \bigcup\{\{\alpha_h\} \cup a_h : h \in H'\}$ is a subset of λ^+ of cardinality $\leq \lambda$, hence we can find $\langle \varepsilon_\alpha : \alpha \in b\rangle$ such that $\langle \text{Rang}(\eta_\alpha \restriction [\varepsilon_\alpha, \theta)) : \alpha \in b\rangle$ is a sequence of pairwise disjoint subsets of λ. Let us define a $g : \lambda^+ \to \theta$ such that $\alpha \in b \Rightarrow g(\alpha) = \varepsilon_\alpha$. Now if $h \in H'$, let $a_h = \{\beta_\zeta : \zeta < \theta\}$ (increasing with ζ), so

$$h(\beta_\zeta) = \sup\{\varepsilon : \eta_{\alpha_h}(\varepsilon) = \eta_{\beta_\zeta}(\varepsilon)\}$$

so $h(\beta_\zeta) \leq \max\{\varepsilon_{\alpha_h}, \varepsilon_{\beta_\zeta}\} = \max\{g(\alpha_h), g(\beta_\zeta)\}$.
So choose $\xi = g(\alpha_h)$ and we get the desired conclusion.
To finish we use part (1). $\square_{4.3}$

4.3A Claim. *1) If there is an H as in (a), (b) of 4.2(1) which is $(< \mu) - \sigma$-free not $\lambda - \sigma$-free <u>then</u> there is a $\lambda' \in [\mu, \lambda] \cap SQ_{\theta,\sigma}$. Similarly for 4.2(2), 4.2(3).*
2) If $pp_{\Gamma(\theta)}(\lambda) > \lambda^+$, $\lambda >$ cf$(\lambda) = \theta$ (or just $(*)_\lambda$ of 4.0) and $\lambda \geq \sigma$ <u>then</u> $SQ_{\theta,\sigma} \cap [\lambda^+, \lambda^\theta] \neq \emptyset$.

Proof. 1) Straightforward.
2) This follows from 4.3(2) and 4.1B(1). $\square_{4.3A}$

4.4 Claim. *1) The following implications hold for any λ:*

$$(a) \Rightarrow (b) \Leftrightarrow (b)^+ \Leftrightarrow (c) \Leftrightarrow (c)^+ \Rightarrow (d),$$

where

- (a) $\lambda \in SQ_{\aleph_0}$
- (b) *There is a $(< \lambda)$-CWH not λ-CWH first countable space.*
- $(b)^+$ *There is a space like in (b), which is in addition $(< \lambda)$-metrizable.*
- (c) *There is a $(< \lambda) -{}^* CWN$ not ${}^*CWN$ first countable space with λ points.*
- $(c)^+$ *There is a space like in (c), which is in addition $(< \lambda)$-metrizable.*
- (d) $\lambda \in SQw_{\aleph_0}$.

2) $\lambda \in SQd_{\theta,\sigma} \Rightarrow \lambda \in SQ_{\theta,\sigma} \Rightarrow \lambda \in SQw_{\theta,\sigma} \Rightarrow [\lambda, \lambda^\theta] \cap SQd_{\theta,\sigma} \neq \emptyset$ for $\sigma \leq \theta^+$.
3) $\lambda \in SP_{\theta,\sigma} \Rightarrow \lambda \in SPw_{\theta,\sigma} \Rightarrow [\lambda, \lambda^\theta] \cap SPd_{\theta,\sigma} \neq \emptyset$ for $\sigma \leq \theta^+$.
4) We can replace $\aleph_0$, "first countable Hausdorff topological spaces" <u>by</u> $\theta, \mathcal{T}_\theta$ respectively (of course, $\lambda > \theta$).

Proof. 1),4)
(a) implies (b), $(b)^+$, (c), $(c)^+$.

So assume H exemplifies that $\lambda \in SQ_\theta$; without loss of generality $\mathrm{Dom}(h) \subseteq \lambda$ for $h \in H$. We can use the space

$$X = \{y_i : i < \lambda\} \cup \{z_h : h \in H\} \cup \{x_{h,i} : h \in H \text{ and } i \in \mathrm{Dom}(h)\},$$

and for $\zeta < \theta$ let

$$u_\zeta[z_h] = \{z_h\} \cup \{x_{h,i} : i \in \mathrm{Dom}(h), (\zeta,\zeta) \notin C\ell(h(i))\},$$

$$u_\zeta[y_i] = \{y_i\} \cup \{x_{h,i} : h \in H, i \in \mathrm{Dom}(h), (\zeta,\zeta) \notin C\ell(h(i))\}$$

and $x_{h,i}$ is isolated.

Suppose $H' \subseteq H$, $|H'| < \lambda$ and let
$X[H'] = \{y_i : i < \lambda\} \cup \{z_h : h \in H'\} \cup \{x_{h,i} : h \in H, i \in \mathrm{Dom}(h)\}$.
Let $g : \lambda \to \theta$ be such that for every $h \in H'$, for some $\zeta[h] < \theta$ we have

$$i \in \mathrm{Dom}(h) \Rightarrow (g(i), \zeta[h]) \in C\ell(h(i)).$$

Let us choose for $t \in X[H']$ a neighborhood v_t:

<u>if</u>	$t = x_{h,i}$	<u>then</u>	$v_t = \{x_{h,i}\}$
<u>if</u>	$t = y_i$	<u>then</u>	$v_t = u_{g(i)}[y_i]$
<u>if</u>	$t = z_h$	<u>then</u>	$v_t = u_{\zeta[h]}[z_h]$.

Now

$$\langle v_{y_i} : i < \lambda\rangle\hat{}\langle v_{z_h} : h \in H'\rangle\hat{}\langle v_{x_{h,i}} : i < \lambda, h \in H \text{ and } x_{h,i} \notin \bigcup_{j<\lambda} v_{y_j} \cup \bigcup_{h \in H'} v_{z_h}\rangle$$

is a partition of $X[H']$ to pairwise disjoint open sets. In each basic open set there is at most one point which is not isolated, and if so it has a neighborhood base consisting of a decreasing sequence of (open) sets of length θ.
This suffices to show that X is $(< \lambda)$-metrizable when $\theta = \aleph_0$ and as required generally (for 4)).
[Why? Suppose $X' \subseteq X$ and $X = \bigcup_{i<i(*)} U_i$, where each U_i is open, U_i for $i < i(*)$ are pairwise disjoint and for every $i < i(*)$ at most one $x_i \in U_i$ is not isolated, and x_i has a countable neighborhood base. If $i < i(*)$ and $x_i \in U_i$ is non-isolated, let $\{u^i_n : n < \omega\}$ be a neighborhood for x_i, and without los of generality we have $u^i_{n+1} \subseteq u^i_n \subseteq U_i$ for $n < \omega$. For $x \in X'$ let $i(*)$ be the $i < i(*)$ such that $x \in U_i$. Now define

$$d(x,y) = \begin{cases} 1 & \text{if} \quad i(x) \neq i(y) \\ 0 & \text{if} \quad x = y \\ \mathrm{Min}\{1/n : x, y \in u^i_n\} & \text{if} \quad x \neq y \text{ but } i(x) = i(y) \end{cases}$$

and check.]

As for showing that X is not CWH (hence not metrizable and not CWN), note that $\{y_i : i < \lambda\} \cup \{z_h : h \in H\}$ is a discrete subspace.
If it is separated, we have a sequence of pairwise disjoint neighborhoods:

$$\langle u_{g(i)}[y_i] : i < \lambda \rangle\hat{}\langle u_{\zeta(h)}[z_h] : h \in H \rangle.$$

But H is not free (in the sense of Definition 4.2(1)) and we get a contradiction.

$\underline{(b)^+ \Rightarrow (b)}$.

Trivial.

$\underline{(b) \Rightarrow (b)^+}$.

Let X exemplify clause (b), so without loss of generality $|X| = \lambda$. Let Y be a discrete subspace of cardinality λ which cannot be separated. Let X^+ be the topology on the set of points of X generated by basic open sets of X and $\{\{x\} : x \in X \backslash Y\}$.

Now X^+ is not $\lambda - CWH$ (Y still exemplifies it). But X^+ is $(< \lambda)$-metrizable as:

If $Z \subseteq X, |Z| < \lambda$, then we can find a sequence $\langle u_z : z \in Z \cap Y \rangle$ of pairwise disjoint open sets, and in $X \restriction u_z$, every point is isolated except z, which has a neighborhood basis of cardinality $\aleph_0$, and every $x \in Z \setminus \bigcup_{z \in Z \cap Y} u_z$ is isolated.

As noted above, this is enough.

$\underline{(b)^+ \Rightarrow (c)^+}$

Trivial (as $(< \lambda)$-metrizable $\Rightarrow (< \lambda) -^* CWN$).

$\underline{(c)^+ \Rightarrow (c)}$

Trivial.

$\underline{(c) \Rightarrow (b)^+}$

If $X, \langle Y_i : i < \alpha \rangle$ exemplify clause (c) in (1) with $\langle u_\zeta(y) : \zeta < \theta \rangle$ a decreasing neighborhood basis of y, we can get another example X' to the third clause, as follows.
We are, without loss of generality, assuming that $|X| = \lambda$. Now we define a topological space X':

$$X' = \bigcup_{i<\alpha} Y_i \cup \Big\{ x_{y,z,\zeta,\xi} : \text{for some } i \neq j < \alpha, y \in Y_i, \\ z \in Y_j, u_\zeta[y] \cap u_\xi[z] \neq \emptyset \Big\}$$

with the neighborhood bases for $y, z \in \bigcup_{i<\alpha} Y_i$ given by

$$u'_\varepsilon[t] = \{t\} \cup \Big\{ x_{y,z,\zeta,\xi} : x_{y,z,\zeta,\xi} \in X', \text{ and:} \\ t = y \wedge \varepsilon \leq \zeta \text{ or } t = z \wedge \varepsilon \leq \xi \Big\}$$

and each $x_{y,z,\zeta,\xi}$ is isolated; note that for $t \in \bigcup_{i<\alpha} Y_i, \langle u'_\varepsilon[t] : \varepsilon < \theta\rangle$ is decreasing with intersection $\{t\}$, so $X' \in \mathcal{T}_\theta$ is a Hausdorff space.

Clearly $Y =: \bigcup_{i<\alpha} Y_i$ is discrete. Assume that $\langle u'_{\varepsilon(y)}[y] : y \in Y\rangle$ is a sequence of pairwise disjoint open sets. Then let

$$U_i = \bigcup\{u_{\varepsilon(y)}[y] : y \in Y_i\}.$$

So in X, U_i is an open set (as a union of open sets),

$$Y_i \subseteq U_i \text{ as } y \in u_{\varepsilon(y)}[y].$$

Therefore, there are i, j such that

$$U_i \cap U_j \neq \emptyset \text{ and clearly}$$

$$\begin{aligned} i \neq j \;\&\; U_i \cap U_j \neq \emptyset &\Rightarrow \exists y \in Y_i \exists z \in Y_j (u_{\varepsilon(y)}[y] \cap u_{\varepsilon(z)}[z] \neq \emptyset) \\ &\Rightarrow x_{y,z,\varepsilon(y),\varepsilon(z)} \text{ is well defined} \\ &\Rightarrow \text{ in } X' \text{ we have that } u'_{\varepsilon(y)}[y] \cap u'_{\varepsilon(z)}[z] \neq \emptyset. \end{aligned}$$

This is a contradiction.

So we conclude that Y cannot be separated in X', so X' is not $\lambda - CWH$.

Next, assume that $Z \subseteq X', |Z| < \lambda$, so in $X, \langle Y_i \cap Z : i < \alpha, Y_i \cap Z \neq \emptyset\rangle$ can be separated, say by $\langle U_i : i < \alpha, Y_i \cap Z \neq \emptyset\rangle$. So for $y \in Y \cap Z$, there is an $\varepsilon(y)$, such that $u'_{\varepsilon(y)}[y] \subseteq U_i$ (the isolated points in $X' \cap Z \backslash Y$ can be taken care of easily so we ignore them).

Now, if $y \neq z \in Y \cap Z$ then:

(i) if $(\exists i)(y, z \in Y_i)$ then $u'_{\varepsilon(y)}[y] \cap u'_{\varepsilon(z)}[z] = \emptyset$.

(ii) If $y \in Y_i, z \in Y_j, i \neq j$, and $u'_{\varepsilon(y)}[y] \cap u'_{\varepsilon(z)}[z] \neq \emptyset$,
then $x_{y,z,\varepsilon(y),\varepsilon(z)}$ exists, so
$\emptyset \neq u_{\varepsilon(y)}[y] \cap u_{\varepsilon(z)}[z] \subseteq U_i \cap U_j$ which is a contradiction.

That X' is $(< \lambda)$-metrizable now follows as in $(b) \Rightarrow (b)^+$.

<u>$(c) \Rightarrow (d)$</u>.

Assume that X is a first countable $(< \lambda) - {}^*CWN$ not $\lambda - {}^*CWN$-space, without loss of generality with the set of points λ, so there is a sequence $\langle Y_i : i < \lambda\rangle$ of pairwise disjoint subsets of X such that $Y_i \neq \emptyset$, Y_i is clopen in $X \restriction (\bigcup_{j<\lambda} Y_j)$ and $\langle Y_i : i < \lambda\rangle$ cannot be separated. For $y \in Y =: \bigcup_{i<\lambda} Y_i$ let $\bar{u}[y] = \langle u_\zeta[y] : \zeta < \theta\rangle$ be a neighborhood basis of the topology for y, and without loss of generality $\varepsilon < \zeta < \theta \Rightarrow u_\zeta[y] \subseteq u_\varepsilon[y]$. Let

$$H = \Big\{(h, \bar{u}) : \text{for some } i < \lambda \text{ and for some } y \in Y_i,$$
$$(h, \bar{u}) = (h_y, \bar{u}_y), \text{ which means :}$$
$$\mathrm{Dom}(h) = \bigcup_{j \neq i} Y_j, C\ell(h(z)) = \{(\zeta, \xi) \in \theta \times \theta : u_\zeta[y] \cap u_\xi[z] = \emptyset\}$$
$$\text{and } \bar{u} \text{ is } \langle u_\zeta(y) \cap \mathrm{Dom}(h) : \zeta < \theta \rangle \Big\}.$$

Note that $h(z)$ is uniquely determined by $C\ell(h(z))$ (since we know that $u(z)$ is a pie). Now we check that H exemplifies SQw_θ, i.e. the clauses in 4.2(3). Clauses (a), (b) are immediate.
As for clause (c), let $H' \subseteq H, |H'| < \lambda$, and, without loss of generality, let $Z' \subseteq \bigcup\{\mathrm{Dom}(h) : (h, \bar{u}) \in H\}$ with $|Z'| < \lambda$. Let

$$Y' =: \{y : y \in \bigcup_{i<\lambda} Y_i, \text{ and } y \in Z' \text{ or } (h_y, \bar{u}_y) \in H'\},$$

so $|Y'| < \lambda$; we can find
$X' \subseteq X, |X'| \leq |Y'| + \theta < \lambda$ such that $Y' \subseteq X'$, and for every $y, z \in Y'$ and $\zeta < \theta, \xi < \theta$, we have $u_\zeta[y] \cap u_\xi[z] \neq \emptyset \Rightarrow u_\zeta[y] \cap u_\xi[z] \cap X' \neq \emptyset$. As $|X'| < \lambda$ we know that X' (i.e. $X \restriction X'$) is CWN, and $\langle Y_i \cap X' : i < \alpha \rangle$ is a discrete sequence of closed sets in X' hence there is a function $g : Y' \to \theta$ such that

(*) if $i < j < \lambda, y \in Y' \cap Y_i, z \in Y' \cap Y_j$, then
$u_{g(y)}[y] \cap u_{g(z)}[z] = \emptyset$ (intersecting with X' is immaterial).

Hence by the choice of g

(**) if $i \neq j$ $(i < \lambda, j < \lambda)$, $y \in Y' \cap Y_i$, $z \in Y' \cap Y_j$
then $(g(y), g(z)) \in C\ell(h_y(z))$.

This is enough.

We are left with proving that H is not free, so suppose $f, g : Y \to \theta$ satisfies

$\otimes$ for every $y \in Y$,
for every $z \in \mathrm{Dom}(h_y)$ we have $(g(y), f(z)) \in C\ell(h_y(z))$,

so without loss of generality $f = g$.
For $i < \lambda$ let

$$U_i = \bigcup\{u_{g(y)}[y] : y \in Y_i\}.$$

So U_i, being the union of open sets is open.
If $i < j, y \in Y_i, z \in Y_j$ then

$$u_{g(y)}[y] \cap u_{g(z)}[z] \neq \emptyset \Rightarrow (g(y), g(z)) \in C\ell(h_y(z))$$
$$\Rightarrow (g(y), f(z)) = (g(y), g(z)) \in C\ell(h_y(z)).$$

Contradiction, by the choice of f and g.
So $u_{g(y)}[y] \cap u_{g(z)}[z] = \emptyset$, as $y \in Y_i, z \in Y_j$ were arbitrary, $U_i \cap U_j = \emptyset$.
We conclude that $\langle Y_i : i < \lambda \rangle$ can be separated, which is a contradiction.
2) We prove each implication
(A) $\lambda \in SQd_{\theta,\sigma} \Rightarrow \lambda \in SQ_{\theta,\sigma} \Rightarrow \lambda \in SQw_{\theta,\sigma}$. Obvious.
(B) $\lambda \in SQw_{\theta,\sigma} \Rightarrow SQd_{\theta,\sigma} \cap [\lambda, \lambda^\theta] \neq \emptyset$ when $\sigma \leq \theta^+$.

Now without loss of generality $\sigma \in \{1, \theta\}$ and the reader can think of $\sigma = 1$ only. Assume that H exemplifies $\lambda \in SQw_{\theta,\sigma}$. By the definition, $(h, \bar{u}) \in H \Rightarrow u_\zeta \subseteq \mathrm{Dom}(h)$ & $\bigcap_{\xi<\theta} u_\xi = \emptyset$. Let for each $(h, \bar{u}) \in H$,

$$H^*_{(h,\bar{u})} = \Big\{ f : f \text{ is a function from ordinals to } \mathrm{Pie}(\theta \times \theta) \text{ and for some set } v,\ \mathrm{Dom}(f) = v \subseteq \mathrm{Dom}(h), |v| = \theta, \text{ but } \zeta < \theta \Rightarrow |v \backslash u_\zeta| < \theta, \text{ and } (\forall \alpha \in v)[C\ell(f(\alpha)) \supseteq C\ell(h(\alpha))], \text{ and } f \text{ is simple} \Big\}$$

and $H^* = \bigcup \{ H^*_{(h,\bar{u})} : (h, \bar{u}) \in H \}$.
It is easy to check that H^* satisfies clauses (a) and (b) from 4.2(1) and (e) of 4.2(2) and $|H^*| \in [\lambda, \lambda^\theta]$.

As for clause (c) of 4.2(1), let $H' \subseteq H^*$, $|H'| < \lambda$, let $H' = \{f_j : j < j(*)\}$, $j(*) < \lambda$, and for every $j < j(*)$, $f_j \in H^*_{(h_j,\bar{u}_j)}$ for some $(h_j, \bar{u}_j) \in H$. So $v_j =:$ $\mathrm{Dom}(f_j)$ is as in the definition of $H^*_{(h_j,\bar{u}_j)}$. Define $H'' = \{(h_j, \bar{u}_j) : j < j(*)\}, Y = \bigcup_{j<j(*)} v_j$. Now H'' is a subset of H of cardinality $< \lambda$, $Y \subseteq Ord$ and $|Y| < \lambda$. As H exemplifies $\lambda \in SQw_{\theta,\sigma}$, we can find a $\langle g_i : i < i(*) \rangle$, $i(*) < 1 + \sigma$, $g_i \in {}^\lambda\theta$ and $\langle (H''_i, Y_i) : i < i(*) \rangle$ such that $H'' \times Y = \bigcup_{i<i(*)} H''_i \times Y_i$ and for every $(h_j, \bar{u}_j) \in H''$ for some $i = i(j) < i(*)$ we have

$$(\exists \zeta < \theta)(\exists \xi < \theta)(\forall \alpha \in u_{j,\zeta} \cap Y_i)[(g_i(\alpha), \xi) \in C\ell(h_j(\alpha))].$$

Here $\bar{u}_j = \langle u_{j,\zeta} : \zeta < \theta \rangle$. Let $H_i =: \{f_j : i(j) = i\}$.
Now $\langle g_i : i < i(*) \rangle$, $\langle (H_i, Y_i) : i < i(*) \rangle$ are O.K. for H', too, as $C\ell(f_j(\alpha)) \supseteq C\ell(h_j(\alpha))$ and $|v_j \backslash u_{i,\zeta}| < \theta$.

We are left with clause (d) of 4.2(1), so assume $i(*) < 1 + \sigma$ and $g_i \in {}^\lambda\theta$, H^*_i, Y_i for $i < i(*)$ exemplify H^* is σ-free. Let

$$H_i =: \{(h, \bar{u}) : \neg(\exists \zeta < \theta)(\exists \xi < \theta)(\forall \alpha \in u_\zeta \cap Y_i)[(g_i(\alpha), \xi) \in C\ell(h(\alpha))]\}.$$

By the choice of H we have $H \times \lambda \neq \bigcup_{i<i(*)} H_i \times Y$, but the inclusion $\supseteq$ is obviously true. So there is a pair $((h, \bar{u}), \alpha^*) \in H \times \lambda \setminus \bigcup_{i<i(*)} H_i \times Y_i$. Let $\langle a_i : i < i(*) \rangle$ be a partition of θ to unbounded subsets, and we choose by induction on $\zeta < \theta$, an ordinal $\alpha_\zeta \in u_\zeta$ and $\Upsilon_\zeta < \theta$ such that if $\zeta \in a_i$ then $\alpha_\zeta \in Y_i$ if possible and

$$\Upsilon_\zeta \in \theta \backslash \left[\bigcup_{\xi<\zeta} (g_i(\Upsilon_\xi) \cup \Upsilon_\xi) + 1 \right]$$

$$\text{and } (g_i(\alpha_\zeta), \Upsilon_\zeta) \notin C\ell(h(\alpha_\zeta)).$$

Lastly we define a function f, $\mathrm{Dom}(f) = \{\alpha_\zeta : \zeta < \theta\}$ and $f(\alpha_\zeta)$ is such that

$$C\ell(f(\alpha_\zeta)) = \{(\gamma_1, \gamma_2) : \gamma_1 < \theta, \gamma_2 < \theta, \text{ and } (\gamma_1, \gamma_2) \not\leq (g_i(\alpha_\zeta), \Upsilon_\zeta)\}.$$

Let $v =: \{\alpha_\zeta : \zeta < \theta\}$, so $f \in H^*_{(h,\bar{u})} \subseteq H^*$. Hence for some $j < i$ the pair (f, α^*) belongs to (H_i, Y_i). This contradicts $((h, \bar{u}), \alpha^*) \notin (H_i^*, Y_i)$.
(3) As in (2), $\lambda \in SPd_{\theta,\sigma} \Rightarrow \lambda \in SP_{\theta,\sigma} \Rightarrow \lambda \in SPw_{\theta,\sigma}$ is obvious.
We need to prove that $\lambda \in SPw_{\theta,\sigma} \Rightarrow SPd_{\theta,\sigma} \cap [\lambda, \lambda^\theta] \neq \emptyset$ when $\sigma \leq \theta^+$.
The proof if similar to that of (2). We start with H exemplifying that $\lambda \in SPw_{\theta,\sigma}$. We assume that for each $(h, \bar{u}) \in H, \bar{u}$ is standard. So for $(h, \bar{u}) \in H$, we define

$$H^*_{(h,\bar{u})} = \Big\{ f : f \text{ is a function from ordinals (i.e. from } \lambda) \text{ to } \theta \text{ and } f \text{ is } 1-1,$$
$$\text{and for some set } v,\ \mathrm{Dom}(f) = v \subseteq \mathrm{Dom}(h),$$
$$\text{we have that } |v| = \theta, \text{ but}$$
$$\zeta < \theta \Rightarrow |v \backslash u_\zeta| < \theta, \text{ while } (\forall \alpha \in v) f(\alpha) \leq h(\alpha) \Big\}.$$

Let $H^* = \bigcup \{H^*_{(h,\bar{u})} : (h, \bar{u}) \in H\}$.
Checking that this H^* is as required is similar to (2). For example, to see 4.1(1)d), with $\sigma = 1$ suppose that g exemplifies that H^* is free. By the choice of H, there is an $(h, \bar{u}) \in H$ such that

$$\neg(\exists \zeta < \theta)(\exists \zeta < \theta)(\forall \alpha \in u_\zeta)[h(\alpha) \leq \mathrm{Max}\{g(\alpha), \xi\}].$$

We choose by induction on $\zeta < \theta$, an ordinal $\alpha_\zeta \in u_\zeta \cap Y$ and $\Upsilon_\zeta \in \theta$ such that

$$\Upsilon_\zeta \in \theta \backslash \left[\bigcup_{\xi<\zeta} (g(\Upsilon_\xi) \cup \Upsilon_\xi) + 1 \right]$$

and

$$h(\alpha_\zeta) > \mathrm{Max}\{g(\alpha_\zeta), \Upsilon_\zeta\}.$$

Then we let $f(\alpha_\zeta)$ be such that

$$f(\alpha_\zeta) \leq \mathrm{Max}\{g(\alpha_\zeta), \Upsilon_\zeta\}$$

but

$$f(\alpha_\zeta) \notin \{f(\alpha_\xi) : \xi < \zeta\}.$$

(4) Included in the proof of (1). $\square_{4.4}$

4.5 Claim. *Assume $\lambda \in SP_{\theta,\sigma}$, μ is a strong limit with $cf(\mu) > \theta$, and $2^\mu = \mu^+ > \lambda$.*

Then there is a $\kappa \in [\lambda, \mu^+]$, a regular cardinal such that $\kappa \in SP^+_{\theta,\sigma}$ where

4.6 Definition. 1) $\kappa \in SP^+_{\theta,\sigma}$ means that κ is regular $> \theta$ and we can find an $S \subseteq \{\delta < \kappa : cf(\delta) = \theta\}$ stationary, $\bar{\eta} = \langle \eta_\delta : \delta \in S\rangle$, $\bar{h} = \langle h_\delta : \delta \in S\rangle$, such that

(a) η_δ is a strictly increasing sequence of ordinals of length θ with limit δ
(b) $h_\delta : \mathrm{Rang}(\eta_\delta) \to \theta$ is strictly increasing
(c) $H = \{h_\delta : \delta \in S\}$ is $(< \kappa)$-σ- free not σ-free (in 4.1's sense).

2) $\kappa \in SP^*_{\theta,\sigma}$ if in the above we add:

(d) $\bar{\eta}$ is tree like, i.e. $\eta_{\delta_1}(\varepsilon_1) = \eta_{\delta_2}(\varepsilon_2) \Rightarrow \varepsilon_1 = \varepsilon_2$ & $\eta_{\delta_1} \restriction \varepsilon_1 = \eta_{\delta_2} \restriction \varepsilon_2$.

3) $\kappa \in SP^{\otimes}_{\theta,\sigma}$ if in part (1) we replace (b) by

(b)′ h_δ : Rang $(\eta_\delta) \to \theta$ is constant.

Remark. 1) The assumption "$\mu^+ = 2^\mu$" (in 4.5) is very reasonable because of 4.3(2) (and 4.3A(3) from the topological point of view).
2) Basically the proof of 3.3(2) is a way of getting nicer examples of incompactness.

Proof of 4.5. Use $\diamondsuit_{\{\delta<\mu^+:cf(\delta)=\theta\}}$ and imitate the proof of 3.3(2) (noting that if h fails g then for some $a \subseteq \mathrm{Dom}(h)$, $|a| = h$, $h \restriction a$ is strictly increasing and $h \restriction a$ fails g). $\square_{4.5}$

4.6A Observation. 1) $SP^*_{\theta,\sigma} \subseteq SP^+_{\theta,\sigma} \subseteq SP_{\theta,\sigma}$.
2) If $\langle h_\delta, \eta_\delta : \delta \in S\rangle$, κ satisfies the preliminary requirements and clauses (a), (b) of Definition 4.6 and H is $(< \kappa_1)$-free, $\kappa_1 > \theta$ then for some $\mu \in [\kappa_1, \kappa]$, $\mu \in SP^+_\theta$.
3) Similarly for $SP^*_{\theta,\sigma}$.

Proof. (1) is trivial. For (2) and (3), do like in 2.3. $\square_{4.6A}$

4.6.B Conclusion. For $\lambda > \theta = cf(\theta), \chi = \beth_\chi > \lambda$, the following are equivalent:

(A) for some $\mu \in [\lambda, \chi)$, $\mu \in SP_\theta$
(B) for some $\mu \in [\lambda, \chi)$, $\mu \in SP^+_\theta$.

Proof. By 4.6A(1), (B) $\Rightarrow$ (A), as for (A) $\Rightarrow$ (B), let $\mu = \beth_{\lambda^{+(\theta^+)}}$, if $pp(\mu) > \mu^+$ use 4.3(2) and if $pp(\mu) = \mu^+$ use 4.5. $\square_{4.6.B}$

4.7 Claim. *Assume $\theta = \theta^{<\theta}$.*
Let $\langle h_\delta, \eta_\delta : \delta \in S\rangle$ exemplify $\lambda \in SP^*_\theta$ (even omitting "η_δ converge to δ, η_δ strictly increasing"). Then any θ^+-complete forcing preserves the non-freeness of $\{h_\delta : \delta \in S\}$.

Proof. Instead of the domain of the functions h_δ being a subset of λ, we can assume that it is $T =: \{\eta_\delta \restriction \zeta : \delta \in S$ and $\zeta < \theta$ is a successor ordinal$\}$ we can by (d) of 4.6(2) (identify $\eta_\delta(\zeta)$ with $\eta_\delta \restriction (\zeta+1)$, so $\mathrm{Dom}(h_\delta) = \{\eta_\delta \restriction \zeta : \zeta < \theta$ is a successor ordinal $\}$). Suppose Q is a θ^+-complete forcing notion, $p \in Q$ and $p \Vdash$ "$\underset{\sim}{g} : T \to \theta$ exemplifies $\{h_\delta : \delta \in S\}$ is free ". We now define by induction on $\ell g(\eta) < \theta$ a sequence $\langle p_{\eta,t}, \varepsilon_{\eta,t} : t \in T_\eta\rangle$ for $\eta \in T$ such that:

(α) $T_\eta \subseteq {}^{\ell g(\eta) \geq}\theta$, is closed under initial segments
(β) $t \lhd s \in T_\eta \Rightarrow p_{\eta,t} \leq_Q p_{\eta,s}$
(γ) if $t \in T_\eta$, either $\bigwedge_{\zeta<\theta} t\hat{\ } <\zeta> \in T_\eta$ or $\bigwedge_{\zeta<\theta} t\hat{\ } <\zeta> \notin T_\eta$
(δ) if $t \in {}^{\ell g(\eta)\geq}\theta$, $\ell g(t)$ is a limit ordinal and $(\forall \zeta < \ell g(t))(t \restriction \zeta \in T_\eta)$, then $t \in T_\eta$,
(ε) if $\nu \lhd \eta$ then $T_\nu \subseteq T_\eta$ and $t \in T_\nu \Rightarrow (p_{\eta,t}, \varepsilon_{\eta,t}) = (p_{\nu,t}, \varepsilon_{\nu,t})$
(ζ) if $\ell g(\eta)$ is a limit ordinal then
$T_\eta = \{t : t \in \bigcup_{\nu \lhd \eta} T_\nu$ or $\ell g(t)$ is a limit ordinal and $(\forall s)[s \lhd t \Rightarrow s \in \bigcup_{\nu\lhd\eta} T_\nu]\}$.
(η) assume $\eta = \nu\hat{\ } <\alpha>$ and s is a $\lhd$-maximal element of T_ν, then:
  (a) if $\{\zeta < \theta : p_{\eta,s} \nVdash_Q$ "$\underset{\sim}{g}(\eta) \neq \zeta$"$\}$ is bounded in θ then s is a $\lhd$-maximal element of T_η.
  (b) if $A = \{\zeta < \theta : p_{\eta,s} \nVdash$ "$\underset{\sim}{g}(\eta) \neq \zeta$"$\}$ is unbounded in θ, then for every $\zeta < \theta$, $s\hat{\ } <\zeta>$ is a maximal member of T_η, and $p_{s\hat{\ }<\zeta>}$ forces a value $\varepsilon_{s\hat{\ }<\zeta>} > \ell g(\eta)$ to $\underset{\sim}{g}(\eta)$.

We can carry this definition.

($*$) if $\delta \in S$ then for some $\zeta = \zeta_\delta < \theta$ and $t = t_\delta \in T_{\eta_\delta \restriction \zeta}$ we have: t is a $\lhd$-maximal member of $T_{\eta \restriction \xi}$ for every $\xi \in [\zeta, \theta)$.

[why? otherwise we can construct a $t \in {}^\theta\theta$ such that $(\forall s)[s \lhd t \Rightarrow s \in \bigcup_{\xi<\theta} T_{\eta_\delta \restriction \xi}]$, $t(\varepsilon) > \varepsilon$ and for unboundedly many $\xi < \theta$, for some $s\hat{\ } <\zeta> \lhd t$ we have $s\hat{\ } <\zeta> \in T_{\eta_\delta \restriction (\xi+1)} \setminus T_{\eta_\delta \restriction \xi}$ and $\varepsilon_{s\hat{\ }<\zeta>} > h_\delta(\eta_\delta \restriction (\xi+1))$.
Now $\{p_{\nu,s} : \nu \lhd \eta_\delta, s \lhd t, s \in T_\nu\}$ has an upper bound in Q, say p^*. Then p^* forces for $\underset{\sim}{g}(\eta \restriction (\xi+1))$ a value $> h_\delta(\eta \restriction (\xi+1)), \xi$; this is a contradiction to $p \Vdash$ "$\underset{\sim}{g}$ exemplifies the freeness of $\{h_\delta : \delta \in S\}$"].

For $t \in {}^{\theta>}\theta$ and $\zeta < \theta$ we define $S_{t,\zeta} = \{\delta \in S : \zeta, t$ can serve as ζ_δ, t_δ from $(*)\}$.
Clearly $\{h_\delta : \delta \in S_{t,\zeta}\}$ is free, hence $\{h_\delta : \delta \in S\}$ is $(2^{<\theta})^+$-free. Now as $2^{<\theta} = \theta$,

clearly we can have $S = \bigcup_{\zeta<\theta} S_\zeta$ and $\{h_\delta : \delta \in S_\zeta\}$ is free. Define $h : T \to \theta$ by $h(\eta) = \sup\{h_\zeta(\eta) : \zeta < \ell g(\eta)\}$. This h shows $\{h_\delta : \delta \in S\}$ is free. Contradiction. $\square_{4.7}$

Remark. We now sum up our results; for simplicity we speak on the case $\theta = \aleph_0$, $\sigma = 1$.

4.8 Theorem. *Assume $\lambda < \mu$ and $(\forall \chi < \mu)[\chi^{\aleph_0} < \mu)$ (possibly $\mu = \infty$). Then the following are equivalent:*

(A) *There is a first countable Hausdorff space X such that:*
- (a) *X is $(<\lambda)$-CWH*
- (b) *X is not λ-CWH*
- (c) *X has $<\mu$ points.*

(A)$^+$ *There is a space X like in (B), and in addition*
- (a)$^+$ *X is $(<\lambda)$-metrizable.*

(B) *There is a first countable Hausdorff space X such that:*
- (a) *X is $(<\lambda)-^*$ CWN*
- (b) *X is not $\lambda-^*$ CWN*
- (c) *X has $<\mu$ points.*

(B)$^+$ *There is an X like in (C), and in addition,*
- (a)$^+$ *X is $(<\lambda)$-metrizable.*

(C) *there is a family H of functions, each with domain a countable set of ordinals and range $\subseteq \omega$ such that:*
- (a) *H is $(<\lambda)$-free*
- (b) *H is not free*
- (c) *$|H| < \mu$.*

(C)$^+$ as in (D) and
- *(d) $\bigcup\{Dom(h) : h \in H\} = \lambda' \in [\lambda, \mu)$*
- *(e) each h is one to one.*

(C)$'$ $[\lambda, \mu) \cap SP_{\aleph_0} \neq \emptyset$

(C)$''$ $[\lambda, \mu) \cap SPw_{\aleph_0} \neq \emptyset$

(C)$'''$ $[\lambda, \mu) \cap SPd_{\aleph_0} \neq \emptyset$

(D) *there is a* $H = \langle h_\alpha, < u_{\alpha,n} : n < \omega >: \alpha \in v\rangle$ *with* $u_{\alpha,n+1} \subseteq u_{\alpha,n} \subseteq v$ *and each* h_α *is a function from ordinals to* Pie$(\aleph_0 \times \aleph_0)$ *defined on* $u_{\alpha,0}$ *such that:*

(a) H *is not free in the sense of* SQw

(b) *for* $v' \in [v]^{<\lambda}$, $H \restriction v'$ *is free in the sense of* SQw

(c) $|v| < \mu$.

$(D)'$ $[\lambda, \mu) \cap SQ_{\aleph_0} \neq \emptyset$

$(D)''$ $[\lambda, \mu) \cap SQw_{\aleph_0} \neq \emptyset$

$(D)'''$ $[\lambda, \mu) \cap SQd_{\aleph_0} \neq \emptyset$

4.8A Theorem. *In 4.8 if* $(\forall\kappa < \mu)(\beth_{\theta^+}(\kappa) < \mu)$ *(really* $(\forall\kappa < \mu)(\beth_{\omega_1}(\kappa) < \mu)$ *is O.K., <u>then</u> we can add*

(E) *for some regular* $\kappa \in [\lambda, \mu)$ *we have* $INCWH^2(\kappa)$

$(E)'$ $\lambda \in SP^+_{\aleph_0}$

Proof of 4.8 and 4.8A. By 4.4(1) (the $(b) \Leftrightarrow (b)^+ \Leftrightarrow (c) \Leftrightarrow (c)^+$ part) we know the equivalence of (A), $(A)^+$, (B) $(B)^+$.
By 4.3(1) and 4.1 $(C) \Leftrightarrow (C)'$.
By 4.4(3) we have $(C)' \Rightarrow (C)'' \Rightarrow (C)'''$.
By 4.3A(1) and 4.2 $(D) \Leftrightarrow (D)'$.
By 4.4(2) $(D)' \Rightarrow (D)'' \Rightarrow (D)'''$.
By 4.2B(2) $(D)''' \Rightarrow (C)'$.
By 4.2B(1) $(C)' \Rightarrow (D)'$, $(C)'' \Rightarrow (D)''$, $(C)''' \Rightarrow (D)''$.

Together we get the equivalence of $(C), (D), (C)', (D)', (C)'', (D)'', (C)'', (D)''$.

By 4.4(1), i.e. $(a) \Rightarrow (b) \Rightarrow (d)$ we have $(D)' \Rightarrow (A) \Rightarrow (D)''$, so by the last sentence and the first paragraph we have finished the proof of 4.8. For 4.8A use 4.6B. $\square_{4.8}$

4.9 Fact. Let $\lambda = cf(\lambda) > \theta = cf(\theta)$. The following statements, (A) and (B), are equivalent:

$(A) = (A)_{\lambda,\theta}$ There is $H = \bigcup_{i<\lambda} H_i$ such that:

(α) H_i is increasing continuous

(β) H_i is a family of functions h to θ, Dom(h) is a set of θ ordinals, h is one to one

(γ) each H_i is free, but H is not free, in the sense of SP.

$(B) = (B)_{\lambda,\theta}$. Let $X = X_{\lambda,\theta} =: {}^{\lambda}\theta$

$$F = F_{\lambda,\theta} =: \Big\{ f : f \text{ a partial function from } X \text{ to } \theta, |\mathrm{Dom}\ f| = \theta \text{ and } (\forall^* i < \lambda)(\forall^* \eta \in \mathrm{Dom}(f))[f(\eta) \leq \eta(i)] \text{ and } f \text{ is one to one} \Big\}.$$

Then there is no $G : X \to \omega$ such that

$$f \in F \Rightarrow (\forall^* \eta \in \mathrm{Dom}(f))[f(\eta) \leq G(\eta)].$$

Proof. (A) $\Rightarrow$ (B).

Let H, H_i $(i < \lambda)$ exemplify (A), let $A = \bigcup\{\mathrm{Dom}(h) : h \in H\}$, and let $g_i : A \to \theta$ exemplify "H_i is free".

We define an equivalence relation E on $A : \alpha E \beta \Leftrightarrow \bigwedge_{i<\lambda} g_i(\alpha) = g_i(\beta)$. If for some $h \in H$ and α, the set $(\alpha/E) \cap \mathrm{Dom}(h)$ has cardinality θ, choose $i < \lambda$ such that $h \in H_i$, and g_i cannot satisfy the requirement. Hence $|(\alpha/E) \cap \mathrm{Dom}(h)| < \theta$ for all α and h.

Let $h^{\otimes}$ be a function with domain $\mathrm{Dom}(h)$, $h^{\otimes}(\alpha) = \sup\{h(\beta) : \beta \in \alpha/E\}$, for all $h \in H$. Now $H' =: \{h^{\otimes} : h \in H\}$, $H_i'' =: \{h^{\otimes} : h \in H_i'\}$ exemplify (A) too. So without loss of generality E is the equality on A.

Next for each $\alpha \in A$ let $\eta_\alpha \in {}^{\lambda}\theta (= X)$ be defined by $\eta_\alpha(i) = g_i(\alpha)$, so $\alpha \neq \beta \Rightarrow \eta_\alpha \neq \eta_\beta$. For $h \in H$ let $\mathrm{Dom}(h) = \{\alpha_{h,\zeta} : \zeta < \theta\}$ be an enumeration such that $\langle h(\alpha^*_{h,\zeta}) : \zeta < \theta \rangle$ is increasing and not eventually constant. For $h \in H$ let

$$\mathcal{P}_h = \{a \subseteq \theta : \langle h(\alpha^*_{h,\zeta}) : \zeta \in a \rangle \text{ is strictly increasing}\}$$

and let the function f_h be defined by:

$$\mathrm{Dom}(f_h) = \{\eta_{\alpha_{h,\zeta}} : \zeta < \theta\} \text{ and } f_h(\eta_{\alpha_{h,\zeta}}) = h(\alpha_{h,\zeta}).$$

Now

$(*)$ $h \in H \wedge a \in \mathcal{P}_h \Rightarrow f_h \restriction a \in F.$

[Why? Let $i(*) = \mathrm{Min}\{i : h \in H_i\}$ (well defined as $H = \bigcup_{i<\lambda} H_i$), so $i \in [i(*), \lambda)$ implies $h \leq^* (g_i \restriction \mathrm{Dom}(h))$. So for some $\zeta(*) < \theta$, for every $\zeta \in [\zeta(*), \theta)$ we have $h(\alpha_{h,\zeta}) \leq g_i(\alpha_{h,\zeta})$, but $f_h(\eta_{\alpha_{h,\zeta}}) = h(\alpha_{h,\zeta})$ and $g_i(\alpha_{h,\zeta}) = \eta_{\alpha_{h,\zeta}}(i)$ so: for every $i < \lambda$ large enough for all but $< \theta$ members $\eta = \eta_{\alpha_{h,\zeta}}$ of $\mathrm{Dom}(f_h)$, $f_h(\eta) = h(\alpha_{h,\zeta}) \leq g_i(\alpha_{h,\zeta}) = \eta_{\alpha_{h,\zeta}}(i) = \eta(i)$ as required].

So assume G is a function from X to ω such that

$(**)$ $f \in F \Rightarrow (\forall^* \eta \in \mathrm{Dom}(f))[f(\eta) \leq G(\eta)]$

and we should get a contradiction. Let us define $g \in {}^A\theta$ by $g(\alpha) = G(\eta_\alpha)$. So for $h \in H$ assume $b = \{\zeta < \theta : h(\alpha_{h,\zeta}) > G(\eta_{\alpha_{h,\zeta}})\}$ is unbounded, so there is $a \subseteq b$, $a \in \mathcal{P}_h$. So we have $f_h \restriction a \in F$ hence by $(*) + (**)$ for some $\zeta(*) < \theta$, $\zeta \in [\zeta(*), \theta) \cap a \Rightarrow f_h(\eta_{\alpha_{h,\zeta}}) \le G(\eta_{\alpha_{h,\zeta}})$. But $f_h(\eta_{\alpha_{h,\zeta}}) = h(\alpha_{h,\zeta})$, and $g(\alpha_{h,\zeta}) = G(\eta_{\alpha_{h,\zeta}})$ so

$$\zeta \in [\zeta(*), \theta) \cap a \Rightarrow h(\alpha_{h,\zeta}) \le g(\alpha_{h,\zeta}).$$

So g shows that H is free, contradiction. We have proved (B).

<u>$(B) \Rightarrow (A)$</u>

The demand $A = \bigcup_{h \in H} \mathrm{Dom}(h) \subseteq \mathrm{Ord}$ is immaterial, so let $A = X$ and $H = F_{\lambda,\theta}$. Lastly for $i < \lambda$ let $g_i : A \to \theta$ be given by $g_i(\eta) = \eta(i)$, and

$$H_i = \{f \in F : \text{ for every } j \in [i, \lambda) \text{ we have } (\forall^* \eta \in \mathrm{Dom}(f))[f(\eta) \le \eta(j)\}.$$

$\square_{4.9}$

4.10 Conclusion. $INCWH(\lambda)$ implies $(B)_{\lambda,\theta}$ of Fact 4.9 implies

$$(\exists \mu)[\lambda \le \mu \le 2^\lambda \ \& \ INCWH(\mu)].$$

4.11 Remark. It is well known that

(∗) if there is a real valued measure m on $P(\lambda)$
$G(f) = \mathrm{Min}\{n : m(f^{-1}(\{n\}) > 0\}$,

then G contradicts $(B)_{\lambda,\aleph_0}$.

Also, it is consistent that $SP_{\aleph_0} \subseteq (2^{\aleph_0})^+$. This follows from the consistency of the PMEA (Product Measure Extension Axiom) and Fact 4.9.
The consistency of PMEA is due to Kunen. See [Fl84] for an exposition.

§5 More on freeness

5.1 Definition. For $\ell \in \{0, 1, 2, 3, 4\}$ and regular cardinal θ we define

$$SP^\ell_{\theta,\sigma} = \Big\{ \lambda : \text{there is a family } H \text{ such that:}$$

(a) every $h \in H$ is a partial function from ordinals to θ

(b) $h \in H \Rightarrow |\mathrm{Dom}(h)| = \theta$

(c) every $H' \subseteq H$ of cardinality $< \lambda$ is free in the sense of $P^\ell_{\theta,\sigma}$

(d) H is not free in the sense of $P^\ell_{\theta,\sigma} \Big\}$, where H' is σ-free in the sense of $P^{\ell}_{\theta,\sigma}$ if $H' = \bigcup_{i<i(*)} H'_i$ for some $i(*) < 1+\sigma$, and each H'_i is free in the sense of P^{ℓ}_{θ}, which means:

(0) if $\ell = 0$, there is a function g from ordinals to θ such that

$$(\forall h)(\exists \xi < \theta)[h \in H'_i \Rightarrow (\forall \alpha \in \mathrm{Dom}(h))[h(\alpha) \leq g(\alpha) \vee h(\alpha) \leq \xi]].$$

(So $SP^0_{\theta,\sigma} = SP_{\theta,\sigma}$).

(1) if $\ell = 1$, there is a function g from ordinals to θ, such that

$$(\forall h)[h \in H'_i \Rightarrow |\{\alpha : h(\alpha) > g(\alpha)\}| < \theta].$$

(2) if $\ell = 2$, there is a g like in (1), and

$$(\forall h)[h \in H'_i \Rightarrow h \text{ one to one}].$$

(3) if $\ell = 3$, there is a g like in (1), and

$$(\forall h)[h \in H'_i \Rightarrow h \text{ is constant function}].$$

(4) if $\ell = 4$, there is a g like in (1), and

$$(\forall h)[h \in H'_i \Rightarrow h \text{ is constant or } h \text{ is one to one}].$$

If $\sigma = 1$, we omit it from the notation.

5.2 Claim. *If $\lambda \in SP^1_{\theta,\sigma}$, then $[\lambda, \lambda + 2^\theta] \cap SP^4_{\theta,\sigma} \neq \emptyset$. (Here $\sigma \leq \theta^+$).*

Proof. We divide the proof into two cases.

Case 1 $\sigma = 1$. Let H exemplify $\lambda \in SP^1_\theta$. Let

$$H^{\oplus} \underset{\text{def}}{=} \Big\{ h \restriction A : h \in H \,\&\, A \in [\mathrm{Dom}(h)]^\theta \;\&\, (h \restriction A \text{ is constant } \underline{\text{or}}\; h \restriction A \text{ is one to one}) \Big\}.$$

So

(α) $|H^{\oplus}| \leq \lambda + 2^\theta$.

(β) Let $H' \subseteq H^{\oplus}$ with $|H'| < \lambda$. Hence

$$H' = \{h_j \restriction A_j : j < j^* < \lambda\}$$

for some $H'' = \{h_j : j < j^*\} \subseteq H$ and $A_j \in [\mathrm{Dom}(h_j)]^\theta$, for $j < j^*$.

Hence H'' is free in the sense of P^1_θ. Let g exemplify this. Hence for every $h_j \in H''$, we have

$$|\{\alpha : h_j(\alpha) > g(\alpha) \,\&\, \alpha \in \mathrm{Dom}(h_j)\}| < \theta.$$

In particular, g exemplifies that H' is free in the sense of P^4_θ.

(γ) Assume that $H^{\oplus}$ is free in the sense of P^4_θ.

Before we proceed, an easy observation.

Subclaim (A). If h is a function from ordinals to θ, then at least one of the following holds:

(i) there is a subset A of $\mathrm{Dom}(h)$ such that $h \restriction A$ is a constant.

(ii) there is a subset A of $\mathrm{Dom}(h)$ such that $h \restriction A$ is one to one.

[Why? If not (i), then $|\mathrm{Rang}(h)| = \theta$, by the regularity of θ.]

Assume that g exemplifies that $H^{\otimes}$ is free in the sense of P^4_θ. For every $h \in H^{\otimes}$ let

$$B_h = \{\alpha \in A : h(\alpha) > g(\alpha)\}.$$

If $|B_h| = \theta$ let $A \subseteq B_h$ be such that $|A| = \theta$, $h \restriction A$ is one to one or constant, but $h \restriction A \in H^{\oplus}$ and $\bigwedge_{\alpha \in A} h \restriction A(\alpha) > g(\alpha)$, contradiction. So $h \in H \Rightarrow |B_h| < \theta$ as required.

Case 2 $\sigma = \theta^+$.

Let H exemplify $\lambda \in SP^1_{\theta,\theta^+}$. For $h \in H$, let $\mathrm{Dom}(h) = \{\alpha^h_\zeta : \zeta < \theta\}$ be an increasing enumeration.

As before we define

$$H^{\oplus} \underset{\mathrm{def}}{=} \Big\{ h \restriction A : h \in H \,\&\, A \in [\mathrm{Dom}(h)]^\theta \;\&\, (h \restriction A \text{ is constant } \underline{\text{or}} \; h \restriction A \text{ is one to one}) \Big\}.$$

It is easily seen that

(a) $|H^{\oplus}| \leq 2^\theta + \lambda$

(b) $H^{\oplus}$ is $(< \lambda) - \theta^+$-free in the sense of P^4_θ.

Fact (B). $H^{\oplus}$ is not θ^+-free in the sense of P^4_θ

Proof of Fact (B). Suppose otherwise, so without loss of generality $H^{\oplus} = \bigcup_{i<\theta} H^{\oplus}_i$, and each $H^{\oplus}_i$ is free in the sense of P^4_θ. Let this be witnessed by g_i.

without loss of generality, $\mathrm{Dom}(g_i) = \bigcup_{h \in H} \mathrm{Dom}(h)$, as $|H| = \lambda$.

Also without loss of generality,

$$i < j < \theta \Rightarrow \bigwedge_{\beta \in \bigcup_{h\in H} \mathrm{Dom}(h)} g_i(\beta) < g_i(j).$$

[Why? Since we can replace $\langle g_i : i < \theta\rangle$ by $\langle g'_i : i < \theta\rangle$ defined by

$$g'_i(\beta) = \sup[\{g_j(\beta) + 1 : j < i\} \cup \{g_i(\beta)\}].]$$

Let

$$H_i \overset{\mathrm{def}}{=} \{h \in H : (\exists \xi < \theta)[\zeta \geq \xi \Rightarrow h(\alpha^h_\zeta) \leq g_i(\alpha^h_\zeta)]\},$$

for $i < \theta$.

Hence, each H_i is free in the sense of P^1_θ, so it suffices to show that $H = \bigcup_{i<\theta} H_i$. So suppose $h \in H \setminus \bigcup_{i<\theta} H_i$. Let

$$A_i \overset{\text{def}}{=} \{\alpha^h_\zeta : \zeta < \theta \,\&\, h(\alpha^h_\zeta) > g_i(\alpha^h_\zeta)\},$$

for $i < \theta$. So, as $h \notin \bigcup_{i<\theta} H_i$, for every $i < \theta$ we have $|A_i| = \theta$. Also, by the choice of $\langle g_i : i < \theta\rangle$, we have

$$i < j < \theta \Rightarrow A_i \supseteq A_j.$$

Hence, we can find a set $A \in [\mathrm{Dom}(h)]^\theta$ such that for all $i < \theta$ we have $|A \setminus A_i| < \theta$. There is a subset $B \subseteq A$ such that $|B| = \theta$ and $h \restriction B \in H^\oplus$. Hence $h \restriction B \in H_i$ for some $i < \theta$. But $|\{\alpha \in B : h(\alpha) < g_i(\alpha)\}| < \theta$, in contradiction with the choice of g_i.

This finishes the proof of the second case. Since, as remarked in 4.1A, the σ-freeness for $\sigma < \theta$ is equivalent to 1-freeness, we have finished the proof. $\square_{5.2}$

5.3 Observation. 1) For $\sigma > \theta^+$, we have $SP^3_{\theta,\sigma} = \emptyset$.

[Why? Suppose $\lambda \in SP^3_{\theta,\sigma}$ for some $\sigma > \theta^+$, and this is exemplified by a family H. Let

$$H_i \overset{\text{def}}{=} \{h \in H : h \text{ is constantly } i \text{ on its domain}\}.$$

Hence $H = \bigcup_{i<\theta} H_i$.

But each H_i is free in the sense of P^3_θ, as exemplified by $g_i \equiv i + 1$.]

2) For $\sigma > \theta$, for every λ we have $\lambda \in SP^2_{\theta,\sigma}$ iff $\lambda \in SP^4_{\theta,\sigma}$.

[Why? Certainly $SP^2_{\theta,\sigma} \subseteq SP^4_{\theta,\sigma}$. Suppose H exemplifies that $\lambda \in SP^4_{\theta,\sigma}$. Let

$$H^\oplus \overset{\text{def}}{=} \{h \in H : h \text{ is one to one}\}.$$

Then by (1) we know $|H^\oplus| = \lambda$ and $H^\oplus$ is not σ-free in the sense of P^4_θ. However, each $H' \subseteq H^\oplus$ is σ-free in the sense of P^4_θ, so $H^\oplus$ exemplifies that $\lambda \in SP^2_{\theta,\sigma}$.]

3) Observation 4.1A(0) now means that $SP^3_{\theta,\sigma}$ is the same as $SPd_{\theta,\sigma}$.

5.4 Claim.

(1) $\lambda \in SP^3_{\theta,\sigma} \Rightarrow \lambda \in SP^1_{\theta,\sigma}$

(2) $\lambda \in SP^2_{\theta,\sigma} \Rightarrow \lambda \in SP^1_{\theta,\sigma}$

(3) $\lambda \in SP^4_{\theta,\sigma} \Rightarrow \lambda \in SP^1_{\theta,\sigma}$

Proof. (1)-(3) Obvious from the definition. $\square_{5.4}$

5.5 Claim. $\lambda \in SP^4_{\theta,\sigma} \Leftrightarrow \lambda \in SP^2_{\theta,\sigma} \vee \lambda \in SP^3_{\theta,\sigma}$.

Proof. $\Leftarrow$ Follows from the definition.

$\Rightarrow$ Let H exemplify $\lambda \in SP^4_{\theta,\sigma}$, and let

$$H^1 \overset{\text{def}}{=} \{h \in H : h \text{ is one to one}\} \text{ and}$$

$$H^2 \overset{\text{def}}{=} \{h \in H \, h \text{ is a constant}\}.$$

So both H^1, H^2 are $(< \lambda) - \sigma$-free in the sense of P^1_θ.

If one of H^1, H^2 is not $\lambda - \sigma$-free in the sense of P^1_θ, we are done. Otherwise let $H^\ell = \bigcup_{i<i_\ell(*)} H^\ell_i$ for $\ell \in \{1,2\}$ and $i_\ell(*) < 1 + \sigma$ such that H^ℓ_i is free in the sense of P^1_θ, as exemplified by g^ℓ_i.

If $\sigma = 1$ let g be

$$g_0(\alpha) \overset{\text{def}}{=} \text{Max}\{g^1_0(\alpha), g^2_0(\alpha)\}.$$

It exemplifies H is free.

If $\sigma > 1$ without loss of generality $\sigma = \theta^+$ and let $i(*) = i_1(*) + i_2(*)$

$$H_i = \begin{cases} H^1_i & \underline{\text{if}} \quad i < i_1(*) \\ H^2_{i-i_1(*)} & \underline{\text{if}} \quad i \in [i_1(*), i_1(*) + i_2(*)) \end{cases}$$

$$g_i = \begin{cases} g^1_i & \underline{\text{if}} \quad i < i_1(*) \\ g^2_i & \underline{\text{if}} \quad i \in [i_1(*), i_1(*) + i_2(*)). \end{cases}$$

Then $H = \bigcup_{i<i(*)} H_i$ and each H_i is free, as exemplified by g_i. $\square_{5.5}$

5.6 Claim. $\lambda \in SP^3_\theta \Rightarrow [\lambda, \lambda^\theta] \cap SP^2_\theta \neq \emptyset$.

Proof.

Let H exemplify $\lambda \in SP^3_\theta$, and let for $\varepsilon < \theta$,

$$H_\varepsilon \overset{\text{def}}{=} \{h \in H : h \text{ is constantly } \varepsilon\}.$$

Let $\text{Dom}(h) = \{\alpha^h_\zeta : \zeta < \theta\}$ be an enumeration with no repetitions, for $h \in H$, and let

$$A \overset{\text{def}}{=} \{\alpha^h_\zeta : h \in H \ \& \ \zeta < \theta\}.$$

We define

$$G \overset{\text{def}}{=} \Big\{ (\beta, \bar{h}) : \bar{h} = \langle h_\zeta ; \zeta \in W \rangle \text{ for some } W = W_{(\beta,\bar{h})} \in [\theta]^\theta \text{ we have}$$
$$\text{(i) } h_\zeta \in H_\zeta \text{ for } \zeta \in W$$
$$\text{(ii) } \forall \zeta \in W \ [\beta \in \{\alpha^{h_\zeta}_\varepsilon : \varepsilon \in (\zeta, \theta)\}]$$
$$\text{(iii) } \langle \text{Min}\{\varepsilon : \alpha^{h_\zeta}_\varepsilon = \beta\} : \zeta \in W \rangle \text{ is strictly increasing} \Big\}.$$

For each $(\beta, \bar{h}) \in G$, we define a function $f_{(\beta,\bar{h})}$ such that

$$\text{Dom}(f_{(\beta,\bar{h})}) = \text{Rang}(\bar{h}) = \{h_\zeta : \zeta \in W_{(\beta,\bar{h})}\}, \text{ and}$$

$$f_{(\beta,\bar{h})}(h_\zeta) = \text{the unique } \varepsilon \text{ such that } \beta = \alpha^{h_\zeta}_\varepsilon.$$

Let $F \overset{\text{def}}{=} \{f_{(\beta,\bar{h})} : (\beta, \bar{h}) \in G\}$.

Remark. Our aim is to use F to exemplify that $[\lambda, \lambda^\theta] \cap SP^2_\theta \neq \emptyset$. However, if $f \in F$, then the domain of f is not a set of ordinals, but a subset of H. This does not matter, as $|H| = \lambda$.

Fact (a). Each $f_{(\beta,\bar{h})}$ is one to one, in fact $f_{(\beta,\bar{h})}(h_\zeta)$ is strictly increasing in ζ.

[Why? Suppose $\zeta_1 < \zeta_2 \in W_{(\beta,\bar{h})}$. By the last clause in the definition of G, we know that $f_{(\beta,\bar{h})}(h_{\zeta_1}) < f_{(\beta,\bar{h})}(h_{\zeta_2})$.]

Fact (b). F is not free in the sense of P^1_θ.

Proof of Fact (b). Suppose otherwise, and let g witness this. We define $g^\oplus$ on $\bigcup\{\mathrm{Dom}(h) : h \in H\}$ by

$$g^\oplus(\beta) \stackrel{\mathrm{def}}{=} \sup\{\varepsilon(h) : h \in H \,\&\, \beta \in \{\alpha^h_\zeta : \zeta \in (g(h), \theta)\}\}.$$

Subfact 1. $g^\oplus(\beta) < \theta$.

[Why? Otherwise we can find for some $\beta \in \mathrm{Dom}(g^\oplus)$, a sequence $\langle h'_\zeta : \zeta < \theta\rangle$ in H such that

$$\varepsilon(h'_\zeta) > \zeta \,\&\, \beta \in \{\alpha^{h'_\zeta}_\xi : \xi \in (g(h'_\zeta), \theta)\}.$$

By thinning out, we can find a sequence $\langle h_\zeta : \zeta \in W\rangle$ for some $W \in [\theta]^\theta$, such that for $\zeta \in W$ we have

$$\varepsilon(h_\zeta) = \zeta \,\&\, \beta \in \{\alpha^{h_\zeta}_\xi : \xi \in (g(h_\zeta), \theta)\}.$$

Hence $(\beta, \bar{h}) \stackrel{\mathrm{def}}{=} \langle h_\zeta : \zeta \in W\rangle) \in G$, and so $f_{(\beta,\bar{h})} \in F$. However, for every $\zeta \in W$ we have $f_{(\beta,\bar{h})}(h_\zeta) > g(h_\zeta)$, in contradiction with the choice of g.]

Subfact 2. For $h \in H$, for every $\zeta < \theta$ large enough, we have $g^\oplus(\alpha^h_\zeta) \geq h(\alpha^h_\zeta)$ $(= \varepsilon(h))$.

[Why? Suppose $\zeta > g(h)$, hence

$$\alpha^h_\zeta \in \{\alpha^h_\xi : \xi \in (g(h), \theta)\},$$

so

$$g^\oplus(\alpha^h_\zeta) \geq \varepsilon(h) = h(\alpha^h_\zeta),$$

by the definition of $g^\oplus$.]

Hence we proved Fact (b). $\square_{(b)}$

Fact (c). F is $(< \lambda)$-free in the sense of P^1_θ.

Proof of Fact (c). Let $F' \subseteq F$ with $|F'| < \lambda$. Let $F' = \{f_{(\beta_i,\bar{h}_i)} : i < i(*) < \lambda\}$.

Now let

$$H' \stackrel{\mathrm{def}}{=} \{h_\zeta : \zeta \in W(\beta_i, h_i) \,\&\, i < i(*)\}.$$

Hence $H' \in [H]^{<\lambda}$ (as $\lambda > \theta$). Let $g^\oplus$ be a function which exemplifies that H' is free in the sense of P^3_θ. We claim that g below shows that F' is free.

For $h \in H$ we let $\mathrm{Dom}(h) = \{\alpha^h_\zeta : \zeta < \theta\}$ be an increasing enumeration. Then we let

$$g(h) \stackrel{\text{def}}{=} \mathrm{Min}\{\xi < \theta : \varepsilon \geq \xi \Rightarrow g^\oplus(\alpha^h_\varepsilon) \geq h(\alpha^h_\varepsilon)\}.$$

Note: $g(h)$ is well defined by the choice of $g^\oplus$.
So, let $f_{(\beta,\bar{h})} \in F'$, and let $W = W(\beta, \bar{h})$. Let

$$A \stackrel{\text{def}}{=} \{\zeta < \theta : f_{(\beta,\bar{h})}(h_\zeta) > g(h_\zeta)\}.$$

If $\zeta \in A$, then $f_{(\beta,\bar{h})}(h_\zeta) > g(h_\zeta)$, so $g^\oplus(\alpha^{h_\zeta}_{f_{(\beta,\bar{h})}(h_\zeta)}) \geq h_\zeta(\alpha^{h_\zeta}_{f_{(\beta,\bar{h})}(h_\zeta)})$ by the definition of $g(h_\zeta)$. In other words, $g^\oplus(\beta) \geq \zeta$. Hence $A \subseteq g^\oplus(\beta) + 1$, and so $|A| < \theta$. $\square_{5.6}$

5.7 Claim. $\lambda \in SQd_{\theta,\sigma} \Rightarrow [\lambda, \lambda^\theta) \cap SP^1_{\theta,\sigma} \neq \emptyset$.

Proof. Let $H = \{h_j : j < \lambda\}$ exemplify that $\lambda \in SQd_{\theta,\sigma}$. Let us enumerate $\mathrm{Dom}(h_j) = \{\alpha^j_\zeta : \zeta < \theta\}$, as in clause (e) of 4.2(2). Hence

$$C\ell(h_j(\alpha^j_\zeta)) = \{(\varepsilon_1, \varepsilon_2) : \varepsilon_1 < \theta \,\&\, \varepsilon_2 < \theta \,\&\, \neg[(\varepsilon_1, \varepsilon_2) \leq (\beta^j_\zeta, \gamma^j_\zeta)]\}$$

for some $\langle \gamma^j_\zeta : \zeta < \theta \rangle$ which is strictly increasing and $\gamma^j_\zeta > \bigcup_{\xi<\zeta} \beta^j_\xi$. Let $h^\oplus_j$ be the function with $\mathrm{Dom}(h^\oplus_j) = \{\alpha^j_\zeta : \zeta < \theta\}$ and defined by $h^\oplus_j(\alpha^j_\zeta) = \beta^j_\zeta + 1$. Let $H^\oplus = \{h^\oplus_j : j < \lambda\}$.

Suppose that $H^\oplus = \bigcup_{i<i(*)} H^\oplus_i$ for some $i(*) < 1 + \sigma$, and each $H^\oplus_i$ is free in the sense of P^1_θ, and let g_i $(i < i(*))$ exemplify this.

Hence for every $j < \lambda$ we have that $h_j \in H^\oplus_i \Rightarrow \{\zeta : h^\oplus_j(\alpha^j_\zeta) = \beta^j_\zeta + 1 > g_i(\alpha^j_\zeta)\}$ is bounded in θ. In particular, there is an ordinal $\xi < \theta$ such that

$$\zeta \in [\xi, \theta) \Rightarrow (g_i(\alpha^j_\zeta), g_i(\alpha^j_\zeta)) \in C\ell(h_j(\alpha^j_\zeta)).$$

This contradicts the assumption that H is not λ-free in the sense of $Qd_{\theta,\sigma}$.

Now suppose that $H' \subseteq H^\oplus$ with $|H'| < \lambda$. Let

$$H'' \stackrel{\text{def}}{=} \{h_j : h^\oplus_j \in H' \text{ (and } j < \lambda \text{ of course)}\},$$

hence $H'' \subseteq H$ with $|H''| < \lambda$. So $H'' = \bigcup_{i<i(*)} H''_i$ for some $i(*) < 1 + \sigma$, and each H''_i is free in the sense of Qd_θ.
Let this be exemplified by g_i, for $i < i(*)$.
For $i < i(*)$ let $H'_i \stackrel{\text{def}}{=} \{h^\oplus_j : h_j \in H''_i\}$, so $H' = \bigcup_{i<i(*)} H'_i$. Suppose $h^\oplus_j \in H'_i$. If $h^\oplus_j(\alpha^j_\zeta) > g_i(\alpha^j_\zeta)$, then $\beta^j_\zeta + 1 > g_i(\alpha^j_\zeta)$.
Let $\xi_j < \theta$ be such that

$$(\forall \alpha \in \mathrm{Dom}(h_j))[(g_i(\alpha), \xi_j) \in C\ell(h_j(\alpha))]$$

Let $\xi < \theta$ be such that $\zeta \geq \xi \Rightarrow \gamma^j_\zeta > \xi_j$. Hence $g_i(\alpha^j_\zeta) > \beta^j_\zeta$, so $g_i(\alpha^j_\zeta) \geq \beta^j_\zeta + 1 = h_j(j_\zeta)$. Hence H'_i is free in the sense of P^1_θ. $\square_{5.7}$

5.8 Remark.

We have now finally proved 4.2B(2) (i.e. $\lambda \in SQd_{\theta,\sigma} \Rightarrow [\lambda, \lambda^\theta] \cap SP_{\theta,\sigma} \neq \emptyset$), as:
By 5.7 $\lambda \in SQd_{\theta,\sigma} \Rightarrow [\lambda, \lambda^\theta] \cap SP^1_{\theta,\sigma} \neq \emptyset$.
By 5.2 $[\lambda, \lambda^\theta] \cap SP^1_{\theta,\sigma} \Rightarrow [\lambda, \lambda^\theta] \cap SP^4_{\theta,\sigma} \neq \emptyset$.
By 5.5 and 5.6, $[\lambda, \lambda^\theta] \cap SP^4_{\theta,\sigma} \neq \emptyset \Rightarrow [\lambda, \lambda^\theta] \cap SP^2_{\theta,\sigma} \neq \emptyset$.
By 4.1A(0), $[\lambda, \lambda^\theta] \cap SP_{\theta,\sigma} \neq \emptyset$.

REFERENCES

[Fl84] W.G. Fleissner. The Normal Moore Space Conjecture. In *Handbook of Set–Theoretic Topology*, pages 733–760. 1984.

[FoLa88] M. Foreman and R. Laver. Some Downward Transfer Properties for λ_2. *Advances in Mathematics*, **67**:230–238, 1988.

[Sh 108] Saharon Shelah. On successors of singular cardinals. In *Logic Colloquium '78 (Mons, 1978)*, volume 97 of *Stud. Logic Foundations Math*, pages 357–380. North-Holland, Amsterdam-New York, 1979.

[Sh 111] Saharon Shelah. On power of singular cardinals. *Notre Dame Journal of Formal Logic*, **27**:263–299, 1986.

[JShS 320] Istvan Juhasz, Saharon Shelah, and Lajos Soukup. More on countably compact, locally countable spaces. *Israel Journal of Mathematics*, **62**:302–310, 1988.

[Sh 351] Saharon Shelah. Reflecting stationary sets and successors of singular cardinals. *Archive for Mathematical Logic*, **31**:25–53, 1991.

[Sh 355] Saharon Shelah. $\aleph_{\omega+1}$ has a Jonsson Algebra. In *Cardinal Arithmetic*, volume 29 of *Oxford Logic Guides*, chapter II. Oxford University Press, 1994.

[Sh 371] Saharon Shelah. Advanced: cofinalities of small reduced products. In *Cardinal Arithmetic*, volume 29 of *Oxford Logic Guides*, chapter VIII. Oxford University Press, 1994.

[Sh 400] Saharon Shelah. Cardinal Arithmetic. In *Cardinal Arithmetic*, volume 29 of *Oxford Logic Guides*, chapter IX. Oxford University Press, 1994.

[Sh 430] Saharon Shelah. Further cardinal arithmetic. *Israel Journal of Mathematics*, **accepted**.

Institute of Mathematics
The Hebrew University
Jerusalem, Israel

Rutgers University
Department of Mathematics
New Brunswick, NJ USA

Contemporary Mathematics
Volume **192**, 1996

CARDINAL INVARIANTS ASSOCIATED WITH HAUSDORFF CAPACITIES

JURIS STEPRĀNS

ABSTRACT. Let $\lambda(X)$ denote Lebesgue measure. If $X \subseteq [0,1]$ and $r \in (0,1)$ then the r-Hausdorff capacity of X is denoted by $H^r(X)$ and is defined to be the infimum of all $\sum_{i=0}^{\infty} \lambda(I_i)^r$ where $\{I_i\}_{i\in\omega}$ is a cover of X by intervals. The r Hausdorff capacity has the same null sets as the r-Hausdorff measure which is familiar from the theory of fractal dimension. It is shown that, given $r < 1$, it is possible to enlarge a model of set theory, V, by a generic extension $V[G]$ so that the reals of V have Lebesgue measure zero but still have positive r-Hausdorff capacity.
AMS Classification: 04A15, 28A12
Key words: Hausdorff measure, generic extension

1. INTRODUCTION

If $r \in [0,1]$ then for any set $X \subseteq [0,1]$ the r-Hausdorff capacity of X is denoted by $H^r(X)$ and is defined to be the infimum of all t such that there is a cover of X by intervals, $X \subseteq \bigcup_{i=0}^{\infty} I_i$, such that $t = \sum_{i=0}^{\infty} \lambda(I_i)^r$. This notion may be familiar from its use along the way to defining r-Hausdorff measure. Given $\beta > 0$, $H^r_\beta(X)$ is defined, for any set $X \subseteq [0,1]$, to be the infimum of all t such that there is a cover of X by intervals, $X \subseteq \bigcup_{i=0}^{\infty} I_i$, such that $t = \sum_{i=0}^{\infty} \lambda(I_i)^r$ and such that the length of each interval I_i is less than β. The r-Hausdorff measure of a set X is then defined to be the supremum of $H^r_\beta(X)$ as β ranges over all positive real numbers. Although the topic of this paper if r-Hausdorff capacity rather than r-Hausdorff measure, this is of little consequence since the zero sets in both cases are the same. The crucial difference between the two is that, while the r-Hausdorff measure is countably additive, the r-Hausdorff capacity is only subadditive if $r \in (0,1)$. A proof of the fact that H^r is actually a capacity can be found in [6] on page 90. For more details on Hausdorff measure consult [6], [2] or [3].

For the rest of this paper let r be a fixed real number such that $0 < r < 1$. Let $\lambda(X)$ denote the Lebesgue measure of any measurable set $X \subseteq [0,1]^n$. It will be shown that it is possible to generically extend an arbitrary model of set theory so that the ground model reals have Lebesgue measure 0 but still have positive r-Hausdorff measure. If this process could be iterated ω_2 times and the ground model satisfied the Continuum Hypothesis then it would yield a model where every set of size $\aleph_1$ has

Research on this paper was partially supported by NSERC.

Lebesgue measure zero yet there is a set of size $\aleph_1$ which has positive r-Hausdorff capacity. This raises the possibility of defining a new, and possibly interesting, class of cardinal invariants using the obvious extension of Hausdorff capacity to $\mathbb{R}^n$. Define $\mathcal{H}(n,r)$ to be the ideal of all subsets of $[0,1]^n$ whose r-Hausdorff capacity is zero and consider the associated cardinal invariants such as $\text{cov}(\mathcal{H}(n,r))$, $\text{cof}(\mathcal{H}(n,r))$ and $\text{add}(\mathcal{H}(n,r))$. The main question left unanswered in this paper can be stated as follows: For any $n \in \omega$ and $r < n$ it is consistent that $\text{unif}(L) \neq \text{unif}(\mathcal{H}(n,r))$?

This is related to the following question posed by P. Komjath.

Question 1.1. *Suppose that every set of size $\aleph_1$ has Lebesgue measure zero. Does it follow that the union of any set of $\aleph_1$ lines in the plane has Lebesgue measure zero?*

To see the relationship between this question and r-Hausdorf capacity consider that it is easy to find countably many unit squares in the plane such that each line passes through either the top and bottom or the left and right sides of at least one of these squares. It is therefore possible to focus attention on all lines which pass through the top and bottom of the unit square. For any such line L there is a pair (a,b) such that both the points $(a,0)$ and $(1,b)$ belong to L. If the mapping which send a line L to this pair (a,b) is denoted by β then it is easy to see that β is continuous and that if $S \subseteq [0,1]^2$ is a square of side ϵ then the union of $\beta^{-1}S$ has measure ϵ while S itself has measure ϵ^2. In other words, the Lebesgue measure of the union of $\beta^{-1}X$ is no larger than the 1-Hausdorff capacity of X for any $X \subseteq [0,1]^2$. Hence, if the answer to Question 1.1 was negative this would imply that the conjecture is true. While this was the motivation for studying the problem of Hausdorff capacity, it may be that the notion of Haudsorff capacity is actually more central than Komjath's question itself.

Since much of the material in this paper is quite technical, a few remarks concerning its organization are in order. Section 2 contains the general definition of a large class of partial orders whose splitting nodes are large with respect to some ideal. The definition of property KP(r) isolates a subset of these partial orders for which a preservation property can be proved. A key hypothesis in the preservation theorem is that the ideals used to define the degree of branching of the trees be Σ^1_1. The remaining results in Section 2 are designed to deal with some of the difficulties arising from this requirement. Section 3 contains the main preservation result: Forcing with a partial order which consists of trees whose branching is large with respect to an ideal which satisfies property KP(r) preserves the outer Lebesgue measure of the ground model. The main point of this paper is that it is possible to find such an ideal which will also make the ground model have zero Hausdorff measure. In Section 4, the main concept needed to define this ideal is introduced in Definition 4.1. The reader should read the remarks following this definition to understand its motivation because the remaining results of the section are purely technical lemmas designed to overcome the difficulty that stems from the requirement that the ideal be Σ^1_1. These technical results continue in Section 5 after the definition of the ideals to be used. The key result of this section is Lemma 5.6 which establishes that the ideals satisfy KP(r). What has not yet been established by this point is that the ideals are proper. If they are not proper then there are no sets of positive measure and so the partial orders defined are empty. Hence Section 6 contains the crucial result that these ideals are indeed proper. This section contains extensions of results originally obtained in [7]. The main theorem is then established in Section 7.

The notation used in this paper is standard except, possibly, for the following. If $1 \leq n \leq m$ then define $\pi_n : [0,1]^m \to [0,1]^n$ by

$$\pi_n(x_1, x_2, \ldots, x_m) = (x_1, x_2, \ldots, x_n).$$

If $X \subseteq [0,1]^m$ then $\pi_n(X)$ will denote the image of X under the mapping π_n. If $A \subseteq [0,1]^d$ and $1 \leq n < d$ then, for any $x \in [0,1]^n$, the notation A_x will be used to denote $\{y \in [0,1]^{d-n} : \pi_n(x,y) = x \text{ and } (x,y) \in A\}$.

2. A General Class of Forcing Partial Orders

This section will be devoted to examining a generalization of Superperfect forcing obtained by insisting that on a dense set of nodes the splitting is into a set of positive measure with respect to some ideal. Such generalizations have been considered by various authors. Throughout this section the term ideal will always refer to a proper ideal on ω which contains all finite sets. In later sections ideals will be constructed on countable sets other than ω, but it will simplify notation to ignore this for now.

Definition 2.1. *Let $\mathcal{I} = \{\mathcal{I}_n\}_{n\in\omega}$ be a sequence of ideals. The partial order $\mathbb{P}(\mathcal{I})$ will be defined to consist of trees $T \subseteq {}^{\omega}\omega$ such that for every $t \in T$ one of the following two alternatives holds:*

- $|\{n \in \omega : t \wedge n \in T\}| = 1$
- $\{n \in \omega : t \wedge n \in T\} \in \mathcal{I}^+_{|t|}$

If $\{n \in \omega : t \wedge n \in T\} \in \mathcal{I}^+_{|t|}$ then t will be said to be a branching node of T and the set of branching nodes will be denoted by $B(T)$. Define $\mathbb{P}(\mathcal{I})$ to consist of all T such that for every $t \in T$ there is $s \in B(T)$ such that $t \subseteq s$. The ordering on $\mathbb{P}(\mathcal{I})$ is inclusion.

It is left to the reader to verify that $\mathbb{P}(\mathcal{I})$ is proper and, indeed, that it satisfies Axiom A. A standard argument works.

Suppose now that $T \in \mathbb{P}(\mathcal{I})$. Then the root of T is the unique minimal member of $B(T)$ and is denoted by root(T). If $t \in B(T)$ then the set of successors of t is denoted by $\text{succ}_T(t)$ and is defined by $\text{succ}_T(t) = \{n \in \omega : t \wedge n \in T\}$. The branching height of t will be denoted by branching-height(t) and is defined to be $|\{s \subseteq t : s \in B(T)\}|$ — so branching-height(root(T)) = 1. A subset $S \subseteq T$ will be said to be a *subtree* if it is closed under taking initial segments. The tree *generated* by $X \subseteq T$ is simply the set of all initial segments of members of X. Observe that succ_S can be defined for any subtree, regardless of whether or not $S \in \mathbb{P}(\mathcal{I})$. A subset $S \subseteq T$ will be said to be a *full subtree* of T if and only if for every $t \in S$ either t is a maximal member of S or $\mathrm{succ}_T(t) = \mathrm{succ}_W(t)$. If $t \in T$ then $T\langle t\rangle$ is defined to be the subtree of T consisting of all $s \in T$ such that either $s \subseteq t$ or $t \subseteq s$. If $S \subseteq T$ is a subtree then define the *interior* of S to be the set of all non maximal elements of $B(T) \cap S$ and denote this by $\mathrm{int}(S)$ — the dependence on T will suppressed.

If $T \in \mathbb{P}(\mathcal{I})$ then define a function Ψ on $B(T)$ to be *approximating* if $\Psi(t) \subseteq [0,1]$ is a finite union of rational intervals for each $t \in B(t)$ and it is monotone in the sense that $\Psi(t) \subseteq \Psi(s)$ if $t \subseteq s$. If $T \in \mathbb{P}(\mathcal{I})$, $x \in [0,1]$ and Ψ on $B(T)$ is approximating then define $R(T, \Psi, x)$ to be the tree generated by $\{t \in B(T) : x \notin \Psi(t)\}$.

Definition 2.2. *An ideal $\mathcal{I}$ will be said to satisfy $KP(r)$ if and only if for all*

- $\theta < 1$
- $X \in \mathcal{I}^+$
- *functions F from X to the Borel subsets of $[0,1]$ satisfying that $H^r(F(x)) \leq \theta$ for each $x \in X$*
- $\epsilon > 0$

there is some $Y \subseteq [0,1]$ as well as $Z \subseteq [0,1]$ such that

- $H^r(Y) \leq \theta$
- $\lambda(Z) < \epsilon$
- $\{x \in X : y \notin F(x)\} \in \mathcal{I}^+$ *for every $y \in [0,1] \setminus (Y \cup Z)$*

Well founded trees will play an important role in the following discussion but the standard equivalance between well founded trees and trees with rank functions is not as convenient a slight modification of this notion. If $T \in \mathbb{P}(\mathcal{I})$ and $S \subseteq T$ then the *standard rank* of S will denote the rank of $S \cap B(T)$ when this is considered as a tree under the inherited ordering. Later on, a different rank function will be introduced and it should not be confused with the standard rank.

If $T \in \mathbb{P}(\mathcal{I})$ and $W \subseteq T$ is a full subtree then $W' \subseteq W$ will be said to be *large* if:

- $\mathrm{root}(W) \in W'$
- if $t \in W' \setminus B(T)$ then t is not maximal in W'
- if $t \in \mathrm{int}(W) \cap W'$ then $\mathrm{succ}_{W'}(t) \in \mathcal{I}^+_{|t|}$.

Lemma 2.1. *Suppose that*

- *$\mathcal{I} = \{\mathcal{I}_n : n \in \omega\}$ is a sequence of ideals, each satisfying $KP(r)$*
- *$T \in \mathbb{P}(\mathcal{I})$ and $W \subseteq T$ is a well founded full subtree of standard rank β*
- *Ψ is an approximating function on $B(T) \cap W$*
- $\theta < 1$
- *$H^r(\Psi(t)) < \theta$ for any $t \in B(T) \cap W$*

then, there is some $x \in [0,1]$ such that $R(W, \Psi, x)$ is a large subtree of W.

Proof: It will be shown by induction on $\beta \in \omega_1$ that the following, stronger condition holds:

Q(β): If $s \in T$ and $W \subseteq T\langle s\rangle$ is a well founded full subtree of standard rank β, $\theta < 1$, $\epsilon > 0$ and Ψ is an approximating function on $B(T) \cap W$ such that $H^r(\Psi(t)) < \theta$ for any $t \in W \cap B(T)$ then, there is some $Y \subseteq [0,1]$ and $Z \subseteq [0,1]$ such that

- $R(W\langle t\rangle, \Psi, x)$ is a large subtree of $W\langle t\rangle$ for each $x \in [0,1] \setminus (Y \cup Z)$
- $H^r(Y) \leq \theta$
- $\lambda(Z) < \epsilon$.

First notice that this implies the lemma by choosing $s = \mathrm{root}(T)$ and $\epsilon < 1 - \theta$ because $\lambda(Y) \leq H^r(Y)$ and so, if $\lambda(Z) < \epsilon$ then $[0,1] \setminus (Y \cup Z)$ is not empty. If $\beta = 0$ the statement is vacuous and Q(1) is implied by KP(r). Now assume that Q(γ) has been established for all $\gamma \in \beta$. If W is a well founded subtree of $T\langle s\rangle$, $\theta < 1$, $\epsilon > 0$ and Ψ is an approximating function on $W \cap B(T)$ then for each $n \in \mathrm{succ}_T(s)$ the standard rank of $W\langle s \wedge n\rangle$ is less than β. Moreover, $H^r(\Psi(t)) < \theta$ for any $t \in W\langle s \wedge n\rangle \cap B(T)$. It is therefore possible to apply the induction hypothesis to $T\langle s \wedge n\rangle$ for each $n \in \mathrm{succ}_T(s)$ to find Y_n and Z_n such that

- $H^r(Y_n) \leq \theta/2^{n+1}$
- $\lambda(Z_n) < \epsilon/2^{n+1}$
- if $x \in [0,1] \setminus (Y_n \cup Z_n)$ then $R(W\langle s \wedge n\rangle, \Psi, x)$ is a large subtree of $W\langle s \wedge n\rangle$.

Now let $X = \mathrm{succ}_T(s) \in \mathcal{I}^+_{|s|}$. Choose a function F on X such that $F(d) \supseteq Y_d$ and $F(d)$ is a G_δ such that $H^r(F(d)) = H^r(Y_d) \leq \theta/2$ for each $d \in X$. It follows from KP(r) that there are $Y' \subseteq [0,1]$ and $Z' \subseteq [0,1]$ such that

- $H^r(Y') \leq \theta/2$
- $\lambda(Z') < \epsilon/2$
- $\{d \in X : x \notin F(d)\} \in \mathcal{I}^+_{|s|}$ for each $x \in [0,1] \setminus (Y' \cup Z')$

Now let $Z = Z' \cup \bigcup\{Z_n : n \in X\}$, $Y = Y' \cup \bigcup\{Y_n : n \in \omega\}$ and note that $\lambda(Z) < \epsilon$ and, by the subadditivity of H^r, $H^r(Y) \leq \theta$. Hence, in order to verify that Q(β) holds it suffices to show that if $x \in [0,1] \setminus (Y \cup Z)$ and $t \in R(W, \Psi, x) \cap \mathrm{int}(W)$ then $\mathrm{succ}_{R(W,\Psi,x)}(t) \in \mathcal{I}^+_{|t|}$. If $t = s$ this follows from the application of KP(r) and the fact that $x \notin Y' \cup Z'$. In every other case it follows from the use of the induction hypothesis because $t \supseteq s \wedge n$ for some n and, therefore, $x \notin Y_n \cup Z_n$ implies that $\mathrm{succ}_{R(W,\Psi,x)}(t) \in \mathcal{I}^+_{|t|}$. ■

For the remainder of this section fix a sequence of ideals $\mathcal{I} = \{\mathcal{I}_n : n \in \omega\}$ and $T \in \mathbb{P}(\mathcal{I})$. For $t \in B(T)$ and $n \in \mathrm{succ}_T(t)$ define $t \oplus n$ to be the least $s \in B(T)$ such that $t \wedge n \subseteq s$. If $X \subseteq T$ is a subtree then a rank function ρ_X will be defined on $B(T) \cap X$ by bar induction. To begin, define $\rho_X(t)$ to be 0 if there is some $t' \subseteq t$ such that $t' \in B(T)$ and $\mathrm{succ}_X(t') \in \mathcal{I}_{|t'|}$. Define $\rho_X(t)$ to be the least ordinal β such that there is some $A \in \mathcal{I}_{|t|}$ such that $\rho_X(t \oplus n)$ is defined for each $n \in \mathrm{succ}_X(t) \setminus A$

and $\rho_X(t \oplus n) \in \beta$ for any such n. The rank of X is defined to be the rank of its root, provided this is defined. A subtree $X \subseteq T$ will be defined to be *small* if $\rho_X(t)$ is defined for all $t \in B(T) \cap X$ and $\rho_X(t) > 0$ if and only if $t \in \mathrm{int}(X)$.

Lemma 2.2. *If $X \subseteq T$ is a subtree and there is some $t \in X \cap B(T)$ for which $\rho_X(t)$ is not defined then X contains a member of $\mathbb{P}(\mathcal{I})$.*

Proof: This is standard. Let S be the subtree of T generated by all $t \in X \cap B(T)$ such that $\rho_X(t)$ is not defined. Notice that if $t \in S$ then

$$\{n \in \mathrm{succ}_X(t) : \rho_X(t \oplus n) \text{ is not defined}\} = \mathrm{succ}_X(t)$$

and this belongs to $\mathcal{I}^+_{|t|}$. Hence $S \in \mathbb{P}(\mathcal{I})$ provided that it is not empty. The hypothesis of the lemma guarantees that this is not the case. ■

For any subtree $W \subseteq T$ and any function $\theta \in \prod_{w \in \mathrm{int}(W)} \mathcal{I}_{|w|}$ define

$$W^\theta = \{w \in W : (\forall n \in \omega)(w(n) \notin \theta(w \restriction n))\}$$

or, in other words, W^θ is obtained by throwing away $\mathcal{I}_{|t|}$ many successors, determined by θ, of each $t \in \mathrm{int}(W)$. If W and X are subtrees of T define $W \prec X$ if and only if there exists $\theta \in \prod_{w \in \mathrm{int}(W)} \mathcal{I}_{|w|}$ such that for every $\theta' \in \prod_{x \in \mathrm{int}(X)} \mathcal{I}_{|x|}$ there is a one-to-one function $G : W^\theta \cap B(T) \to (X^{\theta'} \cap B(T)) \setminus \{\mathrm{root}(X)\}$ which is order preserving in the sense that if $t \subseteq s$ then $G(t) \subseteq G(s)$.

For any small subtree $Y \subseteq T$ a function $\theta \in \prod_{y \in \mathrm{int}(Y)} \mathcal{I}_{|y|}$ will be said to be *a witness to the rank* of Y if and only if for each $t \in \mathrm{int}(Y)$

$$\rho_{Y,T}(t \oplus n) \in \rho_{Y,T}(t)$$

for each $n \in \mathrm{succ}_Y(t) \setminus \theta(t)$.

Lemma 2.3. *Let W and X be small subtrees of T of rank α and β respectively. If $\alpha \in \beta$, then for any $\theta \in \prod_{w \in int(W)} \mathcal{I}_{|w|}$ which is a witness to the rank of W and any $\theta \in \prod_{x \in int(X)} \mathcal{I}_{|x|}$ there is a one-to-one function $G_{\theta,\theta'} : W^\theta \cap B(T) \to X^{\theta'} \cap B(T)$ which is order preserving such that $G_{\theta,\theta'}(root(W)) \neq root(X)$. Moreover, $G_{\theta,\theta'}$ is continuous in the variable θ' in the sense that the mapping $\theta' \mapsto G_{\theta,\theta'}$ is a continuous function from $\prod_{x \in int(X)} \mathcal{I}_{|x|}$ to ${}^{W^\theta \cap B(T)}(X^{\theta'} \cap B(T))$ where $\mathcal{I}_n$ is considered as a subspace of 2^ω.*

Proof: Suppose that $\alpha \in \beta$ and W and X are small subtrees of T of rank α and β respectively. Let $\theta \in \prod_{w \in \mathrm{int}(W)} \mathcal{I}_{|w|}$ be a witness to the rank of W. For every $\theta' \in \prod_{x \in \mathrm{int}(X)} \mathcal{I}_{|x|}$ a function $G_{\theta,\theta'} : W^\theta \cap B(T) \to X^{\theta'} \cap B(T)$ can be defined by induction on the branching height of nodes of $W^\theta \cap B(T)$. The induction hypothesis will be that $\rho_X(G_{\theta,\theta'}(t)) \geq \rho_W(t)$. Define $G_{\theta,\theta'}(\mathrm{root}(W)) = \mathrm{root}(X) \oplus m$ where m is the least integer such that $m \in \mathrm{succ}_X(\mathrm{root}(X)) \setminus \theta'(\mathrm{root}(X))$ and $\rho_X(\mathrm{root}(X) \oplus m) \geq \alpha$.

Such an m must exist because $\rho_X(\mathrm{root}(X)) = \beta > \alpha$ and $\theta'(\mathrm{root}(X)) \in \mathcal{I}_{|\mathrm{root}(X)|}$. If t and $t \oplus n$ are both in $W^\theta \cap B(T)$ and $G_{\theta,\theta'}(t)$ and $G_{\theta,\theta'}(t \oplus i)$ are defined for $i \in n$ then define $G_{\theta,\theta'}(t \oplus n) = G_{\theta,\theta'}(t) \oplus k$ where k is the least integer such that

$$k \in \mathrm{succ}_X(G_{\theta,\theta'}(t)) \setminus (\theta'(G_{\theta,\theta'}(t)) \cup \{G_{\theta,\theta'}(t \oplus i)(|G_{\theta,\theta'}(t)|) : i \in n\})$$

and $\rho_X(G_{\theta,\theta'}(t \oplus k)) \geq \rho_W(t \oplus n)$. The reason such a k exists is that, by the induction hypothesis, $\rho_X(G_{\theta,\theta'}(t)) \geq \rho_W(t)$ and hence $\rho_X(G_{\theta,\theta'}(t) > \rho_W(t \oplus n)$ because θ is a witness to the rank of W and $t \oplus n \in W^\theta$. Since $\mathcal{I}_{|G_{\theta,\theta'}(t)|}$ contains all finite subsets it must be that there is some

$$k \in \mathrm{succ}_X(G_{\theta,\theta'}(t)) \setminus (\theta'(G_{\theta,\theta'}(t)) \cup \{G_{\theta,\theta'}(t \oplus i)(|G_{\theta,\theta'}(t)|) : i \in n\})$$

such that $\rho_X(G_{\theta,\theta'}(t \oplus k) \geq \rho_W(t) \oplus n)$. Since the inductive hypothesis is preserved, this construction can be carried out for all nodes in W^θ. Obviously $G_{\theta,\theta'}$ is a one-to-one, order preserving function. By construction, $G_{\theta,\theta'}(\mathrm{root}(W)) \neq \mathrm{root}(X)$.

The continuity of $G_{\theta,\theta'}$ in the variable θ' follows from the minimal choice of the integer k such that

$$k \in \mathrm{succ}_X(G_{\theta,\theta'}(t)) \setminus (\theta'(G_{\theta,\theta'}(t)) \cup \{G_{\theta,\theta'}(t \oplus i)(|G_{\theta,\theta'}(t)|) : i \in n\})$$

and $\rho_X(G_{\theta,\theta'}(t \oplus k) \geq \rho_W(t \oplus n)$. An open neighbourhood of $G_{\theta,\theta'}$ in

$$^{W^\theta \cap B(T)}(X^{\theta'} \cap B(T))$$

is specified by a restriction of $G_{\theta,\theta'}$ to a finite subset. Given a finite subset $a \subseteq W^\theta$ it is possible to find a finite set $b \subseteq X^{\theta'}$ such that if $t \in a$, $G_{\theta,\theta'}(t) = s \oplus m$ and $i \in m \setminus \theta'(s)$ then $s \oplus i$ also belongs to b. Let $M \in \omega$ be such that the range of each $t \in b$ is contained in M. It is easy to check that if $\theta'' \in \prod_{x \in \mathrm{int}(X)} \mathcal{I}_{|x|}$ is such that $\theta''(t) \cap M = \theta'(t) \cap M$ for each $t \in b$ then $G_{\theta,\theta''} \restriction a = G_{\theta,\theta'} \restriction a$. ■

Lemma 2.4. *If W and X are small subtrees of T of rank α and β respectively, then $W \prec X$ if and only if $\alpha \in \beta$.*

One direction is an immediate consequence of Lemma 2.3 because a witness to the rank of W can always be found. For the other, it will be shown by induction on α that if $\alpha \leq \beta$ then $X \not\prec W$. For $\alpha = 0$ this is trivial so assume that the assertion has been established for all $\alpha' \in \alpha$ and suppose that $X \prec W$. This means that there is some $\theta \in \prod_{x \in \mathrm{int}(X)} \mathcal{I}_{|x|}$ such that for every $\theta' \in \prod_{w \in \mathrm{int}(W)} \mathcal{I}_{|w|}$ there is a one-to-one function $G : X^\theta \cap B(T) \to W^{\theta'} \cap B(T)$ which is order preserving such that $G(\mathrm{root}(X)) \neq \mathrm{root}(W)$. Let θ' be a witness to the rank of W and let the function G from $X^\theta \cap B(T)$ to $W^{\theta'} \cap B(T) \setminus \{\mathrm{root}(W)\}$ be one-to-one and order preserving. It must be that $\rho_W(G(\mathrm{root}(X))) \in \rho_W(\mathrm{root}(W)) = \alpha \leq \beta$. Therefore, it is possible to find $m \in \mathrm{succ}_X(\mathrm{root}(X)) \setminus \theta(\mathrm{root}(X))$ such that $\rho_X(\mathrm{root}(X) \oplus m) \geq$

$\rho_W(G(\text{root}(X)))$. Obviously G, $\theta \restriction X\langle\text{root} \oplus m\rangle$ and $\theta' \restriction W\langle G(\text{root}(X))\rangle$ establish that $X\langle\text{root} \oplus m\rangle \prec W\langle G(\text{root}(X))\rangle$ which contradicts the induction hypothesis. ■

An ideal $\mathcal{I}$ will be said to be Σ^1_1 if it is a Σ^1_1 subset of 2^ω under the natural identification. The next lemma shows that the relation $\prec$ is Σ^1_1 provided that each of the ideals of $\mathcal{I} = \{\mathcal{I}_n\}_{n\in\omega}$ is Σ^1_1. This will require the full conclusion of Lemma 2.3 since the obvious calculation only shows that $\prec$ is Σ^1_3.

Lemma 2.5. *If each of the ideals of $\mathcal{I} = \{\mathcal{I}_n\}_{n\in\omega}$ is Σ^1_1 then the relation $\prec$ defined from them is Σ^1_1.*

Proof: Since each $\mathcal{I}_n$ is Σ^1_1 it is possible to choose continuous functions $f_n : {}^\omega\omega \to 2^\omega$ such that $\mathcal{I}_n$ is the image of f_n. First it will be shown that $W \prec X$ if and only if there is some $\theta \in \prod_{w\in\text{int}(W)} \mathcal{I}_{|w|}$ and a continuous function

$$G : \prod_{x\in\text{int}(X)} {}^\omega\omega \to {}^{W^\theta \cap B(T)}\theta \cap B(T)(B(T) \cap X)$$

such that

(1) for all $\theta' \in \prod_{x\in\text{int}(X)} {}^\omega\omega$, if t belongs to $W^\theta \cap B(T)$ then $G(t) = s \oplus m$ for some $s \in B(T) \cap X$ and some $m \in \text{succ}_X(s)(t) \setminus f_{|s|}(\theta'(s))$

(2) $G(\theta')$ is order preserving for all $\theta' \in \prod_{x\in\text{int}(X)} {}^\omega\omega$

(3) $G(\theta')$ is one-to-one for all $\theta' \in \prod_{x\in\text{int}(X)} {}^\omega\omega$

(4) $G(\theta')(\text{root}(W)) \neq \text{root}(X)$ for all $\theta' \in \prod_{x\in\text{int}(X)} {}^\omega\omega$

Assuming $W \prec X$, it is possible to use Lemma 2.3 to define $G(\theta') = G_{\theta,\mu(\theta')}$ where $\mu(\theta')(x) = f_{|x|}(\theta'(x))$ for each $x \in \text{int}(X)$. Note that G is continuous because of the final sentence of Lemma 2.3 and the continuity of μ, which is a consequence of the continuity of each f_n. The other direction of the equivalence is clear because each f_n is onto $\mathcal{I}_n$.

Hence it suffices to check that the clauses (1) - (4) are equivalent to an arithmetic statement. Since T and $B(T)$ can be used as parameters, the only problematic part is the use of the quantifiers

$$\text{for all } \theta' \in \prod_{x\in\text{int}(X)} {}^\omega\omega .$$

However, the continuity of G allows these to be replaced by quantifiers over approximations to θ'. In particular, it suffices to choose a countable dense subset $C \subseteq \prod_{x\in\text{int}(X)} {}^\omega\omega$ and replace each instance of

$$\text{for all } \theta' \in \prod_{x\in\text{int}(X)} {}^\omega\omega$$

by "for all $\theta' \in C$". ■

Lemma 2.6. *For all α such that $1 \leq \alpha \in \omega_1$ and $t \in B(T)$ there is a well founded full subtree W of standard rank α such that root$(W) = t$ and if $W' \subseteq W$ is any large subtree then $\rho_{W'}(t) = \alpha$.*

Proof: Proceed by induction on α. The case $\alpha = 1$ is trivial so assume the assertion has been established for all $\alpha' \in \alpha$. First suppose that $\alpha = \beta + 1$. Given $t \in B(T)$ choose for each $n \in \mathrm{succ}_W(t)$ a well founded full subtree W_n of standard rank β such that $\mathrm{root}(W_n) = t \oplus n$ and if $W' \subseteq W_n$ is any large subtree then $\rho_{W'}(t \oplus n) = \beta$. Let $W = \cup_{n \in \mathrm{succ}_W(t)} W_n$.

If α is a limit let $\{\beta_n\}_{n\in\omega}$ converge to α from below. Given $t \in B(T)$ choose for each $n \in \mathrm{succ}_W(t)$ a well founded full subtree W_n of standard rank β_n such that $\mathrm{root}(W_n) = t \oplus n$ and if $W' \subseteq W_n$ is any large subtree then $\rho_{W'}(t \oplus n) = \beta_n$. Let $W = \cup_{n \in \mathrm{succ}_W(t)} W_n$. Since $\mathcal{I}_{|t|}$ contains all finite sets this works. ■

3. The preservation Theorem

Theorem 3.1. *If $\mathcal{I} = \{\mathcal{I}_n : n \in \omega\}$ is a sequence of Σ^1_1 ideals satisfying KP(r) such that each $\mathcal{I}_n$ contains all finite sets and G is $\mathbb{P}(\mathcal{I})$ generic over V then $H^r([0,1]\cap V) = 1$ in $V[G]$.*

Proof: Suppose the theorem false — in other words, there is some $\theta < 1$ and $\{J_n\}_{n=0}^{\infty}$, a name for a sequence of intervals with rational endpoints, as well as a condition $T \in \mathbb{P}(\mathcal{I})$ such that

$$T \Vdash \text{“}([0,1] \cap V) \subseteq \bigcup_{n=0}^{\infty} J_n\text{”}$$

and $T \Vdash$ “$\sum_{n=0}^{\infty} \lambda(J_n)^r < \theta$”. By thinning down T it may be assumed that if $t \in B(T)$ and branching-height$_T(t) = n$ then $T\langle t\rangle \Vdash$ “$J_i = J(t,i)$” for some interval $J(t,i)$ with rational endpoints, for each $i \in n$. Let $\Psi(t) = \cup\{J(t,i) : i \in |t|\}$ for each $t \in B(T)$.

If there is some $x \in [0,1]$ such that $R(T,\Psi,x)$ contains some $S \in \mathbb{P}(\mathcal{I})$ then it follows that

$$S \Vdash \text{“}x \notin \bigcup_{n=0}^{\infty} J_n\text{”}$$

contradicting the fact that $S \subseteq T$. Hence by Lemma 2.2 it follows that $\rho_{R(T,\Psi,x)}(t)$ is defined for all $t \in R(T,\Psi,x) \cap B(T)$ and therefore

$$T_x = R(T,\Psi,x) \setminus \{t \in R(T,\Psi,x) : (\exists t' \subsetneq t)\rho_{R(T,\Psi,x)}(t) = 0\}$$

is a small subtree for each $x \in [0,1]$. Notice that “$\rho_{R(T,\Psi,x)}(t) = 0$” is a Σ^1_1 statement because each $\mathcal{I}_n$ is Σ^1_1. Hence $\{T_x : x \in [0,1]\}$ is a Σ^1_1 set.

Since the relation $\prec$ defined on $\{T_x : x \in [0,1]\}$ is also Σ^1_1 by Lemma 2.5, it follows from the Kunen-Martin Theorem [5] and Lemma 2.4 that there is some $\alpha \in \omega_1$ such that the rank of T_x is less than α for each $x \in [0,1]$. Use Lemma 2.6 to find $W \subseteq T$, a well founded full subtree of T, of standard rank α such that if $W' \subseteq W$ is any large subtree then the rank of W' is α. Observe that Ψ is an approximating function on $B(T) \cap W$ such that $H^r(\Psi(t)) < \theta$ for any $t \in W \cap B(T)$. It follows from Lemma 2.1 that there is some $x \in [0,1]$ such that $R(W, \Psi, x)$ is a large subtree of W. It follows that the rank of $R(W, \Psi, x)$ is at least α and this is a contradiction because it implies that the rank of T_x is at least α. ■

The reasonable conjecture at this point is that the conclusion of Theorem 3.1 holds for the countable support iteration of the partial order $\mathbb{P}(\mathcal{I})$. A proof of this would require modifying the preservation technology of Judah-Shelah [4, 1] which was originally developed to show that certain iterations preserve that the ground model reals have positive Lebesgue measure.

4. The Relation Ξ

Sets with positive r-Hausdorff capacity may have measure zero but this type of set will not play an important role in the following discussion. One would like to be able to say that if $\lambda(X) > 0$ then $H^r(X)$ can be calculated from $\lambda(X)$ or, at the very least, one would might hope for some relationship between $H^r(X)$ and $\lambda(X)$. There are easy counterexamples to this though. Let X be such that $H^r(X) = h > 0$ and $\lambda(X) = 0$ and then $\lambda(X \cup [0,a]) = a$ and note that there is obviously no connection between $H^r(X \cup [0,a])$ and a when a is much smaller than h. This sort of example is eliminated by introducing a relation on sets which, roughly speaking, calculates the infimum of $H^r(X \setminus Z)$ as Z ranges over sets of small measure. The relation, which is introduced in the next definition, extends this to all dimensions.

Definition 4.1. *If X and Y are subsets of $[0,1]$ then define the relation $\Xi_{\delta,\epsilon}(X,Y)$ to hold if and only if for every set Z, if $\lambda(Z) < \epsilon$ then $H^r(X \cap Y \setminus Z) > H^r(Y) - \delta$. If X and Y are subsets of $[0,1]^{n+1}$ then define the relation $\Xi_{\delta,\epsilon}(X,Y)$ to hold if and only if*

$$\Xi_{\delta,\epsilon}(\{x \in \pi_1(Y) : \Xi_{\delta,\epsilon}(X_x, Y_x)\}, \pi_1(Y)).$$

The relation $\Xi_{\delta,\epsilon}$ on $[0,1]^n$ can be considered as a crude substitute for an integral when $n > 1$. In fact, one might be tempted to define a better approximation to an integral in the following way. Define $\Xi'_\epsilon(A) = \inf\{H^r(A \setminus Z) : Z \subseteq [0,1] \text{ and } \lambda(Z) < \epsilon\}$ for $A \subseteq [0,1]$. If $A \subseteq [0,1]^{n+1}$ then define

$$\Xi'_\epsilon(A) = \sup\{\delta : \Xi'_\epsilon(\{x \in [0,1] : \Xi'_\epsilon(A_x) \geq \delta\}) \geq \delta\}$$

by induction on n. Notice however that the inequality $\Xi'_\epsilon(A) > \Xi'_\epsilon(B) - \delta$ implies, but is not equivalent to $\Xi_{\delta,\epsilon}(A,B)$ even if $A \subseteq B$. The point is that if X and Y

are subsets of $[0,1]^{n+1}$ and $x \in \pi_1(Y)$ then it is possible that $\Xi_{\delta,\epsilon}(X_x, Y_x)$ holds even though $x \notin \pi_1(X)$.

It is now possible to provide motivation for much the following technical material. The idea is that one would like to be able to use Definition 4.1 to define the ideals used to construct the partial orders as follows: Given a closed set $C \subseteq [0,1]$, a continuous function $f : C \to [0,1]$ and $\delta > 0$ let $X_{C,f,\delta}$ consist of all finite unions of rational intervals U such that for all $\epsilon > 0$ the relation $\Xi_{\delta,\epsilon}(f^{-1}U, C))$ fails and, then, let the ideal be generated by all sets $X_{C,f,\delta}$. The problem with this definition is that it is not clear that the ideal so defined is Σ^1_1 The material leading up to the more complicated Definition 5.2 is intended to rectify this problem. This section begins this process by collecting some facts about the Ξ relation.

Lemma 4.1. *If A and B are subsets of $[0,1]^n$ and $\Xi_{\delta,\epsilon}(A,B)$ holds then for any $Z \subseteq [0,1]^n$ such that*

$$\lambda(Z) < \left(\frac{\epsilon}{2}\right)^n$$

$\Xi_{\delta,\epsilon/2}(A \setminus Z, B)$ also holds.

Proof: This is easily proved using induction on n and a simple application of Fubini's Theorem. ■

The next two lemmas show that the relation Ξ could have been defined from the top down rather than from the bottom up.

Lemma 4.2. *If $d \geq 1$, A and B are subsets of $[0,1]^{d+1}$ and $\Xi_{\delta,\epsilon}(A', \pi_d(B))$ holds and $\Xi_{\delta,\epsilon}(A_x, B_x)$ holds for each $x \in A'$ then $\Xi_{\delta,\epsilon}(A,B)$ holds as well.*

Proof: Proceed by induction on d noting that the case $d = 1$ follows from the definition. ■

The proofs of the next four lemmas are easy and left to the reader.

Lemma 4.3. *If $d \geq 1$, A and B are subsets of $[0,1]^{d+1}$ and $\Xi_{\delta,\epsilon}(A,B)$ holds then so does $\Xi_{\delta,\epsilon}(\{x \in \pi_d(B) : \Xi_{\delta,\epsilon}(A_x, B_x)\}, \pi_d(B)\}$.*

Lemma 4.4. *If A and B are closed subsets of $[0,1]^{d+1}$, $\delta > 0$ and $\epsilon > 0$ then*

$$\{x \in \pi_d(B) : \Xi_{\delta,\epsilon}(A_x, B_x)\}$$

is a Borel set.

Lemma 4.5. *If $\Xi_{\delta,\epsilon}(A,B)$ holds and $A \subseteq B$ then $\Xi_{\delta,\epsilon}(A,A)$ holds as well.*

Lemma 4.6. *If $\Xi_{\delta,\epsilon}(A,B)$ holds, $\delta' > \delta$ and $\epsilon' < \epsilon$ then $\Xi_{\delta',\epsilon}(A,B)$ and $\Xi_{\delta,\epsilon'}(A,B)$ hold as well.*

Lemma 4.7. *If ϵ, δ_1 and δ_2 are greater than 0, $\Xi_{\delta_1,\epsilon}(A,B)\}$ and $\Xi_{\delta_2,\epsilon}(B,C)\}$ both hold and $B \subseteq C$ then $\Xi_{\delta_1+\delta_2,\epsilon}(A \cap B), C\}$ also holds.*

Lemma 4.8. *If $D \subseteq [0,1]$ is a set such that $\Xi_{\delta,\epsilon}(D,D)$ holds and $\delta < H^r(D)$ then there is some $\bar{\epsilon} > 0$ such that $\Xi_{\delta-\bar{\epsilon},\epsilon/2}(D,D)$ holds as well.*

Proof: Let $\bar{\epsilon} > 0$ be such that $\bar{\epsilon} < \min\{H^r(D) - \delta, r^{\frac{1}{1-r}}, \epsilon\}$. Suppose that that $\lambda(Z) < \epsilon/2$ and that $\{I_i : i \in \omega\}$ is a cover of $D \setminus Z$. By taking a tail of the sequence it is possible to find $i_0 \in \omega$ such that $\sum_{i=i_0+1}^{\infty} \lambda(I_i) < \bar{\epsilon}/2 \leq \sum_{i=i_0}^{\infty} \lambda(I_i)$ because, if $\sum_{i=0}^{\infty} \lambda(I_i) \leq \bar{\epsilon}/2 < \epsilon/2$ then $\lambda(Z \cup (\bigcup_{i\in\omega} I_i)) < \epsilon$ and so $Z \cup (\bigcup_{i\in\omega} I_i)$ can not possibly be a cover of D. Let J be an initial subinterval of I_{i_0} such that $\sum_{i=i_0+1}^{\infty} \lambda(I_i) + \lambda(J) = \bar{\epsilon}/2$ and note that $\lambda(J) > 0$. It follows that $\lambda(Z \cup J \cup \bigcup_{i>i_0} I_i) < \epsilon$ and so $\lambda(I_{i_0} \setminus J)^r + \sum_{i\in i_0} \lambda(I_i)^r > H^r(D) - \delta$ and hence, using the fact that $r < 1$,

$$\lambda(I_{i_0} \setminus J)^r + \lambda(J) + \sum_{i\neq i_0} \lambda(I_i)^r \geq$$

$$\lambda(I_{i_0} \setminus J)^r + \lambda(J) + \sum_{i\in i_0} \lambda(I_i)^r + \sum_{i=i_0+1}^{\infty} \lambda(I_i) > H^r(D) - \delta + \bar{\epsilon}/2$$

Now notice that

$$\sum_{i\in\omega} \lambda(I_i)^r = \lambda(I_{i_0} \setminus J)^r + \lambda(J) + \sum_{i\neq i_0} \lambda(I_i)^r - (\lambda(I_{i_0} \setminus J)^r + \lambda(J) - \lambda(I_{i_0})^r)$$

$$> H^r(D) - \delta + \bar{\epsilon}/2 - (\lambda(I_{i_0} \setminus J)^r + \lambda(J) - \lambda(I_{i_0})^r).$$

Hence all that has to be shown is that $\lambda(I_{i_0} \setminus J)^r + \lambda(J) - \lambda(I_{i_0})^r \leq 0$.

To see this, define for $a > 0$

$$F_a(x) = (a-x)^r + x - a^r$$

and observe that $\frac{d}{dx}F_a(x) = 1 - \frac{r}{(a-x)^{1-r}}$ and notice that this is negative if $x < a < \bar{\epsilon}$. Moreover, $F_a(0) = 0$ and so F_a is negative on the interval $(0,a)$ if $a < \bar{\epsilon}$. Because $0 < \lambda(J) < \lambda(I_{i_0}) < \bar{\epsilon} < r^{\frac{1}{1-r}}$ it follows that if $\lambda(J) < \lambda(I_{i_0})$ then $\lambda(I_{i_0} \setminus J)^r + \lambda(J) - \lambda(I_{i_0})^r = F_{\lambda(I_{i_0})}(\lambda(J)) \leq 0$. If $\lambda(J) \geq \lambda(I_{i_0})$ then the result is obvious. ■

Definition 4.2. *A subset $X \subseteq [0,1]$ will be said to be elementary if and only if*

$$X = [p_0, q_0] \cup [p_1, q_1] \cup \ldots \cup [p_k, q_k]$$

where p_i and q_i are rational numbers such that $p_i < q_i < p_{i+1}$ for each $i \in k$. A subset $X \subseteq [0,1]^{n+1}$ is elementary if and only if there is an elementary set $[p_0, q_0] \cup [p_1, q_1] \cup \ldots \cup [p_k, q_k] \subseteq [0,1]$ such that

$$X = \bigcup_{i=0}^{k} [p_i, q_i] \times X_i$$

and each $X_i \subseteq [0,1]^n$ is elementary.

Lemma 4.9. *If $U \subseteq [0,1]^n$ is open and $X \subseteq [0,1]^n$ is closed then $\Xi_{\delta,\epsilon}(U,X)$ if and only if for every $\bar{\epsilon} < \epsilon$ there is an elementary $Y \subseteq U$ such that $\Xi_{\delta,\bar{\epsilon}}(Y,X)$.*

Proof: To begin, suppose that for every $\bar{\epsilon} < \epsilon$ there is an elementary $Y \subseteq U$ such that $\Xi_{\delta,\epsilon}(Y,X)$. Let $Y_k \subseteq U$ be such that $\Xi_{\delta,\frac{k\epsilon}{k+1}}(Y_k,X)$ and let $Y = \cup_{k>0} Y_k$. It will be shown by induction on n that $\Xi_{\delta,\epsilon}(U,X)$.

In particular, it will be shown by induction on n that if $Y_m \subseteq [0,1]^n$ are sets such that

- each Y_m is an elementary subset of U
- $\Xi_{\delta,\epsilon(m)}(Y_m,X)$
- $Y_m \subseteq Y_{m+1}$
- $\epsilon(m) \leq \epsilon(m+1)$
- $\lim_{m\to\infty} \epsilon(m) = \epsilon$

then $\Xi_{\delta,\epsilon}(\cup_{m\in\omega} Y_m, X)$. To begin the induction note that the case $n = 1$ is easy. Now suppose that the assertion has been established for n and that $\epsilon(m) > 0$ and X and Y_m are subsets of $[0,1]^{n+1}$ for $m \in \omega$. Suppose that $Z \subseteq [0,1]$ is such that $\lambda(Z) < \epsilon$ and $A = \{x \in \pi_1(X) : \Xi_{\delta,\epsilon}(U_x, X_x)\}$. Note that A is measurable by the induction hypothesis and Lemma 4.4 and $H^r(A \setminus Z \leq H^r(\pi_1(X)) - \delta$ and define

$$W_m = \{x \in \pi_1(X) : \Xi_{\delta,\epsilon(m)}((Y_m)_x, X_x)\}$$

for each $m \in \omega$. Note that if $\epsilon(m) > \lambda(Z)$ then $\lambda(W_m \setminus A) \geq \epsilon(m) - \lambda(Z)$ because, otherwise, $\lambda(Z \cup (W_m \setminus A)) < \epsilon(m)$ and so $H^r(W_m \setminus ((W_m \setminus A) \cup Z)) > H^r(\pi_1(X)) - \delta$ contradicting that $W_m \setminus ((W_m \setminus A) \cup Z) \subseteq A \setminus Z$. Let j be such that $\epsilon(j) > \lambda(Z)$. Then $\{W_m \setminus A : m > j\}$ is a family of measurable sets — the measurability follows from the measurability of A and Lemma 4.4 and the fact that each W_m is elementary, and hence, closed — each of measure at least $\epsilon(j) - \lambda(Z) > 0$. Hence there is some $x \in [0,1]$ such that there are infinitely many $m \in \omega$ such that $x \in W_m \setminus A$. Therefore are there infinitely many $m \in \omega$ such that $\Xi_{\delta,\epsilon(m)}((Y_m)_x, X_x)$ and, by the induction hypothesis, it follows that $\Xi_{\delta,\epsilon}(U_x, X_x)$ holds because $(Y_m)_x \subseteq (Y_{m+1})_x \subseteq U_x$, each $(Y_m)_x$ is elementary. This yields a contradiction.

Conversely, suppose that $\Xi_{\delta,\epsilon}(U,X)$ and let $\bar{\epsilon} < \epsilon$. Proceed by induction on n. The case $n = 1$ is easy since Y can be chosen to be a finite union of intervals such that $\lambda(X \setminus Y) < \epsilon - \bar{\epsilon}$. Therefore suppose that $n \geq 1$, $U \subseteq [0,1]^{n+1}$ and $X \subseteq [0,1]^{n+1}$. Using the induction hypothesis, it follows that for each $x \in \pi_1(X)$, if $\Xi_{\delta,\epsilon}(U_x, X_x)$ then there is an elementary set $Y_x \subseteq U_x$ such that $Xi_{\delta,\bar{\epsilon}}(Y_x, X_x)$. Moreover, since U is open it follows that if $Y_x \neq \emptyset$ then there is an open interval J_x containing x such that $J_x \times Y_x \subseteq U$. If $Y_x = \emptyset$ then $H^r(\emptyset) > H^r(X_x) - \delta$ and there is an open cover $X \subseteq \cup I_i$ such that $\sum \lambda(I_i)^r < \delta$ since X is closed there is an interval J_x containing x

such that $\Xi_{\delta,\epsilon}(\emptyset, X_z)$ holds for all $z \in J_x$. Since $\Xi_{\delta,\epsilon}(U, X)$ holds it must be the case that

$$\Xi_{\delta,\epsilon}(\cup\{J_x : \Xi_{\delta,\bar{\epsilon}}(Y_x, X_x)\}, \pi_1(X))$$

must hold. From the case $n = 1$ it is possible to find an elementary set $Z \subseteq \cup\{J_x : \Xi_{\delta,\epsilon}(Y_x, X_x)\}$ such that $\Xi_{\delta,\bar{\epsilon}+(\epsilon-\bar{\epsilon})/2}(Z, \pi_1(X))$. Since Z can be covered by finitely many intervals J_x it is possible to obtain $Z' = \cup_{i\in j} J_i$ such that

- $\lambda(Z \setminus Z') < (\epsilon - \bar{\epsilon})/2$
- $J_i \cap J_{i'} = \emptyset$ if $i \neq i'$
- for each $i \in j$ there is some $x(i)$ such that $J_i \subseteq J_{x(i)}$

It follows that $\Xi_{\delta,\bar{\epsilon}}(Z', \pi_1(X))$ so let $Y = \cup_{i\in j} J_i \times Y_{x(i)}$. ■

Corollary 4.1. *If U is open and X is closed then the relation $\Xi_{\epsilon,\delta}(U, X)$ is Borel.*

Proof: It follows from Lemma 4.9 that $\Xi_{\epsilon,\delta}(U, X)$ holds if and only if

$$(\forall \bar{\epsilon} < \epsilon)(\exists Y)(Y \text{ is elementary and } \Xi_{\bar{\epsilon},\delta}(Y, X)$$

and from Lemma 4.4 it follows that $\Xi_{\bar{\epsilon},\delta}(Y, X)$ is a Borel statement because Y, being elementary, is closed. ■

Definition 4.3. *If X and Y are subsets of $[0,1]$ then define the relation $\Xi^*_{\delta,\epsilon}(X, Y)$ to hold if and only if for every elementary set Z, if $\lambda(Z) < \epsilon$ then $H^r(X \cap Y \setminus Z) > H^r(Y) - \delta$. If X and Y are subsets of $[0,1]^{n+1}$ then define the relation $\Xi^*_{\delta,\epsilon}(X, Y)$ to hold if and only if*

$$\Xi^*_{\delta,\epsilon}(\{x \in \pi_1(X) : \Xi^*_{\delta,\epsilon}(X_x, Y_x)\}, \pi_1(Y)).$$

Lemma 4.10. *If X and Y are elementary subsets of $[0,1]^n$ then $\Xi_{\delta,\epsilon}(X, Y)$ holds if and only if $\Xi^*_{\delta,\epsilon}(X, Y)$ holds.*

Proof: One direction is clear. For the other, proceed by induction on n. If $n = 1$ then suppose that $\Xi^*_{\delta,\epsilon}(X, Y)$ holds and that $\lambda(Z) < \epsilon$ is such that $H^r(X \cap Y \setminus Z) \leq H^r(Y) - \delta$. It will be shown that there is an elementary Z' such that $\lambda(Z') < \epsilon$ and $H^r(X \setminus Z') \leq H^r(X \setminus Z)$. This clearly suffices.

The existence of Z' will be established by induction on the number of connected components of $X \cap Y$. If $X \cap Y = [a, b]$ is an interval then $H^r(X \cap Y \setminus Z) \geq (b - a - \lambda(Z))^r$ because, if $\{I_i\}_{i\in\omega}$ is a cover of $X \cap Y \setminus Z$ then $\sum_{i=0}^{\infty} \lambda(I_i) \geq b - a - \lambda(Z)$ and hence, since $r < 1$, $\sum_{i=0}^{\infty} \lambda(I_i)^r \geq (\sum_{i=0}^{\infty} \lambda(I_i))^r \geq (b - a - \lambda(Z))^r$. Hence Z' can be chosen to be a sufficiently small rational interval containing $[a, a + \lambda(Z)]$ because then $H^r(X \cap Y \setminus Z') \leq (b - a - \lambda(Z))^r \leq H^r(X \setminus Z)$. Now suppose that $X \cap Y = [a_0, b_0] \cup [a_1, b_1] \cup \ldots \cup [a_k, b_k]$ where $a_i < b_i < a_{i+1} < b_{i+1}$ for each $i \in k$. It is possible to choose open sets $\{U_i^j\}_{i\in\omega}$ such that $X \cap Y \setminus Z \subseteq \cup_{i\in\omega} U_i^j$

and $\sum_{i=0}^{\infty}\lambda(U_i^j)^r < H^r(X\cap Y\setminus Z)+\frac{1}{j+1}$ for each $j\in\omega$. If $j\in\omega$ is such that $(b_0,a_1)\not\subseteq\cup_{i\in\omega}U_i^j$ then it may as well be assumed that $U_i^j\cap U_{i'}^j=\emptyset$ if $U_i^j\cap[a_0,b_0]\neq\emptyset$ and $U_{i'}^j\cap[a_m,b_m]\neq\emptyset$ for some $m>0$. Hence, if there are infinitely many $j\in\omega$ such that $(b_0,a_1)\not\subseteq\cup_{i\in\omega}U_i^j$ then it that follows that $H^r(X\setminus Z)=$

$$H^r([a_0,b_0]\setminus Z)+H^r([a_1,b_1]\cup[a_2,b_2]\cup\ldots\cup[a_k,b_k]\setminus Z)$$

and the induction hypothesis can be used to find elementary Z_0 and Z_1 such that

- $H^r([a_0,b_0]\setminus Z_0)\leq H^r([a_0,b_0]\setminus Z)$
- $H^r(([a_1,b_1]\cup[a_2,b_2]\cup\ldots\cup[a_k,b_k])\setminus Z_1)\leq H^r(([a_1,b_1]\cup[a_2,b_2]\cup\ldots\cup[a_k,b_k])\setminus Z)$
- $\lambda(Z_0)<\epsilon_0$
- $\lambda(Z_1)<\epsilon_1$

where ϵ_0 and ϵ_1 are chosen to be positive so that $\epsilon_0+\epsilon_1\leq\epsilon$ and $\lambda([a_0,b_0]\cap Z)<\epsilon_0$ and $\lambda(([a_1,b_1]\cup[a_2,b_2]\cup\ldots\cup[a_k,b_k])\cap Z)<\epsilon_1$.

Hence it may be assumed that for all but finitely many $j\in\omega$ there is some $b(j)\in\omega$ such that $(b_0,a_1)\subseteq U_{b(j)}^j=(x_j,y_j)$. By restricting attention to an infinite subsequence it may also be assumed that there is some interval $[x,y]$ such that $\lim_{j\to\infty}x_j=x$ and $\lim_{j\to\infty}y_j=y$. It follows that

$$H^r(X\cap Y\setminus Z)=H^r([a_0,x]\setminus Z)+(y-x)^r+H^r(([y,b_1]\cup[a_2,b_2]\cup\ldots\cup[a_k,b_k])\setminus Z)$$

and so the induction hypothesis can be used to find elementary Z_0, Z_1 and Z_2 such that

- $H^r([a_0,x]\setminus Z_0)\leq H^r([a_0,x]\setminus Z)$
- $H^r(([y,b_1]\cup[a_2,b_2]\cup\ldots\cup[a_k,b_k])\setminus Z_1)\leq H^r([y,b_1]\cup[a_2,b_2]\cup\ldots\cup[a_k,b_k]\setminus Z)$
- $\lambda(Z_0)<\epsilon_0$
- $\lambda(Z_1)<\epsilon_1$
- $Z_2=J_x\cup J_y$ and $x\in J_x$ and $y\in J_y$

where ϵ_0, ϵ_1 and ϵ_2 are chosen to be positive so that $\epsilon_0+\epsilon_1+\epsilon_2\leq\epsilon$ $\lambda([a_0,x]\cap Z)<\epsilon_0$, $\lambda(([y,b_1]\cup[a_2,b_2]\cup\ldots\cup[a_k,b_k])\cap Z)<\epsilon_1$ and $\lambda([x,y]\cap Z)<\epsilon_2$. Let $Z=Z_0\cup Z_1\cup Z_2$.

Now suppose that the result has been established for n and that X and Y are elementary subsets of $[0,1]^{n+1}$ and that $\Xi_{\delta,\epsilon}(X,Y)$. In other words,

$$\Xi_{\delta,\epsilon}^*(\{x\in\pi_1(X):\Xi_{\delta,\epsilon}^*(X_x,Y_x)\},\pi_1(Y))$$

holds and, using the induction hypothesis this yields that

$$\Xi_{\delta,\epsilon}^*(\{x\in\pi_1(X):\Xi_{\delta,\epsilon}(X_x,Y_x)\},\pi_1(Y))$$

holds as well. To finish use the case $n=1$ noting that if X and Y are elementary then so is $\{x\in\pi_1(X):\Xi_{\delta,\epsilon}(X_x,Y_x)\}$. ■

Lemma 4.11. *If $K \subseteq [0,1]$ is closed then for all $\mu > 0$ there is a closed subset $K' \subseteq K$ such that $\lambda(K \setminus K') < \mu$ and for each $i \in \omega$ there is $\gamma > 0$ such that $\Xi_{\frac{1}{i+1},\gamma}(K', K')$.*

Proof: First, the following weaker statement will be established: If $K \subseteq [0,1]$ is closed then for all $\mu > 0$, $i \in \omega$ there is $\gamma > 0$ and a closed subset $K' \subseteq K$ such that $\lambda(K \setminus K') < \mu$ and $\Xi_{\frac{1}{i+1},\gamma}(K', K')$. To see this, suppose not and that K, μ and $i \in \omega$ provided a counterexample. Choose inductively open sets A_m such that $\lambda(A_m) < \mu/2^{i+1}$ and

$$H^r(K \setminus (\cup_{j \leq m} A_j)) \leq H^r(K \setminus (\cup_{j \in m} A_j)) - 1/i$$

for each $m \leq i+2$. If it is not possible to do this for some m then it follows that $\Xi_{\frac{1}{i+1},\frac{\mu}{2^{i+1}}}(K \setminus (\cup_{j\in m} A_j), K \setminus (\cup_{j \in m} A_j))$ holds and $\lambda(\cup_{j\in m} A_j) < \mu$. On the other hand, if the induction can be completed then the following inequalities hold:

1 $H^r(K) - \frac{1}{i+1} \geq H^r(K \setminus A_0)$

2 $H^r(K \setminus A_0) - \frac{1}{i+1} \geq H^r(K \setminus (A_0 \cup A_1))$

$\vdots$

m $H^r(K \setminus (\cup_{j \in i+2} A_j) - \frac{1}{i+1} \geq H^r(K \setminus (\cup_{j \leq i+2} A_j))$

and therefore $H^r(K) - \frac{i+2}{i+1} \geq H^r(K \setminus (\cup_{j \leq i+2} A_j)) \geq 0$ contradicting that $K \subseteq [0,1]$ implies that $H^r(K) \leq 1$.

Now use to choose inductively open sets U_i and numbers $\bar{\gamma}(i) > 0$ such that

- $\bar{\gamma}(0) = \mu$
- $\lambda(U_i) < \frac{\bar{\gamma}(j)}{2^{i-j+2}}$ for each $j \leq i \in \omega$
- $\Xi_{\frac{1}{i+1},\bar{\gamma}(i)}(K \setminus (\cup_{j \leq i} U_j), K \setminus (\cup_{j \leq i} U_j))$ holds.

Now let $K' = K \setminus (\cup_{j \in \omega} U_j)$ and note that $\lambda(K \setminus K') < \mu$. Moreover $\Xi_{\frac{1}{i+1},\bar{\gamma}(i)/2}(K', K')$ holds for each $i \in \omega$ because $\lambda(\cup_{m=i+1}^{\omega} U_m) < \bar{\gamma}(i)/2$ and hence,

$$\lambda(K \setminus (\cup_{j \leq i} U_j) \setminus K') < \bar{\gamma}(i)/2.$$

Since $\Xi_{\frac{1}{i+1},\bar{\gamma}(i)}(K \setminus (\cup_{j \leq i} U_j), K \setminus (\cup_{j \leq i} U_j))$ holds by construction, the result follows by setting $\gamma(i) = \bar{\gamma}(i)/2$. ∎

5. The Definition of the Ideals Associated with a Capacity

This section contains the definition of the ideals satisfying KP(r) which will be used to construct the partial orders. Most of the technical concepts have already been introduced but a few more are needed.

Definition 5.1. *A sequence $\{X_i : i \in \omega\}$ of subsets of $[0,1]^d$ will be said to be a normal family if*

- *each X_i is elementary*
- $X_{i+1} \subseteq X_i$
- $\lambda(\pi_n(X_i) \setminus \pi_n(X_{i+1})) < \frac{\lambda(\pi_n(X_i))}{2^{i+2}}$ *for* $n \leq d$
- *for each* $i \in \omega$ *there is* $\beta(i) > 0$ *such that* $\Xi_{\frac{1}{i+1},\beta(i)}(X_j, X_j)$ *holds for all* $j \geq i$.

The family $\{X_i : i \in \omega\}$ *will be said to be of dimension* d. *The function* β *will be called a witness to the normality of the family* $\{X_i : i \in \omega\}$.

Observe that the intersection of any normal family has positive measure. In fact, if $\{X_i : i \in \omega\}$ is a normal family and $X = \cap_{i \in \omega} X_i$ then

$$\lambda(\pi_n(X)) > \frac{(2^{i+1} - 1)\lambda(\pi_n(X_i))}{2^{i+1}}$$

for any $i \in \omega$ and $n \leq d$.

Definition 5.2. *Let* W_n *be the family of all sets* a *which are the interior of an elementary set* A *such that* $\lambda(A) < 2^{-n}$. *Let* $n \in \omega$ *and* $\delta > 0$. *Suppose that* $d \in \omega$, $\mathcal{C} = \{C_i\}_{i \in \omega}$ *is a normal family of dimension* d *and* $f : [0,1]^d \to [0,1]$ *is a continuous function. Define* $\mathfrak{X}(f, \mathcal{C}, \delta)$ *to be the set of all* $a \in W_n$ *such that for every* $\epsilon > 0$ *there are infinitely many* $i \in \omega$ *such that* $\Xi_{\delta,\epsilon}(f^{-1}a, C_i)$ *does not hold. The set* $\mathfrak{X}(f, \mathcal{C}, \delta)$ *will be said to be of dimension* d. *Define* $\mathcal{I}_n^r$ *to be the set of all sets* $Y \subseteq W_n$ *such that there are* $\delta > 0$, $m \geq 1$, *a normal family* $\mathcal{C}$ *of dimension* m *and a continuous function* $f : [0,1]^d \to [0,1]$ *such that* $Y \subseteq \mathfrak{X}(f, \mathcal{C}, \delta)$.

The ideals of Definition 5.2 are defined on the countable set W_n rather than on ω. Theorem 3.1 still applies of course.

Lemma 5.1. *Let* $0 < a < \frac{1}{2^{1/r}+2}$. *If* $A \subseteq [0, a]$ *and* $B \subseteq [1 - a, 1]$ *then* $H^r(A \cup B) = H^r(A) + H^r(B)$.

Proof: Noting that the hypothesis on a implies that $0 < (1 - 2a)^r - 2a^r$ it is possible to choose $\epsilon > 0$ such that $\epsilon < (1-2a)^r - 2a^r$. Since $H^r(A \cup B) \leq 2a^r$ it follows that if $A \cup B \subseteq \cup_{i \in \omega} I_i$ and $\sum_{i \in \omega} \lambda(I_i)^r < H^r(A \cup B) + \epsilon < 2a^r + (1 - 2a)^r - 2a^r$ then none of the intervals I_i contains $[a, 1 - a]$. Since $a < 1/2$ it may as well be assumed that none of the intervals contains $1/2$, or, in other words, that $\{i \in \omega : I_i \cap A \neq \emptyset\}$ is disjoint from $\{i \in \omega : I_i \cap B \neq \emptyset\}$. Hence $H^r(A \cup B) \geq H^r(A) + H^r(B)$. The result follows since $H^r(A \cup B) \leq H^r(A) + H^r(B)$ is true in general. ∎

Lemma 5.2. *Each set* $\mathcal{I}_n^r$ *is closed under finite unions.*

Proof: Let $\mathfrak{X}(f, \{C_i\}_{i\in\omega}, \mu)$ and $\mathfrak{X}(g, \{D_i\}_{i\in\omega}, \rho)$ be any two generators for $\mathcal{I}_n^r$ of dimension d_1 and d_2 respectively. Let d be greater than d_1 and d_2 and define $\bar{C}_i = C_i \times [0,1]^{d-d_1}$ and $\bar{D}_i = D_i \times [0,1]^{d-d_2}$. Define $\bar{f}(x_1, x_2, \ldots x_d) = f(x_1, x_2, \ldots, x_{d_1})$ and $\bar{g}(x_1, x_2, \ldots x_d) = g(x_1, x_2, \ldots, x_{d_2})$.

Next, let a be a rational number such that $0 < a < \frac{1}{2^{1/r}+2}$. Define $\psi_1 : [0,1] \to [0,a]$ by $\psi_1(x) = ax$ and define $\psi_2 : [0,1] \to [1-a, 1]$ by $\psi_2(x) = 1 - ax$. Let $\varphi_i : [0,1]^d \to [0,1]^d$ be defined by $\varphi_i(x_1, x_2, \ldots, x_d) = (\psi_i(x_1), x_2, \ldots, x_d)$ for $i \in \{1,2\}$. Let $B_i = \varphi_1(\bar{C}_i) \cup \varphi_2(\bar{D}_i)$. To see that $\{B_i\}_{i\in\omega}$ is a normal family notice that $\lambda(\pi_n(B_i) - \pi_n(B_{i+1})) =$

$$\int_0^a \varphi_1^{-1}\lambda((\pi_n(\bar{C}_i) - \pi_n(\bar{C}_{i+1}))_x)dx + \int_{1-a}^1 \varphi_1^{-1}\lambda((\pi_n(\bar{D}_i) - \pi_n(\bar{D}_{i+1}))_x)dx$$

and this evaluates to $\lambda(\pi_n(B_i))/2^{i+2}$. It must also be observed that if $\beta_1 : \omega \to (0,1)$ witnesses that $\{C_i\}_{i\in\omega}$ is normal and $\beta_2 : \omega \to (0,1)$ witnesses that $\{D_i\}_{i\in\omega}$ is normal then the function $\beta : \omega \to (0,1)$ defined by $\beta(i) = \min\{a\beta_1(i), a\beta_2(i)\}$ witnesses the normality of $\{B_i\}_{i\in\omega}$. This uses Lemma 5.1. Finally, let h be any continuous extension of $(\bar{f} \circ \varphi_1^{-1}) \cup (\bar{g} \circ \varphi_2^{-1})$ and let $\delta = \min\{a^r\mu, a^r\rho\}$. Clearly $\mathfrak{X}(h, \{B_i\}_{i\in\omega}, \delta) \in \mathcal{I}_n^r$.

It will be shown that $\mathfrak{X}(f, \{C_i\}_{i\in\omega}, \mu)$ is a subset of $\mathfrak{X}(h, \{B_i\}_{i\in\omega}, \delta)$, the proof for $\mathfrak{X}(g, \{D_i\}_{i\in\omega}, \rho)$ being similar. Let $b \in \mathfrak{X}(f, \{C_i\}_{i\in\omega}, \mu)$. This means that for every $\epsilon > 0$ there are infinitely many $i \in \omega$ such that $\Xi_{\mu,\epsilon}((f^{-1}b), C_i)$ fails to hold. Let ϵ and i be fixed such that $\Xi_{\mu,\epsilon}((f^{-1}b), C_i)$ fails. Unraveling the definition of $\Xi_{\mu,\epsilon}$ reveals that

$$H^r(\{x \in \pi_1(C_i) : \Xi_{\mu,\epsilon}((f^{-1}b)_x, (C_i)_x)\} \setminus Z) \leq H^r(\pi_1(C_i)) - \mu$$

for some set Z such that $\lambda(Z) < \epsilon$. From the definition of φ_1 and $\bar{f}$ it follows that

$$H^r(\{x \in \pi_1(\bar{C}_i) : \Xi_{\mu,\epsilon}((h^{-1}b)_{\psi_1(x)}, (B_i)_{\psi_1(x)})\} \setminus Z) \leq H^r(\pi_1(\bar{C}_i)) - \mu$$

and so, observing that $H^r(\psi_1(A)) = a^r H^r(A)$ for any $A \subseteq [0,1]$, it follows that

$$H^r(\{x \in \pi_1(\varphi_1(\bar{C}_i)) : \Xi_{\mu,\epsilon}((h^{-1}b)_x, (B_i)_x)\} \setminus Z') =$$

$$H^r(\{\psi_1(x) : x \in \pi_1(\bar{C}_i) \text{ and } \Xi_{\mu,\epsilon}((h^{-1}b)_{\psi_1(x)}, (B_i)_{\psi_1(x)})\} \setminus Z')$$

$$\leq a^r(H^r(\pi_1(\bar{C}_i)) - \mu) = H^r(\pi_1(\varphi_1(\bar{C}_i))) - \delta$$

where Z' is the image of Z under ψ_1. Notice that $\lambda(Z') = a\lambda(Z) < \lambda(Z) < \epsilon$.

The next thing to notice is that

$$\{x \in \pi_1(\varphi_1(\bar{C}_i)) : \Xi_{\mu,\epsilon}((h^{-1}b)_x, (B_i)_x)\} \supseteq \{x \in \pi_1(\varphi_1(\bar{C}_i)) : \Xi_{\delta,\epsilon}((h^{-1}b)_x, (B_i)_x)\}$$

because $\delta < \mu$. From Lemma 5.1 it follows that $H^r(\pi_1(B_i)) = H^r(\pi_1(\varphi_1(\bar{C}_i)) + H^r(\pi_1(\varphi_2(\bar{D}_i))$. Therefore,

$$H^r(\{x \in \pi_1(B_i) : \Xi_{\delta,\epsilon}((h^{-1}b)_x, (B_i)_x)\} \setminus Z') \leq$$

$$H^r(\{x \in \pi_1(\varphi_1(\bar{C}_i)) : \Xi_{\delta,\epsilon}((h^{-1}b)_x, (\bar{C}_i)_x)\}) + H^r(\pi_1(\varphi_2(\bar{D}_i)) \leq$$

$$H^r(\{x \in \pi_1(\varphi_1(\bar{C}_i)) : \Xi_{\mu,\epsilon}((h^{-1}b)_x, (\bar{C}_i)_x)\}) + H^r(\pi_1(\varphi_2(\bar{D}_i)) \leq$$

$$H^r(\pi_1(\varphi_1(\bar{C}_i))) - \delta + H^r(\pi_1(\varphi_2(\bar{D}_i))) = H^r(\pi_1(B_i)) - \delta$$

or, in other words $\Xi_{\delta,\epsilon}(h^{-1}b, B_i)$ fails provided that $\Xi_{\mu,\epsilon}(f^{-1}b, C_i)$ fails. Since for every $\epsilon > 0$ there are infinitely many $i \in \omega$ such that $\Xi_{\mu,\epsilon}(f^{-1}b, C_i)$ fails it follows that $b \in \mathfrak{X}(h, \{B_i\}_{i\in\omega}, \delta)$. ■

Lemma 5.3. *If the parameters f, $\mathcal{C}$ and δ are given then the statement "$a \in \mathfrak{X}(f, \mathcal{C}, \delta)$" is arithmetic.*

Proof: Let $\mathcal{C} = \{C_i\}_{i\in\omega}$ be a normal family of dimension n. From Definition 5.2 it follows that $a \in \mathfrak{X}(f, \mathcal{C}, \delta)$ if and only if for every $\epsilon > 0$ there are infinitely many $i \in \omega$ such that $\Xi_{\delta,\epsilon}(f^{-1}a, C_i)$ does not hold. Since a is open and f is continuous it follows from Lemma 4.9 that $a \in \mathfrak{X}(f, \mathcal{C}, \delta)$ if and only if

$$(\forall\epsilon > 0)(\forall m \in \omega)(\exists i > m)(\exists\bar{\epsilon} < \epsilon)(\forall Y)(Y \text{ is elementary and}$$
$$Y \subseteq f^{-1}a \Rightarrow \neg\Xi_{\delta,\bar{\epsilon}}(Y, C_i))$$

and, moreover, $\Xi_{\delta,\bar{\epsilon}}(Y, C_i)$ is equivalent to $\Xi^*_{\delta,\bar{\epsilon}}(Y, C_i)$ when Y and C_i are elementary by Lemma 4.10.

Hence it suffices to show that the statement $\Xi^*_{\delta,\bar{\epsilon}}(Y, C_i)$ is arithmetic. Proceed by induction on n. Notice that the statements $\lambda(Z) < \alpha$ and $H^r(Z) > \alpha$ are arithmetic for elementary sets Z. The case $n = 1$ follows immediately and the induction is carried through because of the elementarity of Y and C_i. ■

Lemma 5.4. *If the parameters β and $\mathcal{C}$ are given then the statement "β witnesses the normality of $\mathcal{C}$" is arithmetic.*

Proof: This follows from Lemma 4.10 and the definition of a normal family because it has already been observed in the proof of Lemma 5.3 that the statement $\Xi^*_{\delta,\bar{\epsilon}}(Y, C_i)$ is arithmetic. ■

Corollary 5.1. *The ideals $\mathcal{I}^r_n$ are all Σ^1_1 ideals.*

Proof: From Definition 5.2 it follows that $Y \in \mathcal{I}^r_n$ if and only if there are $\delta > 0$, $m \geq 1$, a normal family $\mathcal{C}$ of dimension m and a continuous function $f : [0,1]^m \to [0,1]$ such that $Y \subseteq \mathfrak{X}(f, \mathcal{C}, \delta)$. Now apply Lemma 5.3 noting that the existence of a normal family can be expressed with a Σ^1_1 statement. ■

Lemma 5.5. *If $A \subseteq B \subseteq [0,1]^d$ and $X \subseteq B$ are such that*

- $\lambda(\pi_n(B) \setminus \pi_n(A)) < (\frac{\epsilon}{d+1})^n$ *for each $n \leq d$*
- $\Xi_{\delta,\epsilon}(X, B)$

then $\Xi_{\delta,\frac{\epsilon}{d+1}}(X, A)$.

Proof: Proceed by induction in d. If $d = 1$ and $\lambda(Z) < \epsilon/2$ then $\lambda((B \setminus A) \cup Z) < \epsilon$. Hence $H^r(X \cap A \setminus ((B \setminus A) \cup Z)) > H^r(B) - \delta$. Since $X \cap A \setminus ((B \setminus A) \cup Z) = X \cap A \setminus Z$ this suffices.

Suppose the lemma is true for d and that $A \subseteq B \subseteq [0,1]^{d+1}$. Let

$$S_n = \{x \in [0,1] : \lambda((\pi_n(B_x) \setminus \pi_n(A_x)) \geq (\frac{\epsilon}{d+2})^n\}$$

for each $n \leq d$. Since $\lambda((\pi_{n+1}(B) \setminus \pi_{n+1}(A)) < (\frac{\epsilon}{d+2})^{n+1}$ it follows that

$$\lambda(S_n) < \epsilon/(d+2)$$

for each $n \leq d$. If Z is such that $\lambda(Z) < \epsilon/(d+2)$ define $Y(Z) = Z \cup (\cup_{n=1}^{d} S_n) \cup (\pi_1(B) \setminus \pi_1(A))$ and note that that $\lambda(Y(Z)) < \epsilon$. Hence

$$H^r(\{x \in \pi_1(B) : \Xi_{\delta,\epsilon}(X_x, B_x)\} \setminus (Y(Z)) > H^r(\pi_1(B)) - \delta$$

and, moreover, if $\Xi_{\delta,\epsilon}(X_x, B_x)$ holds and $x \notin Y(Z)$ then A_x, B_x and X_x satisfy the hypothesis of the lemma for d and, furthermore, $x \in \pi_1(A)$. Therefore

$$H^r(\{x \in \pi_1(A) : \Xi_{\delta,\frac{\epsilon}{d+1}}(X_x, A_x)\} \setminus Z) > H^r(\pi_1(B)) - \delta > H^r(\pi_1(A)) - \delta$$

and this implies that

$$H^r(\{x \in \pi_1(A) : \Xi_{\delta,\frac{\epsilon}{d+2}}(X_x, A_x)\} \setminus Z) > H^r(\pi_1(A)) - \delta$$

Since Z was arbitrary, this means that $\Xi_{\delta,\frac{\epsilon}{d+2}}(X, A)$ holds. ■

Corollary 5.2. *If* $\{C_i\}_{i\in\omega}$ *is a normal family of dimension d then the following are equivalent:*

(1) *There is* $\epsilon > 0$ *such that* $\Xi_{\delta,\epsilon}(X, C_i)$ *holds for all but finitely many* $i \in \omega$.
(2) *There is* $\epsilon > 0$ *such that* $\Xi_{\delta,\epsilon}(X, C_i)$ *holds for infinitely many* $i \in \omega$.
(3) *There is* $\epsilon > 0$ *such that* $\Xi_{\delta,\epsilon}(X, C_i)$ *holds for some* $i \in \omega$ *such that*

$$\lambda(\pi_n(C_i) \setminus \pi_n(\cap_{j\in\omega} C_j)) < (\frac{\epsilon}{d+1})^n$$

for each $n \leq d$.

Proof: To get that (3) implies (1) use Lemma 5.5 noting that if $j > i$ then $\lambda(\pi_n(C_i) \setminus \pi_n(C_j)) < (\frac{\epsilon}{d+1})^n$ for each $n \leq d$ and so, $\Xi_{\delta,\frac{\epsilon}{d+1}}(X, C_j)$ holds. ■

Lemma 5.6. *Each of the ideals* $\mathcal{I}_n^r$ *of Definition* 5.2 *satisfies* $KP(r)$.

Proof: Suppose not. Then there is some $n \in \omega$ such that the ideal $\mathcal{I}_n^r$ does not satisfy KP(r). This means that there exist

- $\theta > 0$
- $X \in \mathcal{I}^+$
- a function F from X to the Borel subsets of $[0,1]$ satisfying that $H^r(F(x)) \leq \theta$ for each $x \in X$
- $\epsilon > 0$

such that for every $Y \subseteq [0,1]$ and $Z \subseteq [0,1]$ such that

- $H^r(Y) \leq \theta$
- $\lambda(Z) < \epsilon$

it must be the case that $\{a \in X : y \notin F(a)\} \in \mathcal{I}$ for some $y \in [0,1] \setminus (Y \cup Z)$. Using Definition 5.2, it is possible to rephrase this as follows: For every $Y \subseteq [0,1]$ and $Z \subseteq [0,1]$ such that $H^r(Y) \leq \theta$ and $\lambda(Z) < \epsilon$ it must be that there is some $y \in [0,1] \setminus (Y \cup Z)$ and there are $\delta > 0$, $m \geq 1$, a normal family $\mathcal{C}$ of dimension m and a continuous function $f : [0,1]^m \to [0,1]$ such that $\{a \in X : y \notin F(a)\} \subseteq \mathfrak{X}(f, \mathcal{C}, \delta)$

Let $\mathcal{E}_m$ be the set of all elementary subsets of $[0,1]^m$ considered to have the discrete topology. It follows that $\prod_{m\in\omega} \mathcal{E}_m$ is homeomorphic to the irrationals. Let $\mathcal{N}_m$ be the subspace of $\prod_{m\in\omega} \mathcal{E}_m$ consisting of all ξ such that $\{\xi(n)\}_{n\in\omega}$ is a normal family and observe that, because it is a closed subspace of $\prod_\omega \mathcal{E}_m$, $\mathcal{N}_m$ is a Polish space. Let $\mathcal{C}([0,1]^m)$ be the space of continuous functions from $[0,1]^m$ to $[0,1]$ with the metric induced by the supremum norm. Let

$$\mathcal{P}_m = \mathcal{C}([0,1]^m) \times \mathcal{N}_m \times (0,1) \times (0,1)^\omega$$

and let $\mathcal{P} = \cup_{m\in\omega} \mathcal{P}_m$ and note that $\mathcal{P}$ is a Polish space. Let Ω be the set of all $(z, g, \xi, \delta, \beta) \in [0,1] \times \mathcal{P}$ such that $\{a \in X : z \notin F(a)\} \subseteq \mathfrak{X}(g, \{\xi(n)\}_{n\in\omega}, \delta)$ and the normality of $\{\xi(n)\}_{n\in\omega}$ is witnessed by β. Because X and F can be coded by reals, the definition of Ω together with Lemma 5.3 and Lemma 5.4 immediately establish that Ω is a Borel subset of the Polish space $[0,1] \times \mathcal{P}$.

It is therefore possible to appeal to the von Neumann Selection Theorem to find a measurable $\Phi : [0,1] \to \mathcal{P}$ such that the domain of Φ is the same as $\pi_1(\Omega)$ and $\Phi \subseteq \Omega$. If x is in the domain of Φ suppose that $\Phi(x) = (g, \xi, \delta, \beta)$ and define $d(x)$ to be the dimension of $\mathfrak{X}(g, \{\xi(n)\}_{n\in\omega}, \delta)$. Then define $\Phi_i^n(x) = \pi_n(\xi(i))$ for each $i \in \omega$ and define $\Phi_\omega^n(x) = \pi_n(\cap_{i\in\omega}\xi(i))$ — if $n > d(x)$ then $\pi_n(\xi(i)) = \xi(i)$. Since $\lim_{i\to\infty} \lambda(\Phi_i^n(x)) = \lambda(\Phi_\omega^n(x))$ for each x in the domain of Φ and $n \in \omega$, it is possible to apply Egerov's theorem countably many times to find a compact set $\bar{K}$ — which is the intersection of a nested sequence of closed sets obtained from the countably many applications of Egerov's theorem — such that

- $\Phi \restriction \bar{K}$ is continuous
- $\Phi_\alpha^n \restriction \bar{K}$ is continuous for each $\alpha \in \omega + 1$
- $\{\lambda(\Phi_i^n(x))\}_{i\in\omega}$ converges uniformly, with respect to the variable x, to $\lambda(\Phi_\omega^n(x))$ on $\bar{K}$
- $\lambda(\bar{K}) > \lambda(\pi(\Omega)) - \epsilon/4$.

Observe that if Z is such that $\lambda(Z) < \epsilon/2$ then $H^r(\bar{K} \setminus Z) > \theta$ because otherwise, it is possible to obtain a contradiction by setting $Y = \bar{K} \setminus Z$ in the definition of KP(r). Now use Lemma 4.11 to find a closed $K \subseteq \bar{K}$ such that $\lambda(\bar{K} \setminus K) < \epsilon/4$ and there exists $\gamma : \omega \to (0,1)$ such that $\Xi_{\frac{1}{i+1},\gamma(i)}(K, K)$ holds for all $i \in \omega$.

Next, the compactness of K implies that there is $m \in \omega$ such that $d(x) \in m$ for each $x \in K$. Furthermore there is $\delta > 0$ such that for every $x \in K$, if $\Phi(x) = (g, \xi, \delta', \beta)$ then $\delta' > \delta$. Since $H^r(K \setminus Z) > \theta$ for each $Z \subseteq [0,1]$ such that $\lambda(Z) < \epsilon/4$ it follows that, by shrinking δ if necessary, it may be assumed that $H^r(K) > \theta + \delta$. Yet another application of compactness yields a function $\beta : \omega \to (0,1)$ such that for each $x \in K$, if $\Phi(x) = (g, \xi, \delta, \beta_x)$ then $\beta_x(i) \geq \beta(i)$ for each $i \in \omega$.

Let $\tau_n = \int_K \lambda(\Phi^n_\omega(x))dx$ for $n \leq m$. Since $\{\Phi^m_i(x)\}_{i\in\omega}$ is a normal family for each x in the domain of Φ it follows from the remarks following Definition 5.1 that

$$\lambda(\Phi^n_i(x)) < \frac{2^{i+1}\lambda(\Phi^n_\omega(x))}{2^{i+1}-1}$$

for each $i \in \omega$ and $n \leq m$. Therefore,

$$\int_K \lambda(\Phi^n_i(x))dx < \frac{2^{i+1}\tau_n}{2^{i+1}-1}$$

and so it is possible to choose an open set L_i such that $K \subseteq L_i$ and

$$\lambda(L_i \setminus K) + \int_K \lambda(\Phi^n_i(x))dx < \frac{2^{i+1}\tau_n}{2^{i+1}-1}$$

for each $n \leq m$ and $H^r(L_i) < H^r(K) + \frac{1}{2i}$ and $\lambda(L_i \setminus K) < \frac{\gamma(i)}{2}$ for each $i \in \omega$. Next, using the continuity of Φ on K, choose a family $\{N_i\}_{i\in\omega}$ such that

- $N_i = [p^i_0, q^i_0] \cup \ldots \cup [p^i_{k(i)}, q^i_{k(i)}]$ is elementary for each i
- $K \cap [p^i_j, q^i_j] \neq \emptyset$ for each $i \in \omega$ and $j \leq k(i)$
- $K \subseteq N_i \subseteq L_{i+1}$
- $N_{i+1} \subseteq N_i$
- $\lambda(N_i \setminus K) < \lambda(K)/2^{i+1}$ (this uses the fact that $\lambda(K) > 0$)
- $\Phi^n_j(x) = \Phi^n_j(y)$ if $j \leq i$ and x and y belong to K and the same component of N_i

Let $C_i = \bigcup_{j=0}^{k(i+2)}[p^{i+2}_j, q^{i+2}_j] \times \Phi^m_{i+2}(z) \times [0,1]^{m-d(x)}$ for $i \in \omega$ where z is chosen arbitrarily from $[p^{i+2}_j, q^{i+2}_j] \cap K$ Then, let $C = \cap_{i\in\omega} C_i$. Observe that $\lambda(\pi_n(C)) = \tau_n$ for $n \leq m$.

Hence, in order to show that $\mathcal{C} = \{C_i\}_{i\in\omega}$ is an $n+1$-dimensional normal family, first observe that if $j \geq i \geq 1$ and $\lambda(Z) < \gamma(2i-1)$ then

$$H^r(N_j \setminus Z) \geq H^r(K \setminus Z) \geq H^r(K) - \frac{1}{2i} \geq H^r(L_i) - \frac{1}{2i} - \frac{1}{2i}$$

and the last expression is at least as large as $H^r(N_j) - \frac{1}{i}$. Hence $\Xi_{\frac{1}{i},\gamma(2i-1)}(N_j, N_j)$ holds for all $j \geq i \geq 1$. Now let $\beta^*(i) = \min\{\gamma(2(i+2)-1), \beta(i+2)\}$. Then $\Xi_{\frac{1}{i},\beta^*(i)}(N_j, N_j)$ holds for all $j \geq i \geq 1$ and so does $\Xi_{\frac{1}{i},\beta^*(i)}(\Phi^m_{j+2}(z), \Phi^m_{j+2}(z))$ because $\beta^*(i) \leq \beta(i+2) \leq \beta_z(i+2)$ for any $z \in K \cap N_{i+2}$. Therefore $\Xi_{\frac{1}{i},\beta^*(i)}(C_j, C_j)$ holds for all $j \geq i$. Hence, in order to show that $\mathcal{C}$ is a normal family it suffices to show that $\lambda(\pi_{n+1}(C_i) \setminus \pi_{n+1}(C_{i+1})) < \frac{\lambda(\pi_{n+1}(C_i))}{2^{i+2}}$ for $n \leq m$. To see this, notice that

$$\lambda(\pi_{n+1}(C_i) \setminus \pi_{n+1}(C_{i+1})) = \int_{N_{i+2}} \lambda(\Phi^n_{i+2}(x))dx - \int_{N_{i+3}} \lambda(\Phi^n_{i+3}(x))dx$$

$$\leq \int_{N_{i+2}} \lambda(\Phi^n_{i+2}(x))dx - \int_K \lambda(\Phi^n_\omega(x))dx$$

$$\leq \lambda((L_{i+2} \setminus K)) + \int_K \lambda(\Phi^n_{i+2}(x))dx - \tau_n \leq \frac{2^{i+3}\tau_n}{2^{i+3}-1} - \tau_n = \frac{\tau_n}{2^{i+3}-1} \leq \frac{\lambda(\pi_n(C_i))}{2^{i+2}}$$

for each $i \in \omega$. Finally, observe that

$$\lambda(\pi_1(C_i) \setminus \pi_1(C_{i+1})) = \lambda(N_i \setminus N_{i+1} \leq \lambda(N_i \setminus K \leq \frac{\lambda(N_i)}{2^{i+1}}$$

for each $i \in \omega$

Now let $f' : C \to [0,1]$ be defined by $f'(x,y) = g(y)$ if $\Phi(x) = (g, \xi, \mu, \zeta)$ and extend f' to a continuous function $f : [0,1]^m \to [0,1]$ arbitrarily. Since $X \notin \mathcal{I}^r_n$ there must be some $a \in X$ such that $a \notin \mathfrak{X}(f, \mathcal{C}, \delta)$. This means that there is some $\epsilon' > 0$ such that $\Xi_{\delta,\epsilon'}(f^{-1}a, C_i)$ holds for all but finitely many $i \in \omega$. In particular,

$$H^r(\{x \in N_i : \Xi_{\delta,\epsilon'}((f^{-1}a)_x, (C_i)_x)\} \setminus Z) > H^r(N_i) - \delta > H^r(K) - \delta > \theta$$

holds for all but finitely many $i \in \omega$ and any Z such that $\lambda(Z) < \epsilon'$. It may, without loss of generality, be assumed that $\epsilon' \leq \epsilon/2$.

Using the uniform convergence of $\{\lambda(\Phi^n_i(x))\}_{i\in\omega}$ it is possible to find $j \in \omega$ such that $\lambda(\Phi^n_i, x) \setminus \Phi^n_\omega(x)) < (\epsilon'/m+1))^n$ for all $x \in K$, $n \leq m$ and $i > j$. Let $i > j$ be such that $\lambda(N_i \setminus K) < \epsilon'$. Since $H^r(F(a)) \leq \theta$ and

$$H^r(\{x \in N_i : \Xi_{\delta,\epsilon'}((f^{-1}a)_x, (C_i)_x)\} \setminus (N_i \setminus K)) > \theta$$

it is possible to choose $y \in K \setminus F(a)$ such that $\Xi_{\delta,\epsilon'}((f^{-1}a)_y, (C_i)_y)$ holds. Observe that if $\Phi(y) = (g, \xi, \delta', \beta')$ then $\xi(n) = (C_n)_y$ for $n > 0$, $g = f_y$ and $\delta < \delta'$. The choice of j guarantees that the hypothesis (3) of Corollary 5.2 is satisfied by i, δ, ϵ', $(f^{-1}a)_y$ and $\{(C_n)_y\}_{n\in\omega}$. It follows that there is some $\epsilon > 0$ such that $\Xi_{\delta,\epsilon}((f^{-1}a)_y, (C_i)_y)$ holds for all but finitely many $i \in \omega$ and hence so does $\Xi_{\delta',\epsilon}((f^{-1}a)_y, (C_i)_y)$. Therefore $a \notin \mathfrak{X}(g, \{\xi(n)\}_{n\in\omega}, \delta')$. This yields a contradiction to the fact that $y \notin F(a)$ and $\Phi(y) = (g, \xi, \delta', \beta')$ implies that $a \in \mathfrak{X}(g, \{\xi(n)\}_{n\in\omega}, \delta')$. ∎

6. The Ideal is Proper

It remains to be shown that the ideals $\mathcal{I}^r_n$ are proper. This will require a careful analysis of the capacity H^r and some generalizations of results from [7]. The key fact about Hausdorff capacity that will be used is that if $B \subseteq E$ is of small Lebesgue measure but evenly distributed throughout E, then $H^r(B)$ will be close to $H^r(E)$. This is made precise in the next lemma whose statement requires the following notation.

Definition 6.1. *For any measurable set $A \subseteq [0,1]$ define $\Delta^i_m(A)$ to be the least real number such that $\lambda(A \cap [0, \Delta^i_m(A)]) = \frac{i\lambda(A)}{m}$.*

Notice that $\Delta^i_m(A)$ is always defined and that if $A = [0,1]$ then $\Delta^i_m(A)$ is nothing more than $\frac{i}{m}$.

Lemma 6.1. *Let $\delta > 0$, $\eta > 0$ and suppose that $E \subseteq [0,1]$ is measurable. If $\Xi_{\delta,\eta}(E,E)$ holds and $\delta < H^r(E)$ then there exists $m \in \omega$ such that if $D \subseteq E$ is any measurable set such that for each $i \in m$*

$$\lambda(D \cap [\Delta^i_m(E), \Delta^{i+1}_m(E)]) \geq \frac{\eta}{m}$$

then $\Xi_{\delta,\frac{\eta}{2m}}(D,E)$.

Proof: Let $m \in \omega$ be so large that the inequality

$$\frac{m^{1-r}\eta^{1+r}}{8 \cdot 2^r} > 1$$

is satisfied. To begin, note that Lemma 4.8 implies that there exists $\bar{\epsilon} > 0$ such that $\Xi_{\delta-\bar{\epsilon},\eta/2}(E,E)$ holds. If $\Xi_{\delta,\frac{\eta}{2m}}(D,E)$ fails then there is some Z such that $\lambda(Z) < \frac{\eta}{2m}$ and an open cover $D \setminus Z \subseteq \bigcup_{i=0}^{\infty} I_i$ such that $\sum_{i=0}^{\infty} \lambda(I_i)^r < H^r(E) - (\delta - \bar{\epsilon})$. Let $B = \{i \in \omega : \lambda(I_i) \geq \frac{1}{2m}\}$ and let $C = \{i \in m : (\forall j \in B)(I_j \cap [\Delta^i_m(E), \Delta^{i+1}_m(E)] \cap E = \emptyset)\}$. Three separate cases, depending on the size of B and C, will be considered.

Case 1 To begin, suppose that $|B| \geq \frac{m\eta}{8}$. Then

$$\sum_{i=1}^{\infty} \lambda(I_i)^r \geq \sum_{i \in B} \lambda(I_i)^r \geq |B|(1/2m)^r \geq \frac{m^{1-r}\mu}{8 \cdot 2^r} > 1$$

Since $\sum_{i=0}^{\infty} \lambda(I_i)^r < H^r(E) - (\delta - \eta/2) < 1$ this is impossible.

Case 2 Suppose now that $|B| < \frac{m\eta}{8}$ and $|C| \leq \frac{m\eta}{4}$. It then follows that if

$$G = \{i \in m : [\Delta^i_m(E), \Delta^{i+1}_m(E)] \cap E \not\subseteq \bigcup_{j=1}^{\infty} I_j\}$$

then $|G| \leq 2 \cdot |B| + |C|$. The reason for this is that if $j \in B$ then there are at most two integers i such that the intervals $[\Delta^i_m(E), \Delta^{i+1}_m(E)]$ intersect I_j but are not contained in I_j — this accounts for the summand $2 \cdot |B|$. All the other intervals

$[\Delta^i_m(E), \Delta^{i+1}_m(E)]$ for $i \in G$ must be disjoint from I_j for every $j \in B$ — this accounts for the other summand $|C|$.

By the assumptions of this case it follows that $2 \cdot |B| + |C| < m\eta/2$ and hence

$$\lambda(\bigcup_{i\in G}[\Delta^i_m(E), \Delta^{i+1}_m(E)] \cap E) < \eta/2$$

Since $\Xi_{\delta-\bar{\epsilon},\eta/2}(E,E)$ holds it may be concluded that $H^r(E \setminus \bigcup_{i\in G}[\Delta^i_m(E), \Delta^{i+1}_m(E)]) > H^r(E) - (\delta - \bar{\epsilon})$. Since $E \setminus (\bigcup_{i\in G}[\Delta^i_m(E), \Delta^{i+1}_m(E)]) \subseteq \bigcup_{i=1}^{\infty} I_i$ this yields a contradiction.

Case 3 Suppose that $|B| < \frac{m\eta}{8}$ and $|C| > \frac{m\eta}{4}$. Let C' be a family of non-consecutive members of C of maximal cardinality — hence, $|C'| \geq |C|/2 > \frac{m\eta}{8}$. Let

$$U_j = \{i \in \omega : I_i \cap [\Delta^j_m(E), \Delta^{j+1}_m(E)] \cap E \neq \emptyset\}$$

for each $j \in C'$ and define $U = \bigcup_{j\in C'} U_j$. Since, for $j \in C$, the sets $[\Delta^j_m(E), \Delta^{j+1}_m(E)] \cap E$ are intersected only by intervals I_i where $i \in \omega \setminus B$, and such intervals I_i are smaller than any $[\Delta^{j+1}_m(E), \Delta^j_m(E)]$, it follows that $U_j \cap U_k = \emptyset$ if k and j are distinct members of C'. Therefore, using the fact that $0 < r < 1$,

$$\sum_{i\in U} \lambda(I_i)^r \geq \sum_{j\in C'} \sum_{i\in U_j} \lambda(I_i)^r \geq$$

$$\sum_{j\in C'} \left(\sum_{i\in U_j} \lambda(I_i)\right)^r \geq \sum_{j\in C'} \lambda(D \cap [\Delta^j_m(E), \Delta^{j+1}_m(E)])^r \geq$$

$$\sum_{j\in C'} (\frac{\mu}{m} - \lambda(Z))^r \geq \frac{m\mu}{8}(\frac{\mu}{2m})^r > 1$$

and once again, as in the first case, this is a contradiction because $D \subseteq [0,1]$. ■

If $X \subseteq [0,1]$ then $F : X \to [0,1]$ will be said to have small fibres if and only if $\lambda(F^{-1}\{x\}) = 0$ for each $x \in [0,1]$. The proof of the Theorem 6.1 and the lemmas preceding it will rely on decomposing an arbitrary continuous function into a piece that has small fibres and a piece which has countable range.

Lemma 6.2. *Let $\mu \in (0,1)$ and suppose that $\{X_i : i \in \omega\}$ is a sequence of mutually independent $\{0,1\}$-valued random variables with mean μ for each $i \in \omega$. Suppose that $C \subseteq [0,1]$ is a measurable set and that for each $j \in n$ the function $F_j : C \to [0,1]$ is measurable with small fibres. For any $\rho > 0$ there is $M \in \omega$ such that for all $m \geq M$ the probability that*

$$\lambda\left(\bigcap_{j\in n}\bigcup_{i\in m} F_j^{-1}[\frac{i}{m}, \frac{i+X_i}{m}]\right) > \frac{\mu^n\lambda(C)}{2}$$

is greater than $1 - \rho$.

Proof: To begin, let $m \in \omega$ be fixed. For any function $\xi \in {}^n m$ define $\theta(\xi) = \lambda(\bigcap_{j\in n} F_j^{-1}[\frac{\xi(j)}{m}, \frac{\xi(j)+1}{m}])$ and let

$$Y(\xi) = \lambda\left(\bigcap_{j\in n} F_j^{-1}[\frac{\xi(j)}{m}, \frac{\xi(j) + X_{\xi(j)}}{m}]\right) = \theta(\xi)\prod_{j\in n} X_{\xi(j)}$$

If $\xi \neq \xi'$ then

$$\lambda\left(\left(\bigcap_{j\in n} F_j^{-1}[\frac{\xi(j)}{m}, \frac{\xi(j)+1}{m}]\right) \cap \left(\bigcap_{j\in n} F_j^{-1}[\frac{\xi(j)}{m}, \frac{\xi(j)+1}{m}]\right)\right) = 0$$

and so $\sum_{\xi\in {}^n m} \theta(\xi) = \lambda(C)$.

Letting $E[Z]$ denote the average value of the random variable Z and $V[Z]$ the variance of Z, it is easy to see that $E[Y(\xi)] = \theta(\xi)\mu^{\sigma(\xi)}$ where $\sigma(\xi)$ represents the cardinality of the range of ξ. Noting that $\sigma(\xi) \leq n$ for all σ, it follows that $E[\sum_{\xi\in {}^n m} Y(\xi)] \geq \mu^n\lambda(C)$. Furthermore,

$$V[\sum_{\xi\in {}^n m} Y(\xi)] = E[(\sum_{\xi\in {}^n m} Y(\xi) - E[Y(\xi)])^2] =$$

$$\sum_{\xi\in {}^n m}\sum_{\xi'\in {}^n m} E[(Y(\xi) - E[Y(\xi)])(Y(\xi') - E[Y(\xi')])].$$

If ξ and ξ' have disjoint ranges then $Y(\xi)$ and $Y(\xi')$ are independent random variables and so

$$E[(Y(\xi) - E[Y(\xi)])(Y(\xi') - E[Y(\xi')])] = E[Y(\xi) - E[Y(\xi)]]E[Y(\xi') - E[Y(\xi')]] = 0$$

while if the ranges of ξ and ξ' are not disjoint then

$$E[(Y(\xi) - E[Y(\xi)])(Y(\xi') - E[Y(\xi')])] =$$

$$E[Y(\xi)Y(\xi')] - E[E[Y(\xi)]Y(\xi')] - E[E[Y(\xi')]Y(\xi)] + E[Y(\xi)]E[Y(\xi')]]$$

$$= E[Y(\xi)Y(\xi')] - E[Y(\xi)]E[Y(\xi')] \leq E[Y(\xi)Y(\xi')]$$

$$= E[\theta(\xi)\prod_{j\in n} X_{\xi(j)}\theta(\xi')\prod_{j\in n} X_{\xi'(j)}] \leq \theta(\xi)\theta(\xi')$$

since $X_i \in \{0,1\}$ for each i. It may be concluded that

$$V[\sum_{\xi\in {}^n m} Y(\xi)] \leq \sum_{j\in n}\sum_{\xi\in {}^n m}\sum_{\substack{\xi'\in {}^n m\\ \xi'(j)=\xi(j)}} \theta(\xi)\theta(\xi') = \sum_{j\in n}\sum_{\xi\in {}^n m}\theta(\xi)\sum_{\substack{\xi'\in {}^n m\\ \xi'(j)=\xi(j)}}\theta(\xi').$$

However, if j is fixed then $\sum_{\substack{\xi'\in {}^n m\\ \xi'(j)=\xi(j)}} \theta(\xi') = \lambda(F_j^{-1}[\frac{\xi(j)}{m}, \frac{\xi(j)+1}{m}])$. Therefore, all that needs to be done is to choose M so large that if $m \geq M$ and $i \in m$ then

$\lambda(F_j^{-1}[\frac{i}{m}, \frac{i+1}{m}]) < \frac{\rho\mu^{2n}\lambda(C)}{4n}$ for each $j \in n$. The reason this suffices is that this implies that

$$V[\sum_{\xi \in^n m} Y(\xi)] \leq \sum_{j \in n} \sum_{\xi \in^n m} \theta(\xi) \frac{\rho\mu^{2n}\lambda(C)}{4n} \leq \frac{\rho\mu^{2n}\lambda(C)^2}{4}$$

and so Chebyshev's Inequality can be applied to conclude that the probability that

$$|\sum_{\xi \in^n m} Y(\xi) - E[\sum_{\xi \in^n m} Y(\xi)]| > \frac{\mu^n \lambda(C)}{2}$$

is less than ρ. Since it has already been established that $E[\sum_{\xi \in^n m} Y(\xi)] \geq \mu^n \lambda(C)$ it follows that the probability that $\sum_{\xi \in^n m} Y(\xi) \geq \mu^n \lambda(C)/2$ is at least $1 - \rho$ as required.

To choose M so large that if $m \geq M$ and $i \in m$ then $\lambda(F_j^{-1}[\frac{i}{m}, \frac{i+1}{m}]) < \frac{\rho\mu^{2n}\lambda(C)}{4n}$ for each $j \in n$, all that is required is compactness and the fact that each F_j has small fibres. Since $F_j^{-1}\{x\} = \bigcap_{k \in \omega} F_j^{-1}[x - 1/k, x + 1/k]$ and $\lambda(F_j^{-1}\{x\}) = 0$ it follows that it is possible to choose a finite cover of $[0,1]$ by open intervals, $\mathcal{C}$, such that if $I \in \mathcal{C}$ then $\lambda(F_j^{-1}I) < \frac{\rho\mu^{2n}\lambda(C)}{4n}$ for each $j \in n$. Hence M must be chosen so large that if $m \geq M$ and $i \in m$ then there is $I \in \mathcal{C}$ such that $[\frac{i}{m}, \frac{i+1}{m}] \subseteq I$. ■

Lemma 6.3. *Suppose that $\delta > 0$, $\mu > 0$, $\eta > 0$ and $k \in \omega$. There is then a real number $\epsilon(\delta, \mu, \eta, k) > 0$ such that if*

- *$\{C_i\}_{i \in k}$ is a family of measurable subsets of $[0,1]$*
- *$F_i : C_i \to [0,1]$ is a measurable function with small fibres for each $i \in k$*
- *$E \subseteq [0,1]$ is a measurable set*
- *$\Xi_{\delta,\eta}(E, E)$ holds and $\delta < H^r(E)$*
- *$\rho > 0$*

then there is $M \in \omega$ such that for all $m > M$ and for any mutually independent, $\{0,1\}$-valued random variables $\{X_i\}_{i \in m}$ with mean μ, the probability that

$$\Xi_{\delta,\epsilon(\delta,\mu,\eta,k)}\left(\bigcap_{i \in k}((F_i^{-1} \bigcup_{j \in m}[\frac{j}{m}, \frac{j + X_j}{m}]) \cup ([0,1] \setminus C_i)), E\right)$$

holds is greater than $1 - \rho$.

Moreover, there is $\theta > 0$ such that if

- *$E' \subseteq [0,1]$ is a measurable set such that $\lambda(E \Delta E') < \theta$*
- *$\{C_i'\}_{i \in k}$ is a family of measurable sets such that $\lambda(C_i \Delta C_i') < \theta$ for each $i \in k$*
- *$\{G_i\}_{i \in k}$ is a family of measurable functions such that*

$$\sup\{|F_i(x) - F_i'(x)| : x \in C \cap C'\} < \theta$$

for each $i \in k$

then the probability that

$$\Xi_{\delta,\epsilon(\delta,\mu,\eta,k)}\left(\bigcap_{i\in k}((G_i^{-1}\bigcup_{j\in m}[\frac{j}{m},\frac{j+X_j}{m}])\cup([0,1]\setminus C_i')),E'\right)$$

holds is still greater than $1-\rho$.

Proof: Let $\alpha = \mu^k/2$ and use Lemma 6.1 to find p such that if $D \subseteq E$ is a measurable set such that for each $i \in p$

$$\lambda(E\cap[\Delta_p^i(E),\Delta_p^{i+1}(E)])\geq\frac{\alpha}{2p}$$

then $\Xi_{\delta,\frac{\alpha}{4p}}(D,E)$ holds. Let $\epsilon(\delta,\mu,\eta,k) = \frac{\alpha}{4p}$. Let $\{P_i : i \in s\}$ enumerate the sets of positive measure which belong to the coarsest partition of E refining each of the partitions $\{[\Delta_p^i(E),\Delta_p^{i+1}(E)]\cap E : i \in p\}$ and $\{C_i\cap E, E\setminus C_i\}$ for $i \in k$. Now use Lemma 6.2 to find $M \in \omega$ such that for all $m \geq M$ the probability that

$$\lambda\left(\bigcap_{j\in k}\left(((F_j\restriction P_n)^{-1}\bigcup_{i\in m}[\frac{i}{m},\frac{i+X_i}{m}])\cup(P_n\setminus C_j)\right)\right)>\alpha\lambda(P_n)$$

is greater than $1-\frac{\rho}{s}$ for each $n \in s$.

Now notice that if $m \geq M$ is fixed then, because each F_i has small fibres, it is possible to find $p(i,j)$ and $q(i,j)$ such that $\frac{i}{m} < p(i,j) < q(i,j) < \frac{i+1}{m}$ and

$$\lambda(F_j^{-1}[\frac{i}{m},\frac{i+1}{m}])-\lambda(F_j^{-1}[p(i,j),q(i,j)])<\frac{\alpha\lambda(P_n)}{2(2k+1)m}$$

for each $j \in k$ and $i \in m$. Now observe that if $\lambda(C_i\Delta C_i') < \frac{\alpha\lambda(P_n)}{2(2k+1)}$ for each $i \in k$ and if $\lambda(E\Delta E') < \frac{\alpha^{\lambda}(P_n)}{2(2k+1)}$ and if $Y_i \in \{0,1\}$ are such that

$$\lambda\left(\bigcap_{j\in k}\left((F_j\restriction P_n)^{-1}\bigcup_{i\in m}[\frac{i}{m},\frac{i+Y_i}{m}]\cup(P_n\setminus C_j)\right)\right)>\alpha\lambda(P_n)$$

then the Lebesgue measure of

$$\bigcap_{j\in k}\left((F_j\restriction P_n)^{-1}\bigcup_{i\in m}[p(i,j),p(i,j)+Y_i(q(i,j)-p(i,j))]\cup(P_n\setminus C_j)\right)$$

is greater than $\dfrac{\alpha\lambda(P_n)}{2}$. Therefore, if $\theta > 0$ is such that

- $\theta < \frac{\alpha\lambda(P_n)}{2(2k+1)}$ for each $n \in s$
- $\theta < p(i,j) - \frac{i}{m}$ for all i and j
- $\theta < \frac{i_1}{m} - q(i,j)$ for all i and j

then if $\{G_i\}_{i\in k}$ is a family of measurable functions such that $\sup\{|F_i(x) - F_i'(x)| : x \in C \cap C'\} < \theta$ for each $i \in k$ then $F_j^{-1}[\frac{i}{m}, \frac{i+1}{m}] \subseteq G_j^{-1}[p(i,j), q(i,j)]$ and hence

$$\lambda\left(\bigcap_{j\in k}\left((G_j \restriction P_n)^{-1}\bigcup_{i\in m}[\frac{i}{m}, \frac{i+Y_i}{m}] \cup (P_n \setminus C_j')\right)\right) > \frac{\alpha\lambda(P_n)}{2}$$

also holds for each $n \in s$.

It now follows that the Lebesgue measure of the set obtained by intersecting the interval $[\Delta_p^i(E'), \Delta_p^{i+1}(E')]$ with

$$\bigcap_{j\in k}\left((G_j \restriction P_n)^{-1}\bigcup_{i\in m}[\frac{i}{m}, \frac{i+Y_i}{m}] \cup ([\Delta_p^i(E'), \Delta_p^{i+1}(E')] \setminus C_j')\right)$$

is greater than

$$\frac{\alpha\lambda([\Delta_p^i(E'), \Delta_p^{i+1}(E')] \cap E')}{2}$$

and the result now follows from Lemma 6.1. ∎

Lemma 6.4. *Let $k \in \omega$ and $\{C_i\}_{i\in k}$ be a family of measurable subsets of $[0,1]$. Let $F_i : C_i \to [0,1]$ be a measurable function for each $i \in k$. Suppose also that $\delta > 0$ and $\eta > 0$. Then, for any $N \in \omega$ there exists $\epsilon > 0$ such that if $\Xi_{\delta,\eta}(E, E)$ holds for some measurable set E then*

$$\Xi_{\delta,\epsilon}(\bigcap_{i\in k}(F_i^{-1}a) \cup ([0,1] \setminus C_i), E)$$

holds for some $a \in W_N$.

Proof: First notice that if $H^r(E) \leq \delta$ then the result follows immediately, so it will be assumed that $H^r(E) > \delta$. For each $i \in k$ let $\{y_j^i : j \in d_i \leq \omega\}$ enumerate all points $y \in [0,1]$ such that $\lambda(F_i^{-1}\{y\}) > 0$. Let $C_i' = C_i \setminus F_i^{-1}\{y_j^i : j \in d_i\}$ and let $F_i' = F_i \restriction C_i'$. Since F_i' has small fibres for each $i \in k$ it follows from Lemma 6.3 that it is possible to choose m so large that if $\{X_i\}_{i\in m}$ are $\{0,1\}$-valued random variables with mean 2^{-N-1} then, letting $\epsilon' = \epsilon(\delta, 2^{-N-1}, \eta, k)$, the probability that

$$\Xi_{\delta,\epsilon'}\left(\bigcap_{i\in k}((F_i')^{-1}\bigcup_{j\in m}[\frac{j}{m}, \frac{j+X_j}{m}]) \cup ([0,1] \setminus C_i'), E\right)$$

holds is at least $3/4$ for any measurable set E such that $\Xi_{\delta,\eta}(E, E)$ holds and $H^r(E) > \delta$. Since the mean of each X_i is 2^{-N-1} it is possible to choose m so large that the probability that

$$\lambda(\bigcup_{j\in m}[\frac{j}{m}, \frac{j+X_j}{m}]) < 2^{-N}$$

is also greater than 3/4. Hence, given E with the required properties, there is $a_0 \in W_N$ such that

$$\Xi_{\delta,\epsilon'}\left(\bigcap_{i\in k}((F_i')^{-1}a_0)\cup([0,1]\setminus C_i'),E\right)$$

holds. Now choose $J \in \omega$ such that $\lambda(\bigcup_{i\in k}\bigcup_{j\geq J} F_i^{-1}\{y_j^i\}) < \epsilon'/2$. It is then easy to find $a \in W_N$ such that $a_0 \cup \{y_j^i : i \in k, j \in J\} \subseteq a$. Let $\epsilon = \epsilon'/2$ and note that it follows that

$$\Xi_{\delta,\epsilon}\left(\bigcap_{i\in k}((F_i)^{-1}a)\cup([0,1]\setminus C_i),E\right)$$

holds because, if

$$Y = (\bigcap_{i\in k}((F_i')^{-1}a_0)\cup([0,1]\setminus C_i'))\setminus\bigcap_{i\in k}((F_i)^{-1}a)\cup([0,1]\setminus C_i)$$

then $Y \subseteq \bigcup_{i\in k}\bigcup_{j\geq J} F_i^{-1}\{y_j^i\}$ and hence $\lambda(Y) < \epsilon'/2$. ■

Lemma 6.5. *Suppose that $k \in \omega$ and $\{C_i\}_{i\in k}$ are measurable subsets of $[0,1]^{d+1}$ and $F_i : C_i \to [0,1]$ are measurable functions such that $(F_i)_x$ has small fibres for each $x \in [0,1]^d$. Let $N \in \omega$, $\delta > 0$ and $\eta > 0$. Then there is $\epsilon > 0$ such that for all closed $E \subseteq [0,1]^{d+1}$ and $\rho > 0$ there is some $a \in W_n$ such that the Lebesgue measure of*

$$\{x \in \pi_d(E) : \Xi_{\delta,\epsilon}(\bigcap_{i\in k}((F_i^{-1}a)\cup([0,1]^{d+1}\setminus C_i))_x, E_x) \text{ or } \neg\Xi_{\delta,\eta}(E_x,E_x) \text{ or } H^r(E_x)\leq\delta\}$$

is at least $\lambda(\pi_d(E))(1-\rho)$.

Proof: Let $\{X_i\}_{i\in\omega}$ be a sequence of mutually independent random variables with mean 2^{-N-1}. Let $\epsilon = \epsilon(\delta, 2^{-N-1}, \eta, k)$ and suppose that $E \subseteq [0,1]^{d+1}$ is closed. Next, choose compact subsets $W_i \subseteq \pi_d(C_i)$ and $V_i \subseteq [0,1]^d \setminus W_i$ for $i \in k$ as well as $E' \subseteq \pi_d(E)$ such that

- $\lambda(\pi_d(E)\setminus E') < \frac{\rho\lambda(\pi_d(E))}{6(1-\rho/2)}$
- $\lambda([0,1]^d \setminus (\bigcap_{i\in k}(W_i \cup V_i))) < \frac{\rho\lambda(\pi_d(E))}{6(1-\rho/2)}$
- $F_i \restriction W_i$ is continuous
- the mapping from $\pi_d(E')$ to $[0,1]$ defined by $x \mapsto \lambda(E_x \cap (p,q))$ is continuous for each pair of rationals p and q such that $0 \leq p < q \leq 1$
- the mapping from $\pi_d(W_i)$ to $[0,1]$ defined by $x \mapsto \lambda((C_i)_x \cap (p,q))$ is continuous for each $i \in k$ and each pair of rationals p and q such that $0 \leq p < q \leq 1$.

An easy application of the Lebesgue Density Theorem shows that a consequence of the last clause is that if $x \in \pi_d(W_i)$ then $\lim_{y\to x}\lambda(W_y \Delta W_x) = 0$. The penultimate clause implies a similar assertion for E'. It is possible to find compact E_1 and E_2, subsets of E', such that

- if $x \in E_1$ then $\Xi_{\delta,\eta}(E_x, E_x)$ holds and $H^r(E_x) > \delta$
- if $x \in E_2$ then $\Xi_{\delta,\eta}(E_x, E_x)$ fails or $H^r(E_x) \leq \delta$
- $\lambda(E' \setminus (E_1 \cup E_2)) < \frac{\rho\lambda(\pi_d(E))}{6(1-\rho/2)}$

because Lemma 4.4 implies that $\{x \in E' : \Xi_{\delta,\eta}(E_x, E_x)\}$ is measurable. Let $Z = E_1 \cap (\bigcap_{i\in k}(W_i \cup V_i))$ and for $x \in Z$ let $K(x) = \{i \in k : x \in \pi_d(W_i)\}$ and notice that $K(x)$ is constant on a neighbourhood of x because the sets V_i and W_i are all compact. If $x \in Z$ and $i \in K(x)$ then $F_i \restriction (W_i)_x$ has small fibres, $H^r(E_x) > \delta$ and $\Xi_{\delta,\eta}(E_x, E_x)$ holds, so it follows from Lemma 6.3 that there is $\theta_x > 0$ and $M_x \in \omega$ such that if $||y - x|| < \theta_x$ and $M \geq M_x$ then the probability that

$$\Xi_{\delta,\epsilon}\left(\left(\bigcap_{i\in K(y)}(F_i^{-1}\bigcup_{j\in m}[\frac{j}{m},\frac{j+X_j}{m}]) \cup ([0,1]\setminus C_i)\right)_y, E_y\right)$$

holds is greater than $1 - \rho^2/2$.

Since Z is compact, it is possible to find a single M such that for all $m > M$ and for any $x \in Z$ the probability that

$$\Xi_{\delta,\epsilon}\left(\left(\bigcap_{i\in K(x)}(F_i^{-1}\bigcup_{j\in m}[\frac{j}{m},\frac{j+X_j}{m}]) \cup ([0,1]\setminus C_i)\right)_y, E_y\right)$$

holds is greater than $1 - \rho^2/2$

Now let $m > M$ be so great that the probability that $\lambda(\bigcup_{j\in m}[\frac{j}{m},\frac{j+X_j}{m}]) < 2^{-N}$ is greater than 2ρ. Define

$$\Gamma(X_0, X_1, \dots, X_m)$$

to be the Lebesgue measure of the set of all $x \in Z$ such that

$$\Xi_{\delta,\epsilon}\left(\left(\bigcap_{i\in K(x)}(F_i^{-1}\bigcup_{j\in m}[\frac{j}{m},\frac{j+X_j}{m}]) \cup ([0,1]\setminus C_i)\right)_x, E_x\right)$$

holds. Note that Corollary 4.1 implies that this set is measurable. The first step is to estimate

$$\alpha_m = \sum_{X_0=0}^{1}\sum_{X_1=0}^{1}\dots\sum_{X_m=0}^{1}\Gamma(X_0, X_1, \dots, X_m)\prod_{i=0}^{m}\mu^{X_i}(1-\mu)^{1-X_i}$$

the average value of $\Gamma(X_0, X_1, \dots, X_m)$. To this end, let

$$\Lambda_x(X_0, X_1, \dots, X_m) \in \{0, 1\}$$

be defined to be 1 if and only if

$$\Xi_{\delta,\epsilon}\left(\left(\bigcap_{i\in K(x)}(F_i^{-1}\bigcup_{j\in m}[\frac{j}{m},\frac{j+X_j}{m}]) \cup ([0,1]\setminus C_i)\right)_x, E_x\right)$$

holds. and observe that α_m is equal to

$$\sum_{X_0=0}^{1}\sum_{X_1=0}^{1}\cdots\sum_{X_m=0}^{1}\left(\int_{x\in Z}\Lambda_x(X_0,X_1,\ldots,X_m)dx\right)\prod_{i=0}^{m}\mu^{X_i}(1-\mu)^{1-X_i}=$$

$$\int_{x\in Z}\left(\sum_{X_0=0}^{1}\sum_{X_1=0}^{1}\cdots\sum_{X_m=0}^{1}\Lambda_x(X_0,X_1,\ldots,X_m)\prod_{i=0}^{m}\mu^{X_i}(1-\mu)^{1-X_i}\right)dx$$

However, notice that

$$\sum_{X_0=0}^{1}\sum_{X_1=0}^{1}\cdots\sum_{X_m=0}^{1}\Lambda_x(X_0,X_1,\ldots,X_m)\prod_{i=0}^{m}\mu^{X_i}(1-\mu)^{1-X_i}$$

is just the probability that

$$\Xi_{\delta,\epsilon}\left((\bigcap_{i\in K(x)}(F_i^{-1}\bigcup_{j\in m}[\frac{j}{m},\frac{j+X_j}{m}])\cup([0,1]\setminus C_i))_x,E_x\right)$$

holds and the choice of m and the fact that $x\in Z$ guarantee that this probability is greater than $1-\rho^2/2$. Hence $\alpha_m\geq\lambda(Z)(1-\rho^2)$.

Now let p be the probability that $\Gamma(X_0,X_1,\ldots,X_m)\geq(1-\rho/2)\lambda(Z)$. Obviously, $p\lambda(Z)+(1-p)(1-\epsilon/2)\lambda(Z)\geq\alpha_m\geq(1-\rho^2)\lambda(Z)$. Solving for p yields that $p\geq 1-2\rho$. Since m was chosen so large that the probability that $\lambda(\cup_{j\in m}[\frac{j}{m},\frac{j+X_j}{m}])<2^{-N}$ is greater than 2ρ, there is at least one $a\in W_N$ such that $\lambda(U)>(1-\rho/2)\lambda(Z)$ where U is the set of all $x\in Z$ such that

$$\Xi_{\delta,\epsilon(\delta,\mu,\eta,k)}\left((\bigcap_{i\in K(x)}(F_i^{-1}a)\cup([0,1]\setminus C_i))_x,E_x\right)$$

holds. Obviously

$$\begin{aligned}\lambda(U\cup E_2)&=\lambda(U)+\lambda(E_2)\geq(1-\rho/2)\lambda(Z)+\lambda(E_2)\\&\geq(1-\rho/2)(\lambda(E_1)-\frac{\rho\lambda(\pi_d(E))}{6(1-\rho/2))})+\lambda(E_2)\\&\geq(1-\rho/2)(\lambda(E_1)+\lambda(E_2))-(1-\rho/2)\frac{\rho\lambda(\pi_d(E))}{6(1-\rho/2)}\geq\end{aligned}$$

$$(1-\rho/2)(\lambda(\pi_d(E))-2\frac{\rho\lambda(\pi_d(E))}{6(1-\rho/2)})-(1-\rho/2)\frac{\rho\lambda(\pi_d(E))}{6(1-\rho/2)}\geq(1-\rho)(\lambda(\pi_d(E)))$$

as required. ∎

Theorem 6.1. *Suppose that* $\{F_i\}_{i\in k}$ *are continuous functions from* $[0,1]^d$ *to* $[0,1]$, $\eta > 0$, $\delta > 0$, $N \in \omega$ *and* $\{A_i\}_{i\in k}$ *are measurable subsets of* $[0,1]^d$. *Then there is* $\epsilon > 0$ *such that for each closed subset* $E \subseteq [0,1]^d$, *if* $\Xi_{\delta,\eta}(E,E)$ *holds then*

$$\Xi_{(d+1)\delta,\epsilon}(\bigcap_{i\in k}((A_i \cap F_i^{-1}a) \cup ([0,1]^d \setminus A_i)), E)$$

also holds for some elementary set $a \in W_N$.

Proof: Note that if $d = 1$ then it is possible to use Lemma 6.5 to find $\epsilon > 0$ such that for all closed sets E, if $\Xi_{\delta,\eta}(E,E)$ holds and $H^r(E) > \delta$ then

$$\Xi_{\delta,\epsilon}(\bigcap_{i\in k}((A_i \cap F_i^{-1}a) \cup ([0,1]^d \setminus A_i)), E)$$

also holds for some elementary set $a \in W_N$ and hence, be Lemma 4.6, so does

$$\Xi_{2\delta,\epsilon}(\bigcap_{i\in k}((A_i \cap F_i^{-1}a) \cup ([0,1]^d \setminus A_i)), E)$$

On the other hand, if $H^r(E) \leq \delta$ then

$$\Xi_{2\delta,\epsilon}(\emptyset E)$$

holds and hence, so does

$$\Xi_{2\delta,\epsilon}(\bigcap_{i\in k}((A_i \cap F_i^{-1}a) \cup ([0,1]^d \setminus A_i)), E)$$

for any $a \in W_n$.

Proceed by induction on d noting that if $d = 1$ then this follows directly from Lemma 6.4. So assume that the lemma has been established for d and that $\{F_i\}_{i\in k}$ are continuous functions from $[0,1]^{d+1}$ to $[0,1]$, $\eta > 0$, $\delta > 0$, $N \in \omega$ and $\{A_i\}_{i\in k}$ are measurable subsets of $[0,1]^{d+1}$. Let $B_i = \{(x,y) \in [0,1]^d \times [0,1] : \lambda((F_i^{-1}\{y\})_x) > 0\}$ and note that

$$B_i = \{(x,y) \in [0,1]^d \times [0,1] : (\exists K \text{ compact})(\lambda(K) > 0 \text{ and } K \subseteq (F_i^{-1}\{y\})_x\}$$

and, because F is continuous, the relation $K \subseteq (F_i^{-1}\{y\})_x$ is Borel. Moreover, so is the statement $\lambda(K) > 0$ when K is compact, and so the set B is Σ^1_1 and hence, measurable. Let B_i^* be the inverse image of B_i under the mapping $(x,y) \mapsto (x, F_i(x,y))$ or, in other words, $(x,y) \in B_i^*$ if and only if $\lambda((F_i^{-1}\{F(x,y)\})_x) > 0$. Since B_i^* is clearly measurable, it follows that so is $C_i = [0,1]^{d+1} \setminus B_i^*$.

Now, for each $i \in k$, let $\{f_i^j : j \in I_i\}$ enumerate a maximal collection of functions such that

- $f_i^j : C_i^j \to [0,1]$ where $C_i^j \subseteq [0,1]^d$ is compact
- f_i^j is continuous
- $f_i^j \subseteq B_i$
- if $x \in C_i^j \cap C_i^{j'}$ then $f_j(x) \neq f_{j'}(x)$

- $\int_{C_i^j} \lambda((F_i^{-1}\{f_j(x)\})_x)dx > 0$.

The first thing to notice is that, for each $i \in k$, such a family must be countable and therefore, $I_i \leq \omega$ without loss of generality. To see this let $E_i^j = \{(x,y) \in [0,1]^d \times [0,1] : F_i(x,y) = f_i^j(x)\}$. If $j \neq j'$ then $E_i^j \cap E_i^{j'} = \emptyset$ and, moreover,

$$\lambda(E_i^j) = \int_{C_i^j} \lambda((F_i^{-1}\{f_j(x)\})_x)dx > 0$$

for any $j \in I_i$. Hence the family of sets E_i^j is countable for each $i \in k$.

Next, it must be shown that

$$\sum_{j\in I_i} \int_{C_i^j} \lambda((F_i)^{-1}\{f_j(x)\})_x)dx = \lambda(B_i^*)$$

so suppose not. Then it must be that $\lambda(B_i^* \setminus \cup_{j\in I_i} E_i^j) > 0$. Since each f_i^j is continuous and $I_i \leq \omega$, it follows that $B_i \setminus (\cup_{j\in I_i} f_i^j)$ is Σ_1^1. Hence it is possible to use the von Neumann selection theorem to find a function f such that the domain of f is $\pi_d(B_i \setminus (\cup_{j\in I_i} f_i^j))$ and f is measurable. Since $\pi_d(B_i \setminus (\cup_{j\in I_i} f_i^j))$ is also equal to $\pi_d(B_i^* \setminus (\cup_{j\in d} E_i^j))$ it must be that $\lambda(\pi_d(B_i \setminus (\cup_{j\in I_i} f_i^j))) > 0$. Hence

$$\int_{\pi_d(B_i \setminus (\cup_{j\in I_i} f_i^j))} \lambda((F_i^{-1}\{f(x)\})_x)dx > 0$$

because $\lambda(F_x^{-1}\{f(x)\}) > 0$ for each $x \in \pi_d(B_i \setminus (\cup_{j\in I_i} f_i^j))$. Finally, by using Lusin's Theorem, it is possible to find a compact set, $D \subseteq \pi_d(B_i \setminus (\cup_{j\in I_i} f_i^j))$ such that $f \restriction D$ is continuous and $\int_D \lambda((F_i^{-1}\{f(x)\}_x)dx > 0$. This contradicts the putative maximality of the family $\{f_i^j : j \in I_i\}$.

Now note that for each $i \in k$ the function $(F_i \restriction C_i)_x$ has small fibres for all x. Applying Lemma 6.5 to $\{F_i \restriction (C_i \cap A_i) : i \in k\}$, δ, $\eta/2$ and $N+1$ it follows that there is some $\epsilon^* > 0$ such that for all $\mu > 0$ and any closed $E \subseteq [0,1]^{d+1}$ there is some $a \in W_{N+1}$ such that the Lebesgue measure of

$$\{x \in \pi_d(E) : \Xi_{\delta,\epsilon^*}((\bigcap_{i\in k}((C_i \cap A_i \cap F_i^{-1}a) \cup ([0,1]^{d+1} \setminus (C_i \cap A_i)))_x, E_x)$$
$$\text{or } \neg\Xi_{\delta,\eta/2}(E_x, E_x) \text{ or } H^r(E_x) \leq \delta\}$$

is at least $(1-\mu)\lambda(E)$.

It is therefore possible to find $K \in \omega$ such that for each $i \in k$

$$\sum_{j\in K} \int_{C_i^j} \lambda((F_i^{-1}\{f_i^j(x)\})_x)dx > \lambda(B_i^*) - \frac{\eta^d \epsilon^*}{2 \cdot 4^d k^2}$$

and so, if S_i is defined to be

$$\{x \in [0,1]^d : \lambda((B_i^* \setminus F_i^{-1}\{f_i^j(x) : j \in K\})_x) \geq \epsilon^*/2k\}$$

then $\lambda(S_i) < \frac{\eta^d}{4^d k}$ for each $i \in k$. Let $U \subseteq [0,1]^d$ be any closed set such that $U \cap S_i = \emptyset$ for each $i \in k$ and $\lambda(U) > 1 - (\eta/4)^d$. Let F_i^j be an arbitrary continuous extension of f_i^j which has domain $[0,1]^d$ and let $A_i^j = \mathrm{dom}(f_i^j)$. It follows from the induction hypothesis that there is $\epsilon' > 0$ such that if $E \subseteq [0,1]^d$ is a closed set such that $\Xi_{\delta,\eta/4}(E, E)$ holds then there is $a \in W_{N+1}$ such that

$$\Xi_{(d+1)\delta,\epsilon'}(\bigcap_{i\in k}\bigcap_{j\in K}((A_i^j \cap (F_i^j)^{-1}a) \cup ([0,1]^d \setminus A_i^j)), E)$$

holds. Let $\epsilon = \min\{\epsilon^*/2, \epsilon'/2, \eta/4\}$.

Now suppose that E is a closed set such that $\Xi_{\delta,\eta}(E, E)$ holds. From the choice of ϵ^* it follows that it is possible to find $a_0 \in W_{N+1}$ such that, letting Z be the set of all $x \in [0,1]^d$ such that

$$\Xi_{\delta,\epsilon^*}((\bigcap_{i\in k}((C_i \cap A_i \cap F_i^{-1}a_0) \cup ([0,1]^{d+1} \setminus (C_i \cap A_i)))_x, E_x)$$

$$\text{or } \neg\Xi_{\delta,\eta/2}(E_x, E_x) \text{ or } H^r(E_x) \leq \delta\}$$

the Lebesgue measure of Z is at least $\lambda(E)(1 - (\epsilon'/2)^d)$. If $\hat{E} = \{x \in \pi_d(E) : \Xi_{\delta,\eta/2}(E_x, E_x)\}$ then $\Xi_{\delta,\eta}(\hat{E}, \pi_d E)$ holds, by Lemma 4.3, because $\Xi_{\delta,\eta}(E, E)$ does. From Lemma 4.4 it follows that $\hat{E}$ is Borel and so there exists a closed set $\bar{E} \subseteq \hat{E}$ such that $\lambda(\hat{E} \setminus \bar{E}) < (\eta/2)^d$. Therefore $\Xi_{\delta,\eta/2}(\bar{E}, \pi_d E)$ holds by lemma 4.1. Another appeal to Lemma 4.1 yields that $\Xi_{\delta,\eta/2}(\bar{E} \cap U, \bar{E})$ and so, from Lemma 4.5 it may be concluded that $\Xi_{\delta,\eta/2}(\bar{E} \cap U, \bar{E} \cap U)$.

The choice of ϵ' guarantees that there is $a_1 \in W_{N+1}$ such that

$$\Xi_{(d+1)\delta,\epsilon'}(\bigcap_{i\in k}\bigcap_{j\in K}((A_i^j \cap (F_i^j)^{-1}a_1) \cup ([0,1]^d \setminus A_i^j)), \bar{E} \cap U)$$

holds. It follows from Lemma 4.1 that so does

$$\Xi_{(d+1)\delta,\epsilon'/2}(Z \cap \bigcap_{i\in k}\bigcap_{j\in K}((A_i^j \cap (F_i^j)^{-1}a_1) \cup ([0,1]^d \setminus A_i^j)), \bar{E} \cap U)$$

and from Lemma 4.7 it follows that

$$\Xi_{(d+2)\delta,\epsilon}(Z \cap \bar{E} \cap U \cap \bigcap_{i\in k}\bigcap_{j\in K}((A_i^j \cap (F_i^j)^{-1}a_1) \cup ([0,1]^d \setminus A_i^j)), \pi_d(E))$$

because $\Xi_{\delta,\epsilon}(\bar{E} \cap U, \pi_d(E))$ holds since $\epsilon \leq \eta/4$. Let $a = a_0 \cup a_1$ and notice that $a \in W_N$.

Using Lemma 4.2, it suffices to show that if

$$x \in Z \cap \bar{E} \cap U \cap \bigcap_{i\in k}\bigcap_{j\in K}((A_i^j \cap (F_i^j)^{-1}a_1) \cup ([0,1]^d \setminus A_i^j))$$

then

$$\Xi_{2\delta,\epsilon}(\bigcap_{i\in k}((A_i \cap F_i^{-1}a) \cup ([0,1]^{d+1} \setminus A_i))_x, E_x)$$

holds. To see that this is so, recall that since $x \in U$ it must be that $\lambda(Y(x)) < \epsilon^*/2$ where

$$Y(x) = \bigcup_{i\in k}(B_i^* \setminus F_i^{-1}\{f_i^j(x) : j \in K\})_x)$$

Moreover, since $x \in Z$ it must be that either

$$\Xi_{\delta,\epsilon^*}(\bigcap_{i\in k}((C_i \cap A_i \cap F_i^{-1}a_0) \cup ([0,1]^{d+1} \setminus (C_i \cap A_i)))_x, E_x)$$

holds or $\Xi_{\delta,\eta/2}(E_x, E_x)$ fails or $H^r(E_x) \leq \delta$. However, since $x \in \bar{E} \subseteq \hat{E}$ it must be that $\Xi_{\delta,\eta/2}(E_x, E_x)$ holds. If $H^r(E_x) \leq \delta$ then $\Xi_{2\delta,\eta}(\emptyset, E_x)$ holds and so, in either case it follows that

$$\Xi_{2\delta,\epsilon^*}((\bigcap_{i\in k}((C_i \cap A_i \cap F_i^{-1}a_0) \cup ([0,1]^{d+1} \setminus (C_i \cap A_i)))_x, E_x)$$

holds. It therefore follows from Lemma 4.1 that

$$\Xi_{2\delta,\epsilon^*/2}((\bigcap_{i\in k}((\cap C_i \cap A_i \cap F_i^{-1}a_0) \cup ([0,1]^{d+1} \setminus (C_i \cap A_i))))_x \setminus Y(x), E_x)\}$$

holds. Therefore it suffices to show that

$$(E \cap C_i \cap A_i \cap F_i^{-1}a_0) \cup (E \setminus (C_i \cap A_i))))_x \setminus Y(x) \subseteq$$
$$((A_i \cap F_i^{-1}a) \cup ([0,1]^{d+1} \setminus A_i))_x \cap E_x$$

for each $i \in k$.

Fix $i \in k$ and suppose that $y \in (E \cap C_i \cap A_i \cap F_i^{-1}a_0) \cup (E \setminus (C_i \cap A_i))_x \setminus Y(x)$. If $y \in E \cap C_i \cap A_i \cap F_i^{-1}a_0$ then $y \in A_i \cap E \cap F_i^{-1}a$. On the other hand, suppose $y \in (E \setminus (C_i \cap A_i))_x \setminus Y(x)$. If $y \in E \setminus A_i$ there is nothing to prove so it may be assumed that $y \in (A_i \setminus C_i)_x \setminus Y(x)$. Then, since $B_i^* \cap E = E \setminus C_i$ it must be that $y \in (B_i^*)_x$ and, since $y \notin Y(x)$, it follows that $y \in F_i^{-1}\{f_i^j(x) : j \in K\}$ and so there is some $m \in K$ such that $F_i(y) = f_i^m(x)$ and, in particular, $x \in A_i^m$. Since

$$x \in \bigcap_{i\in k}\bigcap_{j\in I_K}((A_i^j \cap (F_i^j)^{-1}a_1) \cup ([0,1]^d \setminus A_i^j))$$

it follows that $x \in A_i^m \cap (F_i^m)^{-1}a_1$ and so $F_i(y) = F_i^m(x) \in a_1$. Recalling that $y \in (E \setminus (C_i \setminus A_i))_x$ it follows that $y \in (A_i \cap E \cap F_i^{-1}a)_x$. ■

Corollary 6.1. *For any $n \in \omega$ the ideal $\mathcal{I}_n^r$ is proper.*

Proof: From Lemma 5.2 it suffices to show that that if $\mathfrak{X}(f, \mathcal{C}, \delta')$ is a d-dimensional generator for $\mathcal{I}_n^r$ then $W_n \not\subseteq \mathfrak{X}(f, \mathcal{C}, \delta')$. Let $\beta : \omega \to (0,1)$ witness that $\mathcal{C} = \{C_i\}_{i \in \omega}$ is a normal family. Let m be any integer such that $m > (d+1)/\delta'$. Now apply Theorem 6.1 letting $k = 1$, $\{F_i\}_{i \in k} = \{f\}$, $\eta = \beta(m)$, $\delta = \delta'/(d+1)$, $N = n$ and $\{A_i\}_{i \in k} = \{[0,1]^d\}$, . This yields $\epsilon > 0$ such that for every $i \in \omega$ there is some $a_i \in W_n$ such that $\Xi_{(d+1)\delta,\epsilon}((f^{-1}a_i \cup ([0,1]^d \setminus C_i), C_i)$ holds provided that $\Xi_{\delta,\beta(m)}(C_i, C_i)$ holds. Without loss of generality, it may be assumed that $\epsilon \leq \beta(m)$. This implies that $\Xi_{(d+1)\delta,\epsilon}(f^{-1}a_i, C_i)$ holds provided that $\Xi_{\delta,\beta(m)}(C_i, C_i)$ does. Since $1/m < \delta$ and $\Xi_{1/m,\beta(m)}(C_i, C_i)$ holds for each $i \geq m$ it follows $\Xi_{(d+1)\delta,\epsilon}(f^{-1}a_j, C_j)$ holds for some j such that $\lambda(\pi_n(C_j) \setminus \pi_n(\cap_{i \in \omega} C_i)) < (\frac{\epsilon}{d+1})^n$ for all $n \leq d$. The fact that such a j exists follows from the remark after Definition 5.1. Therefore there is some $\epsilon' > 0$ $\Xi_{\delta',\epsilon'}(f^{-1}a_j, C_i)$ holds for all but finitely many $i \in \omega$ by Corollary 5.2 and hence $a \notin \mathfrak{X}(f, \mathcal{C}, \delta')$. ■

7. The End

The bulk of this paper has been devoted to showing that the r-Hausdorff capacity of the ground model reals is not decreased by forcing with $\mathbb{P}(\mathcal{I})$ when $\mathcal{I} = \{\mathcal{I}_n^r\}_{n \in \omega}$ but it must also be shown that the partial order does what it was designed to do, make the ground model reals have Lebesgue measure zero. This is quite simple.

Theorem 7.1. *If* $\mathcal{I} = \{\mathcal{I}_n^r\}_{n \in \omega}$ *then*

$$1 \Vdash_{\mathbb{P}(\mathcal{I})} \text{“}\lambda(V \cap [0,1]) = 0\text{”}$$

where V *represents the ground model.*

Proof: First notice that if $A \in \mathcal{I}_n^{r+}$ then $[0,1] \subseteq \cup X$ because if $x \in [0,1]$ then, letting $\hat{x} : [0,1] \to [0,1]$ represent the function which is constantly x, it follows that $A \not\subseteq \mathfrak{X}(\hat{x}, \{[0,1]\}_{i \in \omega}, 1/2)$. A standard genericity argument will yield that if $G \in \prod_{n \in \omega} W_n$ is obtained from a $\mathbb{P}(\mathcal{I})$ generic set and $x \in V$ then $x \in G(n)$ for infinitely any n. Since any member of W_n has measure less than 2^{-n}, the result is proved. ■

References

1. T. Bartoszyński, H. Judah, and S. Shelah, *The Cichoń diagram*, J. Symbolic Logic **58** (1993), 401–423.
2. Lennart Carleson, *Selected problems on exceptional sets*, Mathematical Studies, vol. 13, Van Nostrand, New York, 1967.
3. Claude Dellacherie, *Ensembles analytiques, capacites, mesures de Hausdorff*, Lecture Notes in Mathematics, vol. 295, Springer, New York, 1972.

4. H. Judah and S. Shelah, *The Kunen-Miller chart (Lebesgue measure, the Baire property, Laver reals and preservation theorems for forcing)*, J. Symbolic Logic **55** (1990), 909–927.
5. Y. N. Moschovakis, *Descriptive set theory*, Studies in Logic, vol. 100, North-Holland, Amsterdam, 1980.
6. C. A. Rogers, *Hausdorff measures*, Cambridge University Press, London, 1970.
7. T. Salisbury and J. Steprāns, *Hausdorff measure and Lebesgue measure*, preprint.

DEPARTMENT OF MATHEMATICS, YORK UNIVERSITY, 4700 KEELE STREET, NORTH YORK, ONTARIO, CANADA M3J 1P3

Other Titles in This Series

(Continued from the front of this publication)

(See the AMS catalog for earlier titles)